W9-CRB-813

Methods of
DNA and RNA
Sequencing

Methods of DNA and RNA Sequencing

Edited by

Sherman M. Weissman, M.D.

Library of Congress Cataloging in Publication Data

Main entry under title:

Methods of DNA and RNA sequencing.

Includes bibliographical references and index.
1. Nucleotide sequence. 2. Deoxyribonucleic
acid—Analysis. 3. Ribonucleic acid—Analysis.
I. Weissman, Sherman M.
QP625.N89M47   1983   574.87'328   82-22261
ISBN 0-03-059174-0

Published in 1983 by Praeger Publishers
CBS Educational and Professional Publishing
a Division of CBS Inc.
521 Fifth Avenue, New York, New York 10175 U.S.A.

© 1983 by Praeger Publishers

3456789   052   987654321

Printed in the United States of America

LIST OF CONTRIBUTORS

Marvin H. Caruthers
Department of Chemistry
 University of Colorado
 Boulder, CO 80309
 USA

David J. Galas
Département de Biologie Moléculaire
 Université de Genève
 Geneva
 Switzerland

G. Nigel Godson
Biochemistry Department
 New York University Medical
  Center
 New York, N.Y. 10016
 USA

Yoshiyuki Kuchino
Biology Division
 National Cancer Center Research
  Institute
 Tsukiji, Chuo-Ku
 Tokyo, Japan

Allan Marshall Maxam
Sidney Farber Cancer Institute
 Charles A. Dana Cancer Center
 Boston, MA 02115
 USA

Stephen M. Mount
Department of Molecular Biophysics
 and Biochemistry
 Yale University
 New Haven, CT 06510
 USA

Susumu Mishimura
Biology Division, National Cancer
 Center Research Institute
 Tsukiji, Chuo-Ku
 Tokyo, Japan

Debra A. Peattie
Department of Biochemistry
 Stanford University Medical Center
 Stanford, CA 94305
 USA

Erika Randerath
Department of Pharmacology
 Baylor College of Medicine
 Texas Medical Center
 Houston, TX 77030
 USA

Kurt Randerath
Department of Pharmacology
 Baylor College of Medicine
 Texas Medical Center
 Houston, TX 77030
 USA

Albert Schmitz
Départment de Biologie Moléculaire
  Université de Genève
  Geneva
  Switzerland

Michael Smith
Department of Biochemistry
  Faculty of Medicine
  University of British Columbia
  Vancouver, B.C. V6T 1W5
  Canada

Thomas Shenk
Department of Microbiology
  Health Sciences Center
State University of New York
Stony Brook, N.Y. 11794
USA

Joan A. Steitz
Department of Molecular Biophysics
  and Biochemistry
Yale University
New Haven, CT 06510
USA

Sherman M. Weissman
Department of Human Genetics
Yale University School
  of Medicine
New Haven, CT 06510
USA

CONTENTS

2

SYNTHETIC OLIGODEOXYRIBONUCLEOTIDES AS PROBES FOR NUCLEIC ACIDS AND AS PRIMERS IN SEQUENCE DETERMINATION

xi

6

CHARACTERIZATION OF MODIFIED NUCLEOSIDES IN tRNA

Susumu Nishimura & Yoshiyuki Kuchino

10

SIGNALS FOR THE SPLICING OF EUKARYOTIC MESSENGER RNA TRANSCRIPTS

Stephen M. Mount & Joan A. Steitz

PREFACE

The entire structure of a free-living biologic unit is, to a first approximation, encoded in its DNA. Increasingly efficient and universal approaches for cloning discrete fragments of genetic DNA defined by restriction endonuclease cleavage sites have made available in principle each part of even the most complex genomes in large amounts and of sufficient purity for detailed analysis. Development of DNA sequencing techniques based principally on restriction endonuclease-generated fragments and dependent on gel fractionation methods that are capable of resolving the oligonucleotides that differ by 0.2% or less in their chain length have made it possible for many laboratories to determine complete sequences for the genetic material of particular interest to their research. Fortunately for the field, the inventors have made available fairly detailed protocols that make it possible to routinely reproduce the methods with clean results. A particularly exemplary protocol has been provided by Drs. Maxam and Gilbert for the method of chemical sequencing of DNA. One purpose of the present volume has been to update a simple set of such protocols for the more commonly used methods of nucleic acid labeling and sequence analysis, together with a discussion of the principles underlying the methods and some presentation of discussion of common sources of technical difficulties.

The chemical synthesis of deoxynucleotides has become a fundamental tool, both in sequence analysis and in *in vitro* mutagenesis and further characterization of biologically active sequences. Dr. Caruthers has included a chapter providing an update on discussion of the current methods that are applicable for laboratories not specialized in organic chemistry, and Dr. Smith has prepared an extensive discussion of the application of synthetic oligonucleotides to problems of molecular biology. Dr. Godson has reviewed the enzymatic approaches to DNA sequencing, and Dr. Maxam has provided a historical account of the development of chemical methods for DNA sequencing. It has become an increasing challenge to interpret sequences, once available, particularly through the use of valuable chemical methods that have been developed for studying interaction of purified proteins or protein complexes with nucleic acids, and these footprinting methods and extensive discussion of them by Drs. Schmitz and Galas are also included in the present volume.

Finally, a variety of techniques, including the classical techniques of prokaryotic genetics and *in vitro* mutagenesis of genes in functioning animal cells, followed by reintroduction of genes into cellular milieus *in vitro* enzymologic studies, have partly defined at least major subclasses of sequences responsible for transcription initiation and termination of prokaryotes and begun to provide a corresponding definition for eukaryotic sequences involved in these processes, as well as in the uniquely eukaryotic RNA splicing mechanisms. "State of the art" reviews on sequence features involved in initiation and termination of transcription (Dr. Shenk), and RNA splicing (Drs. Mount and Steitz) have been included.

The aim of the editor throughout has been to obtain experts to review their fields with the intention of describing methods in sufficient detail to facilitate ready application in laboratories without previous experience in the field and to provide discussion both of the theoretic underpinnings of the method and of the potential pitfalls. An effort has been made to minimize reduplication of material that has been authoritatively and lucidly reviewed elsewhere; for example, there is no discussion of *in vitro* mutagenesis, or of features of translation initiation sites, since several excellent reviews are available. One major feature of nucleic acid sequence analysis not covered in the present volume is the use of computers, both to expedite handling of the data and to reveal more detailed information than can be seen simply by visual inspection. This is a rapidly burgeoning subject in which many laboratories have their own favorite "data-crunching" systems, and increasing progress has been made in the theoretical aspects of problems related to the detection of potential base-pairing structures, analyses of homology among related sequences, etc. The editor can only refer the interested reader to primary papers such as the excellent collection recently published in *Nucleic Acids Research* (Jan., 1982) and to caution the novice that accessibility to computer analysis of data is almost essential for any extensive sequencing project. It is the hope of the authors that the present set of essays will expedite the work of young or novice investigators and provide a convenient reference or refresher for experienced workers in the field.

Sherman M. Weissman, M.D.

ACKNOWLEDGMENTS

The editor wishes to acknowledge with gratitude the help and patience of Marjorie Veronneau in preparing the manuscript of this book. The effort involved in editing the manuscript was supported by Grant Number CA-16038 from the National Cancer Center, Department of Health and Human Services.

1

New Methods for Chemically Synthesizing Deoxyoligonucleotides

MARVIN H. CARUTHERS

Recent developments in the recombinant DNA field have created a need for sequence-defined deoxyoligonucleotides. These compounds are being used as primers and probes for isolating natural genes (Wallace et al., 1981; Gillam et al., 1977), for experiments involving site-directed mutagenesis (Smith, 1980), and for the synthesis of genes and gene control regions (Agarwal et al., 1970; Khorana, 1979; Itakura et al., 1977; Goedell, Kleid, et al., 1979; Goedell, Heyneker, et al., 1979; Wetzel et al., 1980; Edge et al., 1981). However, until recently the synthesis and isolation of deoxyoligonucleotides has been a difficult and time-consuming task. Ideally, chemical methods should be simple, rapid, versatile, and accessible to the nonchemist. This chapter outlines a synthetic methodology that satisfies all these criteria. The approach has been used successfully by nonchemists in my own laboratory and more recently by other research groups with limited backgrounds in nucleic acid chemistry.

This chapter is divided into three sections. The first section provides a general outline of various approaches that have been used for synthesizing deoxyoligonucleotides. Such an outline is important since the field is highly specialized and not generally familiar to most biologists and biochemists. The second section outlines and discusses our methodology. The final section is a presentation of procedural details useful for those interested in duplicating our approach.

I. GENERAL OUTLINE OF DNA CHEMICAL SYNTHESIS METHODOLOGIES

A large number of synthetic strategies for preparing sequence-defined deoxyoligonucleotides have been developed (Reese, 1978; Amarnath & Broom, 1977). The major challenge is formation of the internucleotide phosphate ester bond. A closely related but secondary problem is the development of amino, phosphorus, and hydroxyl blocking groups that are compatible with the formation of internucleotide bonds. As work in this area has progressed, three chemical methodologies have evolved which differ primarily in relation to the synthesis of the internucleotide phosphodiester bond. More specifically, these methods involve the direct synthesis of the phosphodiesters, the intermediate synthesis of phosphotriesters, and the formation of intermediate phosphite triesters. All three methods can be illustrated using the general scheme depicted in Fig. 1-1.

The diester approach involves the condensation of a nucleoside (I: R_1, p-anisyldiphenylmethyl; R_4, hydrogen) with a nucleotide (II: R_5, phosphate; R_2, aryl or acyl ester) using an activating agent such as dicyclohexylcarbodiimide or arylsulfonyl chlorides in anhydrous pyridine (Khorana, 1968). The product is an appropriately protected phosphate diester (III: R_3, hydrogen). Deoxyoligonucleotides as large as eicosanucleotides have been prepared by these procedures. The synthetic products can be readily purified by diethylaminoethyl cellulose ion exchange chromatography or reverse-

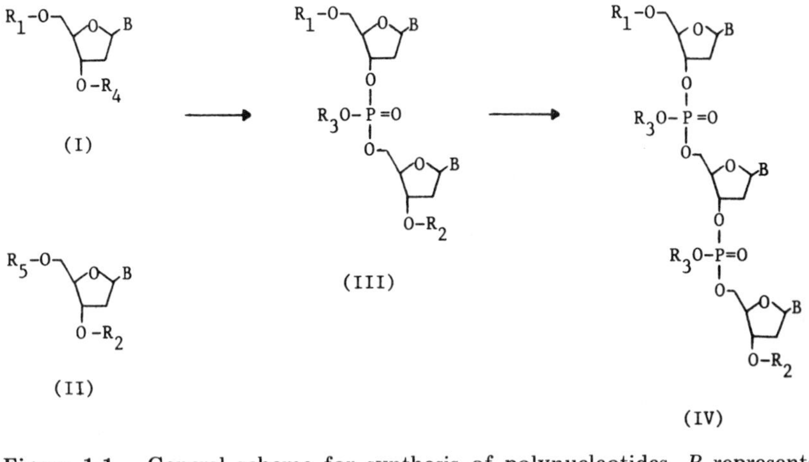

Figure 1-1. General scheme for synthesis of polynucleotides. B represents a protected purine or pyrimidine base. R groups are defined in the text.

phase high-pressure liquid chromatography. Yields per condensation range from 30–60%. This approach has been used to synthesize sequence-defined polynucleotides useful for studies on the genetic code (Khorana et al., 1966), the synthesis of lactose operator and lambda operator DNAs (Goedell et al., 1977; Kawashima, Gadek, & Caruthers, 1977), the synthesis of an alanine transfer RNA gene (Agarwal et al., 1970), and the first total synthesis of a biologically active gene—the *E. coli* tyrosine suppressor III transfer RNA gene (Khorana, 1979). A major advantage of this approach as contrasted to others is that mononucleotide starting materials are readily available and can be easily derivatized for polynucleotide synthesis. The major problem inherent in this approach is the activation of unprotected phosphodiester internucleotide bonds contained within the starting materials and product. This activation by the condensing agent leads to undesirable side reactions that dominate the synthetic procedures as the chain length increases and also reduces the overall yield.

The triester approach involves the synthesis of phosphate triesters as intermediates in polynucleotide synthesis. These triesters are then converted to phosphodiesters. The most attractive feature of this approach is that the highly reactive internucleotide phosphate anions are masked as triesters. The approach was initially developed by Letsinger in his work on polymer supports (Letsinger & Mahadevan, 1965, 1966; Letsinger, Caruthers, & Jerina, 1967) and led to the systematic development of a phosphotriester synthetic methodology (Letsinger & Ogilvie, 1967; Letsinger et al., 1967) which in various modified forms is now widely used in the nucleic acid field. The current most popular synthetic strategy is the following (Broka et al., 1980; Crea et al., 1978). A nucleotide phosphodiester (I: R_1, di-*p*-anisylphenylmethyl; R_4, *p*-chlorophenylphosphate) is condensed with a second nucleotide (II: R_5, hydrogen; R_2, β-cyanoethyl-*p*-chlorophenylphosphate) using an activating agent such as triisopropylbenzenesulfonyltetrazolide (TPSTe). Using the nucleotide phosphodiester in 100% molar excess, the condensation with TPSTe as a coupling reagent is reported to go essentially to completion. The isolation procedure simply involves treatment with benzenesulfonic acid to remove the tritylether (R_1, di-*p*-anisylphenylmethyl) followed by aqueous extraction of starting materials and other ionic components such as triisopropylbenzenesulfonate. Precipitation of the organic phase gives essentially homogeneous dinucleotide (III). This synthesis procedure can then be repeated a second time using the detritylated dinucleotide (III: R_1, hydrogen; R_2, β-cyanoethyl-*p*-chlorophenylphosphate; R_3, *p*-chlorophenyl) and a mononucleotide

(I: R_1, di-p-anisylphenylmethyl; R_4, p-chlorophenylphosphate). The trinucleotide (IV: R_1, di-p-anisylphenylmethyl; R_3, p-chlorophenyl; R_2, β-cyanoethyl-p-chlorophenylphosphate) is then isolated by short-column silica gel chromatography. This trinucleotide is a very versatile reagent for further synthesis. The trityl ether can be removed from one nucleotide (IV: R_1, hydrogen; R_3, p-chlorophenyl; R_2, β-cyano-ethyl-p-chlorophenylphosphate), and the β-cyanoethyl group can be removed using mild base from another trinucleotide (IV: R_1, di-p-anisylphenylmethyl; R_3, p-chlorophenyl; R_2, p-chlorophenyl-phosphate). These two trinucleotides can be joined using TPSTe to form a hexanucleotide. Repetition of this cycle with various hexanucleotides will lead to formation of dodecanucleotides. Thus by synthesizing all 64 possible trinucleotides, the ability to rapidly synthesize any deoxyoligonucleotide is possible. This same basic methodology has been used for synthesizing deoxyoligonucleotides on insoluble polymer supports. Considerable success has been achieved using polyacrylamide (Gait et al., 1980), polyacrylmorpholide (Miyoshi & Itakura, 1979), cellulose (Crea & Horn, 1980), cross-linked polystyrene (Myoshi et al., 1980), and polystyrene grafted on teflon beads (Potapov et al., 1979). Generally each synthesis begins by reacting compound I (R_1, di-p-anisylphenylmethyl; R_4, hydrogen) with a carboxylic acid functional group located on the polymer. The resulting ester anchors the growing deoxyoligonucleotide to the insoluble polymer support. After removal of the trityl ether, synthesis of the deoxyoligonucleotide proceeds by successively adding di- or trinucleotides using the same chemistry as described above for the nonpolymer support phosphotriester approach. Yields per condensation have been reported to be 80–90%. After several cycles, the final product is removed from the support with base. Isolation is completed by ion exchange chromatography after removal of protecting groups. Using presynthesized trinucleotides as condensing units and TPSTe as activating agent, deoxyoligonucleotides containing as many as 28 mononucleotides have been synthesized. A major criticism of this approach is the necessity to use presynthesized di- and trinucleotides as intermediates in order to synthesize deoxyoligonucleotides containing 12 or more mononucleotides. This is necessary because yields are low (80–90%) and reactions require 2–3 hr for completion. The presynthesis of all 16 dinucleotides or all 64 trinucleotides is a costly, time-consuming process.

A recent innovation in oligonucleotide synthesis has been the introduction of the phosphite coupling approach by Letsinger and coworkers (Letsinger et al., 1975; Letsinger & Lunsford, 1976). This approach has been adapted to the synthesis of deoxyoligonucleotides (Matteucci & Caruthers, 1980a, 1981; Caruthers et al., 1980;

Alvarado-Urbina et al., 1981), oligoribonucleotides (Daub & van Tamelen, 1977; Nemer & Ogilvie, 1980a; Ogilvie et al., 1980; Ogilvie & Nemer, 1980), and nucleic acid analogs (Melnick, Finnan, & Letsinger, 1980; Nemer & Ogilvie, 1980b; Burgers & Eckstein, 1978). Generally the approach involves the reaction of a suitably protected nucleoside (I: R_1, di-*p*-anisylphenylmethyl; R_4, hydrogen), a bifunctional phosphitylating agent such as methoxydichlorophosphine or methoxyditetrazoylphosphine, and a second protected nucleoside (II: R_5, hydrogen; R_2, levulinyl). Mild oxidation using iodine in tetrahydrofuran, lutidine, and water generates the natural internucleotide bond. By varying the oxidation procedure, phosphorus analogs such as selenophosphates, imidophosphates, and thiophosphates can be generated. A serious limitation of this methodology has been the instability of the reactive intermediates (nucleoside phosphomonochloridites or monotetrazolides) toward hydrolysis and air oxidation. This problem has recently been solved by the development of a new class of nucleoside phosphites— the nucleoside *N,N*-dimethylaminophosphoramidites (Beaucage & Caruthers, 1981). These compounds are easily prepared by standard organochemical procedures, are stable under normal laboratory conditions to hydrolysis and air oxidation, and are stored as dry, stable powders. Activation by mild acid treatment (amine hydrochlorides, 3-nitrotriazole, or tetrazole) in the presence of a second, appropriately protected nucleoside occurs very rapidly, and the dinucleotide forms in essentially quantitative yield within a few minutes. This approach has two obvious advantages over the others. First, only the four mononucleotide phosphites are required. The synthesis of intermediate dinucleotides or trinucleotides is completely unnecessary. Additionally, the intermediate phosphites can be oxidized to thiophosphates, selenophosphates, imidophosphates, and phosphonates. These analogs could be quite important for studies on various fundamental biological problems and for several potentially useful clinical applications of deoxyoligonucleotides. Our adaptation of this phosphite triester chemistry to deoxyoligonucleotide synthesis on polymer supports is described in the next section.

II. A PROCEDURE FOR THE RAPID CHEMICAL SYNTHESIS OF DEOXYOLIGONUCLEOTIDES

The general synthetic strategy involves adding mononucleotides sequentially to a nucleoside covalently attached to an insoluble polymer support. Reagents, starting materials, and side products are removed simply by filtration. After various additional chemical steps, the next mononucleotide is joined to the growing, polymer-supported deoxyoligonucleotide. At the conclusion of the synthesis, the deoxyoligonucleotide is chemically freed of blocking groups, hydrolyzed from the support, and purified to homogeneity by polyacrylamide gel electrophoresis.

The Support

High-performance liquid chromatography- (HPLC) grade silica gel is used as an insoluble polymer support. Silica gel is a rigid, nonswellable matrix in common organic solvents. Additionally, HPLC-grade silica gel is stable to all the reagents used for deoxyoligonucleotide synthesis and is designed for efficient mass transfer. These features make HPLC-grade silica gel an attractive starting material for deoxyoligonucleotide synthesis. Presently, silica gel preparations purchased from Merck (Fractosil 200 or Fractosil 500) are used. These preparations are easily derivatized and can be readily manipulated in columns, in test tubes, and on sintered glass funnels. Vydac TP-20 was previously used (Matteucci & Caruthers, 1981) but this material is quite flocculent in solutions and tends to plug filtration devices. Other more coarse silica gels, such as the Fractosil preparations, are therefore preferable.

The silica gel support derivatized to contain a deoxynucleoside is prepared using the steps outlined schematically in Fig. 1-2. The initial step involves forming compound 1 by refluxing 3-aminopropyltriethoxysilane with silica gel in dry toluene for 3 hr. The next step is synthesis of compounds 2a–d, which can most easily be accomplished using a procedure modified from published protocols (Gait, Singh, et al., 1980; Chow, Kempe, & Palm, 1981; Caruthers et al., 1982). The complete details for synthesizing these compounds are outlined in the experimental section. When any one of the compounds 2a–d is added to compound 1 in a mixture of dimethylformamide, dioxane, and triethylamine, an intense yellow color rapidly forms, indicating the elimination of p-nitrophenol and the formation of compounds 3a, 3b, 3c, or 3d. Usually, the ratio of reagents is adjusted so that approximately 50 µmol nucleoside per gram silica gel is obtained. Thus 1–2 µmol of a decanucleotide can be produced from 100–200 mg of silica gel. Unreactive amino groups are blocked against further reaction by acylation with acetic anhydride. This overall procedure is quite satisfactory for attaching any nucleoside to the polymeric support. In addition, more con-

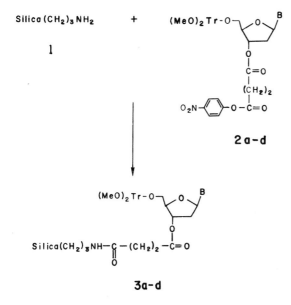

3a-d

Figure 1.2. Synthesis of the polymer support. *B* refers to thymine in 2*a* and 3*a*; to *N*-benzoylcytosine in 2*b* and 3*b*; to *N*-benzoyladenine in 2*c* and 3*c*; and to *N*-isobutyrlguanine in 2*d* and 3*d*. (MeO)$_2$Tr designates the di-*p*-anisylphenylmethyl protecting group.

sistent, reproducible loading is observed than when the previously published method is followed (Matteucci & Caruthers, 1981).

The Synthesis Cycle

The addition of one mononucleotide to compound 3*a*, 3*b*, 3*c*, or 3*d* requires the following four steps: (1) removal of the trityl protecting group with ZnBr$_2$; (2) condensation with the appropriate 5'-*O*-di-*p*-anisylphenylmethyl-3'-methoxy-*N*,*N*-dimethylaminophosphine nucleoside; (3) acylation of unreactive 5'-hydroxyl groups; (4) oxidation of the phosphite triester to the phosphate triester. These steps are summarized in Fig. 1-3. Thus the synthesis proceeds in a 3' to 5' direction by adding one nucleotide per cycle. The individual steps for one synthesis cycle are listed in Table 1-1 and outlined in the following paragraphs.

Zinc bromide in nitromethane completely removes trityl ethers from support-bound deoxyoligonucleotides in 30 min without any associated depurination (Matteucci & Caruthers, 1981, 1980b). The reaction time for this step can be reduced to 5 min using a nitro-

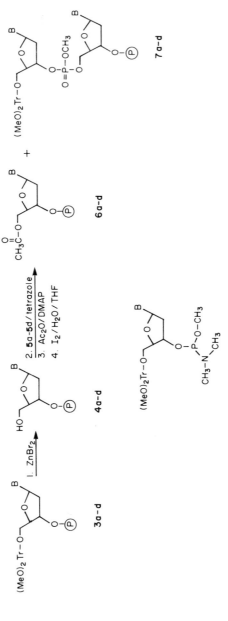

Figure 1-3. Steps in the synthesis of a dinucleotide. *B* refers to thymine in *3a*, *4a*, *5a*, *6a*, and *7a*; to *N*-benzoylcytosine in *3b*, *4b*, *5b*, *6b*, and *7b*; to *N*-benzoyladenine in *3c*, *4c*, *5c*, *6c*, and *7c*; and to *N*-isobutyrlguanine in *3d*, *4d*, *5d*, *6d*, and *7d*. (MeO)₂Tr refers to the di-*p*-anisylphenylmethyl group.

Table 1-1. A Summary of the Chemical Steps for One Synthetic Cycle

Reagent or solvent	Purpose	Time (min)
Satd. $ZnBr_2$ in 5% methanol/ CH_3NO_2	Detritylation	4
CH_3NO_2	Wash	1
Methanol	Wash	1
Acetonitrile	Wash	1
Activated nucleotide in acetonitrile	Add one nucleotide	5
Tetrahydrofuran lutidine/ H_2O (2:2:1)	Wash	3
I_2 Solution	Oxidation	5
Methanol	Wash	1
Tetrahydrofuran	Wash	1
Capping solution	Acylation	5
Methanol	Wash	2
CH_3NO_2	Wash	1

methane/methanol solution (95:5). Addition of methanol simply increases the concentration of $ZnBr_2$ in the saturated solution and therefore increases the detritylation reaction rate. This procedure has been examined extensively since our original publication. For example, Itakura has shown (Kierzek et al., 1981) that 100% debenzoylation of adenine and cytosine amino protecting groups occurs in 6 hr when the N-benzoylnucleoside is exposed to 1 M $ZnBr_2$ in a methanol/chloroform solution (4:1). The conditions as outlined in the experimental section are much less drastic and do not cause debenzoylation even after extensive reaction times (6 hr). An alternative detritylation procedure involves using 2% trichloroacetic acid in methylene chloride. Although this approach initially appeared very promising, Gait has shown through various solution experiments that less depurination is observed with trichloroacetic acid than with other protic acids (Gait, Popov, et al., 1980). However, on silica gel we have observed considerable depurination with trichloroacetic acid. Therefore, at the present time a saturated $ZnBr_2$ solution using nitromethane/methanol (95:5) as solvent is still recommended.

Appropriately protected nucleoside phosphoramidites (5a–d) are used for the sequential addition of mononucleotides. These compounds have several attractive features. They can be synthesized using standard organochemical procedures (Beaucage & Caruthers,

1981) and can be stored as white powders. They are also resistant to aqueous hydrolysis and to oxidation. Consequently, these reagents are stored, measured, and transferred in a manner analogous to any other solid compound. Activation in dry acetonitrile is achieved by the addition of a mild acid. Tetrazole is the activating agent of choice since it is a nonhygroscopic, commercially available material. Other mild acids such as amine hydrochlorides can also be used but are not recommended. Amine hydrochlorides are hygroscopic and would introduce water into the condensation step. The synthesis procedure therefore involves dissolving the nucleoside phosphoramidite and tetrazole in acetonitrile, mixing, and adding the solution to a silica gel support. The condensation reaction is complete within a minute but is usually allowed to proceed for 5 min.

Based on detailed and careful analysis of various condensation reactions, approximately 1-5% of the deoxynucleoside or deoxy-oligonucleotide bound to the support (compounds 4a-d) does not react with the activated nucleotide to form 7a-d. Our present assumption is that these results are due to inaccessible growing deoxyoligo-nucleotides rather than incomplete reaction. Nevertheless, we still prefer to block or cap these unreactive compounds in order to guard against the formation of several deoxyoligonucleotides having hetero-geneous sequences. In the absence of a capping step, the further extension of these failure sequences will lead predominantly to deoxyoligonucleotides having one or two less nucleotides than the expected product. Purification would therefore be difficult. If a capping step is included, the failure sequences will be quite disperse relative to size, with only 1-5% of the products having one nucleo-tide shorter than the preferred deoxyoligonucleotide. This latter case leads to relatively simple purification schemes involving polyacryl-amide gel electrophoresis. Currently, this capping step can best be accomplished using a tetrahydrofuran solution of acetic anhydride and dimethylaminopyridine. The reaction is complete within one to two minutes and does not lead to any detectable side products.

The internucleotide phosphite triester is oxidized to the phos-phate triester using I_2 in 2,6-lutidine, water, and tetrahydrofuran. Oxidation is extremely rapid (1-2 min), and side products are not generated. Attempts to postpone the oxidation until after all con-densation steps have not been encouraging. Several uncharacterized side products are observed.

Several devices have been used to aid in the synthesis of deoxy-oligonucleotides. Initially, our syntheses were completed in manually operated machines (Matteucci & Caruthers, 1980, 1981). The prin-ciples outlined in these machines are now being engineered into a completely automatic and programmable machine. Other devices are also being used; for example, deoxyoligonucleotides have been successfully synthesized in test tubes and on sintered glass funnels. In the former case, each synthesis step is followed by a low-speed centrifugation and decantation to remove solvents and reagents. For the latter case, reagents and solvents are removed simply by filtration. When test tubes or sintered glass funnels are used, six to eight deoxyoligonucleotides can be synthesized simultaneously. The appropriate mononucleotide is added to each sample of silica gel during the synthesis cycle. Otherwise, all the steps in a multiple-synthesis format are the same and can be completed at the same time.

The repetitive addition of mononucleotides to a growing deoxyoligonucleotide attached to silica gel therefore consists of four steps: detritylation, condensation, capping, and oxidation. The cycle, including various washing steps as outlined in Table 1-1, can be completed easily in 30 min. Routinely, compounds containing 18 mononucleotides are synthesized in one day. The chemistry as outlined in the preceding paragraphs has been used for synthesizing deoxyoligonucleotides containing from 6 to 26 mononucleotides.

Isolation of Deoxyoligonucleotides

Once a synthesis has been completed, the deoxyoligonucleotide is freed of protecting groups and isolated by polyacrylamide gel electrophoresis. Silica gel containing the reaction product is first treated with triethylammonium thiophenoxide in dioxane to remove the methyl groups from internucleotide phosphotriesters. This step is followed by treatment with concentrated ammonium hydroxide at $20°C$ for 3 hr to hydrolyze the ester joining the deoxyoligonucleotide to the support. After centrifugation and recovery of the supernatant containing the deoxyoligonucleotide, the N-benzoyl groups from deoxycytosine and deoxyadenosine, and the N-isobutyrl group from deoxyguanosine are removed by warming the concentrated ammonium hydroxide solution at $50°C$ for 12 hr. Finally, the trityl ether is hydrolyzed with 80% acetic acid. This is the preferred detritylating agent after the amino protecting groups have been removed. Depurination is not observed with completely deprotected deoxyoligonucleotides, and, unlike $ZnBr_2$, 80% acetic acid is volatile and easily removed. The reaction mixture containing deprotected deoxyoligonucleotides is then fractionated by electrophoresis on a polyacrylamide gel using the standard tris-borate buffer system. When a standard slab gel is used, as many as eight compounds can be purified

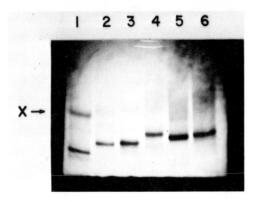

1. d(G-T-T-C-G-T-C-A-G-A-A-G-A-G-T-C)
2. d(A-A-T-T-C-T-A-T-T-A-G-T-C-T-T-T-A)
3. d(C-G-T-A-A-T-A-A-A-G-A-C-T-A-A-T-A-G)
4. d(G-C-G-T-T-C-C-C-T-G-A-C-T-C-T-T-C-T-G-A)
5. d(C-G-A-A-C-C-T-C-C-A-G-G-A-A-C-G-C-C-T-G)
6. d(T-T-A-C-G-C-A-G-G-C-G-T-T-C-C-T-G-G-A-G)

Figure 1-4. Purification of deoxyoligonucleotides synthesized on silica gel. Lanes 1-6 show the results obtained from six syntheses completed on silica gel. Each lane represents a total hydrolysate from silica gel after removal of all deoxyoligonucleotide protecting groups. The sequence of each deoxyoligonucleotide is listed below the gel pattern. X represents xylene cyanol blue which was added to lane 1 as a dye marker. The dye absorbs ultraviolet light and therefore appears as a dark band.

simultaneously. The appropriate deoxyoligonucleotides are visualized using an ultraviolet light and eluted from the gel using standard procedures (Maxam & Gilbert, 1980). An example illustrating this isolation procedure is reproduced in Fig. 1-4. Deoxyoligonucleotides numbered 1 through 6 were simultaneously synthesized on sintered glass funnels. Following removal of all protecting groups, total hydrolysates from silica gel were layered on a polyacrylamide gel and fractionated by electrophoresis. As can be seen from the photograph reproduced in Fig. 1-4, analysis of this gel showed one major band in each lane. Further analysis (not shown) confirms that the major bands correspond to the expected deoxyoligonucleotides whose sequences are also shown as part of Fig. 1-4. Occasionally, a synthesis fails or several additional bands are observed which correspond to significant amounts of various failure sequences. These problems appear to be due to mistakes made during the many steps required to complete manually a large number of simultaneous syntheses and should disappear once a completely automatic machine is available.

Nevertheless, the chemistry is reliable and has so far led to the synthesis of several hundred deoxyoligonucleotides.

III. EXPERIMENTAL SECTION

General Methods

Thiophenol, 4-dimethylaminopyridine, anhydrous $ZnBr_2$, and iodine can be purchased from Aldrich and used without further purification. Reagent-grade acetic anhydride, triethylamine, and t-butylamine are used as received. Common solvents such as dioxane, acetonitrile, nitromethane, and methanol are stored over activated (dried in an oven at $100°C$ for 12 hr) 4A molecular sieves and used without further purification. Dry acetonitrile is obtained by refluxing reagent-grade solvent over CaH_2 for several hours and distilling just prior to usage. 2,6-Lutidine is obtained by refluxing reagent-grade solvent over CaH_2 for 1 hr followed by distillation from 4A molecular sieves. 2,6-Lutidine is stored in the dark. Tetrahydrofuran is dried over sodium-benzophenone and used freshly distilled. Pyridine is distilled first from toluene sulfonyl chloride and then from potassium hydroxide. Toluene is distilled from sodium. Acid-free chloroform is obtained by passing chloroform through an aluminum oxide column just prior to use. 1-H-Tetrazole (Aldrich) is sublimed at 110-$115°C$ at 0.05 mmHg prior to use.

All solution transfers involving dry reagents are completed with clean syringes dried in ovens at $50°C$. When sintered glass funnels or test tubes are used as reaction flasks, the wash cycle immediately preceding the condensation step and the condensation step are completed in an atmosphere of dry nitrogen. This is most easily accomplished by passing a dry nitrogen line through the top of a rubber stopper attached to the top of a test tube or sintered funnel. A second hole in the stopper allows for escape of excess nitrogen. This stopper is removed during all other steps in the cycle. The main reason for this procedure is to ensure that the condensation solution remains dry throughout the synthesis step.

Synthesis of the Support

The initial step is synthesis of the succinilated deoxynucleosides as the p-nitrophenyl esters (compounds 2a, 2b, 2c, and 2d). All four are prepared using the same general procedure. To a solution of 5'-dimethoxytritylnucleoside (5 mmol) in anhydrous pyridine is added 4-dimethylaminopyridine (0.61 g; 5 mmol) and succinic anhydride (6.6 g; 6 mmol). The reaction is monitored by silica gel tlc (acetonitrile/water; 9:1) and is usually complete after 12 hr at 20°C. Occasionally, a second portion of succinic anhydride (0.1 g; 1 mmol) is added. The reaction is next quenched with water (0.1 ml) for 10 min at 20°C. The reaction mixture is evaporated *in vacuo*, and then co-evaporated twice with dry toluene (2 × 20 ml). The residue is redissolved in dichloromethane (40 ml) and the solution is washed successively, once with 10% citric acid (10 ml) and twice with water (2 × 10 ml). The organic solution is dried over anhydrous sodium sulfate and evaporated *in vacuo*. The residue is redissolved in 10 ml dichloromethane (containing 5% pyridine) and the product precipitated into pentane/ether (200 ml; 1:1, v/v). The precipitate is dried *in vacuo* (yield: 70–85%). The succinylated nucleoside (1 mmol) is next dissolved in dioxane (4 ml) containing dry pyridine (0.2 ml), and *p*-nitrophenol (140 mg; 1 mmol) is added. A solution of dicyclohexylcarbodiimide (220 mg) in anhydrous dioxane (1 ml) is added, and the reaction is monitored by tlc (silica gel plate; benzene/dioxane, 3:1). The reaction is virtually complete after 2 hr at room temperature. Dicyclohexylurea is removed by centrifugation, and the supernatant containing the desired product is used directly in the condensation reaction.

Deoxynucleosides are attached to the support using the following general procedure. HPLC-grade silica gel (12 g, Fractosil 200, Merck) is exposed to a 15% relative humidity (saturated aqueous solution of LiCl) for at least 24 hr. The silica is then treated with 3-triethoxysilyl-propylamine (13.8 g, 0.01 M in dry toluene) for 12 hr at 20°C and 18 hr at reflux. It is isolated by centrifugation and then washed successively (3 times each) with toluene, methanol, and 50% aqueous methanol. The silica is shaken with 50% aqueous methanol (200 ml) at 20°C for 18 hr. After isolation by centrifugation, the silica is washed with methanol and ether and dried *in vacuo*. The dried silica is suspended in anhydrous pyridine and treated with trimethylsilyl chloride (15 ml) for 12 hr at 20°C. After isolation by centrifugation, the silica is washed four times with methanol, twice with ether, and then dried *in vacuo*. The dry silica (3 g) is suspended in DMF (5 ml) and a solution of the 5'-dimethoxytritylnucleoside-3'-*p*-nitrophenylsuccinate in dioxane and 1 ml of triethylamine is added. The suspension is shaken at 20°C for 4 hr. Ninhydrin test at this stage indicates the existence of free amino groups on the resin. To cap these groups, acetic anhydride (0.6 ml) is added and the mixture is shaken for another 30 min, after which time a negative ninhydrin test is obtained. The silica is isolated by centrifugation, washed successively (three times each) with DMF, 95% ethanol, dioxane, and ethyl ether, and then dried *in vacuo*. Analysis for the extent of dimethoxytritylnucleoside attached to the support is done spectrophotometrically. An accurately weighed sample of silica (10–15 mg) is treated with 0.1 M *p*-toluenesulfonic acid in acetonitrile, and the optical density of the supernatant obtained after centrifugation is measured at 498 nm (the extinction coefficient of dimethoxytritanol is 7.0×10^4). A typical preparation leads to the following amounts of nucleosides bound to silica gel: $(MeO)_2TrdT$, 62 $\mu mol/g$; $(MeO)_2TrdibG$, 56 $\mu mol/g$; $(MeO)_2TrdbzA$, 65 $\mu mol/g$; $(MeO)_2TrdbzC$, 68 $\mu mol/g$.

Synthesis of Deoxynucleosidephosphoramidites

The careful preparation of compounds 5*a–d* is of critical importance. These compounds are prepared essentially as described previously (41). The synthesis begins with the preparation of chloro-*N,N*-dimethylaminomethoxyphosphine $[CH_3OP(Cl)N(CH_3)_2]$ which is used as a monofunctional phosphitylating agent. A 250-ml addition funnel is charged with 100 ml of precooled ($-78°C$) anhydrous dimethylamine (45.9 g, 1.02 mol). The addition funnel is wrapped with aluminum foil containing Dry Ice in order to avoid evaporation of dimethylamine. This solution is added dropwise at $-15°C$ (ice/acetone bath) over 2 hr to a mechanically stirred solution of methoxydichlorophosphine (47.7 ml, 67.32 g, 0.51 mol) in 300 ml of anhydrous ether in a 1-liter, three-necked round-bottom flask. The addition funnel is removed and the flask is stopped with serum caps tightened with copper wire. The suspension is mechanically stirred for 2 hr at room temperature. The suspension is filtered, and the amine hydrochloride salt is washed with 500 ml anhydrous ether. The filtrate and washings are combined, and ether is distilled at atmospheric pressure. The residue is distilled under reduced pressure. The product is collected at 40–42°C at 13 mmHg and is isolated in 71% yield (51.1 g, 0.36 mol): $d^{25} = 1.115$ g/ml; ^{31}P-N.M.R.; $\delta = -179.5$ ppm ($CDCl_3$) with respect to internal 5% v/v aqueous H_3PO_4 standard; H-N.M.R. doublet at 3.8 and 3.6 ppm $J_{P-H} = 14$ Hz (3H, OCH_3), and two singlets at 2.8 and 2.6 ppm (6H, $N(CH_3)_2$). The mass spectrum showed a parent peak at $m/e = 141$.

Compounds 5*a–d* are prepared by the following procedure. 5′-*O*-Di-*p*-anisylphenylmethyl nucleoside (1 mmol) is dissolved in 3 ml of dry, acid-free chloroform and diisopropylethylamine (4 mmol) in a 10-ml reaction vessel preflushed with dry nitrogen. Using a syringe, $[CH_3OP(Cl)N(CH_3)_2]$ (2 mmol) is added dropwise (30 to 60 sec) to the solution under nitrogen at room temperature. After 15 min the solution is transferred with 35 ml of ethyl acetate into a 125-ml separatory funnel. The solution is extracted four times with an aqueous, saturated solution of NaCl (80 ml). The organic phase is dried over anhydrous Na_2SO_4 and evaporated to a foam under reduced pressure. The foam is dissolved with toluene (10 ml) (5*d* is dissolved with 10 ml of ethyl acetate), and the solution is added dropwise to 50 ml of cold hexanes ($-78°C$) with vigorous stirring. The cold suspension is filtered and the white powder is washed with 75 ml of cold hexanes ($-78°C$). The white powder is dried under reduced pressure and stored under nitrogen. Isolated yields of compounds 5*a–d* are 90–94%. The purity of the products is checked by ^{31}P-N.M.R. Compounds 5*a–d* are characterized as two peaks between -146 and -145 ppm. Various impurities are sometimes observed with peaks between 0 and $+10$ relative to phosphoric acid. These impurities do not appear to inhibit the condensation reactions.

Outline of the Synthesis Cycle

The appropriately derivatized deoxynucleosides attached covalently to silica gel (compound 3*a*, 3*b*, 3*c*, or 3*d*) are treated with a saturated solution of anhydrous $ZnBr_2$ in nitromethane/methanol (95:5) for 4 min. The support is then washed with nitromethane, followed by methanol. Before the condensation step the silica gel is carefully washed several times with dry acetonitrile under a dry inert atmosphere (N_2). Stock solutions of sublimed tetrazole and appropriately protected 2′-deoxynucleoside-3′-*N*,*N*-dimethylaminomethoxyphosphines are prepared in dry acetonitrile and stored over an inert gas atmosphere (N_2). Usually, these stock solutions are sufficient for at least three condensations. For each μmol of deoxynucleoside attached covalently to silica gel, tetrazole (60 μmol) and the deoxynucleoside-3′-*N*,*N*-dimethylaminomethoxyphosphine (20 μmol) are 0.1 M in the condensation mixture. The condensation reaction is completed under an inert gas atmosphere (N_2) and stopped after 5 min. The excess phosphine can be reduced to 5- or 10-fold over support-bound nucleoside if rigorously dry reaction conditions are used. Such conditions are possible in a closed, machine-type device; if test tubes or sintered funnels are used as reaction vessels however, a larger excess of phosphine should be used to ensure that conditions are anhydrous. Immediately following the condensation reaction, the silica gel is washed with a tetrahydrofuran/water/2,6-lutidine solution (2:2:1) for 3 min. Oxidation of trivalent phosphorus to pentavalent phosphate is with a 0.2 M solution of iodine in tetrahydrofuran/water/2,6-lutidine (2:2:1) for 5 min. The silica gel is washed with methanol followed by tetrahydrofuran. Acylation of unreactive hydroxyl groups is completed by adding first a solution of 4-dimethylaminopyridine in dry tetrahydrofuran (2 ml; 6.5% w/v) and then a solution of acetic anhydride in 2,6-lutidine (0.4 ml; 1:1) to the support. After 5 min, this acylation solution is removed and the silica gel is washed with methanol and nitromethane. This step completes one synthesis cycle.

Isolation of Synthetic Deoxyoligonucleotides

After completion of the appropriate synthesis cycle, deoxyoligonucleotides free of protecting groups and side products are isolated using the following procedure. Silica gel containing the deoxyoligonucleotide is first treated with a solution containing thiophenol/dioxane/triethylamine (1:2:2) for 90 min at room temperature. This deprotection step removes the methyl phosphotriester protecting group. The support is next treated with concentrated ammonium hydroxide for 24 hr at 60°C. The liquid phase is evaporated to dryness *in vacuo* and the residue is treated with a solution of *t*-butylamine in methanol (1:1) for 24 hr at 60°C in a screw-cap vial. The reaction mixture is evaporated to dryness *in vacuo* and then loaded on a Sephadex G-50 column. Fractions containing deoxyoligonucleotides are pooled, and the DNA is isolated by ethanol precipitation. The crude DNA pellet is dissolved in formamide and loaded on a 20% denaturing polyacrylamide gel. After electrophoresis, the bands containing deoxyoligonucleotide products on the gel are visualized using an ultraviolet lamp. The absorbance can be enhanced by placing a silica gel thin-layer plate containing a fluorescent dye behind the gel. The gel slice containing the product is eluted and desalted using standard procedures (47). Usually the product is the major UV-light absorbing band on the gel.

ACKNOWLEDGMENTS

This research was supported by grants from the National Institutes of Health (GM21120 and GM25680). M.H.C. also acknowledges support derived from an

NIH Research Career Development Award (1 KO4 GM00076). Several excellent and imaginative graduate and postdoctoral students contributed immeasurably to the development of this chemical methodology. Their names appear in the references cited.

REFERENCES

Agarwal, K. L., Büchi, H., Caruthers, M. H., Gupta, N., Khorana, H. G., Kleppe, K., Kumar, A., Ohtsuka, E., Rajbhandary, U. L., van de Sande, J. H., Sgaramella, V., Weber, H., and Yamada, T. (1970). Total synthesis of the gene for an alanine transfer ribonucleic acid from yeast. *Nature* 227:27-34.

Alvarado-Urbina, G., Sathe, G. M., Liu, W., Gillen, M. F., Duck, P. D., Bender, R., and Ogilvie, K. K. (1981). Automated synthesis of gene fragments. *Science* 214:270-274.

Amarnath, V., and Broom, A. D. (1977). Chemical synthesis of oligonucleotides. *Chem. Rev.* 77:183-217.

Beaucage, S. L., and Caruthers, M. H. (1981). Deoxynucleoside phosphoramidites—A new class of key intermediates for deoxypolynucleotide synthesis. *Tetrahedron Lett.* 22:1859-1862.

Broka, C., Hozumi, T., Arentzen, R., and Itakura, K. (1980). Simplification in the synthesis of short oligonucleotide blocks. *Nucl. Acids Res.* 8:5461-5471.

Burgers, P. M. J., and Eckstein, F. (1978). Synthesis of dinucleoside monophosphorothioates via addition of sulphur to phosphite triesters. *Tetrahedron Lett.* 3835-3838.

Caruthers, M. H., Beaucage, S. L., Efcavitch, J. W., Fisher, E. F., Matteucci, M. D., and Stabinsky, Y. (1980). New chemical methods for synthesizing polynucleotides. *Nucl. Acids Res. Symp. Ser. No. 7* 215-223.

Caruthers, M. H., Stabinsky, Y., Stabinsky, Z., and Peters, M. (1982). New methods for synthesizing promoters and other gene control regions. In *Proceedings of the Symposium on Promoters: Structure and Function*, eds. R. L. Rodriguez and M. J. Chamberlin (Reading, Mass.: Addison-Wesley, Academic Press), pp. 432-451.

Chow, F., Kempe, T., and Palm, G. (1981). Synthesis of oligodeoxyribonucleotides on silica gel support. *Nucl. Acids Res.* 9:2807-2817.

Crea, R., and Horn, T. (1980). Synthesis of oligonucleotides on cellulose by a phosphotriester method. *Nucl. Acids Res.* 8:2331-2348.

Crea, R., Kraszewski, A., Hirose, T., and Itakura, K. (1978). Chemical synthesis of genes for human insulin. *Proc. Natl. Acad. Sci., USA* 75:5765-5769.

Daub, G. W., and van Tamelen, E. E. (1977). Synthesis of oligoribonucleotides based on the facile cleavage of methyl phosphotriester intermediates. *J. Amer. Chem. Soc.* 99:3526-3528.

Edge, M. D., Green, A. R., Heathcliffe, G. R., Neacock, P. A., Schuch, W., Scanlon, D. B., Atkinson, T. C., Newton, C. R., and Markham, A. F. (1981). Total synthesis of a human leukocyte interferon gene. *Nature* 292:756-762.

Gait, M. J., Singh, N., Sheppard, R. C., Edge, M. D., Greene, A. R., Heathcliffe, G. R., Atkinson, T. C., Newton, C. R., and Markham, A. F. (1980). Rapid synthesis of oligodeoxyribonucleotides. IV. Improved solid phase synthesis of oligodeoxyribonucleotides through phosphotriester intermediates. *Nucl. Acids. Res.* 8:1081-1096.

Gait, M. J., Popov, S. G., Singh, M., and Titmas, R. C. (1980). Rapid synthesis of oligodeoxyribonucleotides. V. Further studies in solid phase synthesis of oligodeoxyribonucleotides through phosphotriester intermediates. *Nucl. Acids Res. Symp. Ser. No. 7* 243-257.

Gillam, S., Rattman, F., Jahnke, P., and Smith, M. (1977). Enzymatic synthesis of oligonucleotides of defined sequence: Synthesis of a segment of yeast iso-l-cytochrome *c* gene. *Proc. Natl. Acad. Sci. USA* 74:96-100.

Goeddel, D. V., Heyneker, H. L., Hozumi, T., Arentzen, R., Itakura, K., Yansura, D. G., Ross, M. J., Miozzari, G., Crea, R., and Seeburg, P. (1979). Direct expression in *Escherichia coli* of a DNA sequence coding for human growth hormone. *Nature* 281:544-548.

Goeddel, D. V., Kleid, D. G., Bolivar, F., Heyneker, H. L., Yansura, D., Crea, R., Hirose, T., Kraszewski, A., Itakura, K., and Riggs, A. (1979). Expression in *Escherichia coli* of chemically synthesized genes for human insulin. *Proc. Natl. Acad. Sci. USA* 76:106-110.

Goeddel, D. V., Yansura, D. G., and Caruthers, M. H. (1977). Studies on gene control regions. I. Chemical synthesis of lactose operator deoxyribonucleic acid segments. *Biochem.* 16:1765-1772.

Itakura, K., Hirose, T., Crea, R., Riggs, A. D., Heyneker, H. S., Bolivar, F., and Boyer, H. W. (1977). Expression in *Escherichia coli* of chemically synthesized gene for the hormone somatostatin. *Science* 198:1056-1063.

Kawashima, E., Gadek, T., and Caruthers, M. H. (1977). Studies on gene control regions. IV. Synthesis and biological activity of a λ pseudo operator. *Biochem.* 16:4209-4217.

Khorana, H. G. (1979). Total synthesis of a gene. *Science* 203:614-625.

Khorana, H. G. (1968). Nucleic acid synthesis. *Pure Appl. Chem.* 17:349-381.

Khorana, H. G., Buchi, H., Ghosh, H., Gupta, N., Jacob, T. M., Kossel, H., Morgan, R., Narang, S. A., Ohtsuka, E., and Wells, R. D. (1966). Polynucleotide synthesis and the genetic code. *Cold Spring Harbor Symp. Quant. Biol.* 31:39-49.

Kierzek, R., Ito, H., Bhatt, R., and Itakura, K. (1981). Selective N-deacylation of N,O-protected nucleosides by zinc bromide. *Tetrahedron Lett.* 22:3761-3764.

Letsinger, R. L., and Lunsford, W. B. (1976). Synthesis of thymidine oligonucleotides by phosphite triester intermediates. *J. Amer. Chem. Soc.* 98:3655-3661.

Letsinger, R. L., and Mahadevan, V. (1966). Oligonucleotide synthesis on a polymer support. *J. Amer. Chem. Soc.* 88:5319-5324.

Letsinger, R. L., and Mahadevan, V. (1965). Stepwise synthesis of oligodeoxyribonucleotides on an insoluble polymer support. *J. Amer. Chem. Soc.* 87:3526-3527.

Letsinger, R. L., and Ogilvie, K. K. (1967). A convenient method for stepwise synthesis of oligothymidylate derivatives in large scale quantities. *J. Amer. Chem. Soc.* 89:4801-4803.

Letsinger, R. L., Caruthers, M. H., and Jerina, D. M. (1967). Reactions of nucleosides on polymer supports. Synthesis of thymidylylthymidylylthymidine. *Biochemistry* 6:1379-1388.

Letsinger, R. L., Caruthers, M. H., Miller, P. S., and Ogilvie, K. K. (1967). Oligonucleotide synthesis utilizing β-benzoyl propionyl, a blocking group with a trigger for selective cleavage. *J. Amer. Chem. Soc.* 89:7146-7147.

Letsinger, R. L., Finnan, J. L., Heavner, G. A., and Lunsford, W. B. (1975). Phosphite coupling procedure for generating internucleotide links. *J. Amer. Chem. Soc.* 97:3278-3279.

Matteucci, M. D., and Caruthers, M. H. (1981). Synthesis of deoxyoligonucleotides on a polymer support. *J. Amer. Chem. Soc.* 103:3185-3191.

Matteucci, M. D., and Caruthers, M. H. (1980a). The synthesis of oligodeoxypyrimidines on a polymer support. *Tetrahedron Lett.* 21:719-722.

Matteucci, M. D., and Caruthers, M. H. (1980b). The use of zinc bromide for removal of dimethoxytrityl ethers from deoxynucleosides. *Tetrahedron Lett.* 21:3243-3246.

Maxam, A. M., and Gilbert, W. (1980). Sequencing end-labeled DNA with base-specific chemical cleavages. In *Methods in Enzymology*, eds. L. Grossman and K. Moldave (New York: Academic Press), pp. 499-560.

Melnick, B. P., Finnan, J. L., and Letsinger, R. L. (1980). Oligonucleotide analogues with internucleoside phosphite links. *J. Org. Chem.* 45:2715-2716.

Miyoshi, K., and Itakura, K. (1979). Solid phase synthesis of nonadecathymidylic acid by the phosphotriester approach. *Tetrahedron Lett.* 3635-3638.

Miyoshi, K., Arentzen, R., Huang, T., and Itakura, K. (1980). Solid phase synthesis of polynucleotides. IV. Usage of polystyrene resins for the synthesis of polydeoxyribonucleotides by the phosphotriester method. *Nucl. Acids Res.* 8:5507-5517.

Nemer, M. J., and Ogilvie, K. K. (1980a). The synthesis of oligoribonucleotides. VI. The synthesis of a hexadecamer by a block condensation approach. *Can. J. Chem.* 58:1389-1397.

Nemer, M. J., and Ogilvie, K. K. (1980b). Phosphoramidate analogues of diribonucleoside monophosphates. *Tetrahedron Lett.* 21:4153-4154.

Ogilvie, K. K., and Nemer, M. J. (1980). Silica gel as a solid support in the synthesis of oligoribonucleotides. *Tetrahedron Lett.* 21:4159-4162.

Ogilvie, K. K., Nemer, M. J., Thériault, N., Pon, R., and Seifert, J. (1980). A complete procedure for the chemical synthesis of oligoribonucleotides. *Nucl. Acids Res. Symp. Ser. No. 7* 147-149.

Potapov, V., Veiko, V., Koroleva, O., and Shabarova, Z. (1979). Rapid synthesis of oligodeoxyribonucleotides on a grafted polymer support. *Nucl. Acids Res.* 6:2041-2056.

Reese, C. B. (1978). The chemical synthesis of oligo- and polynucleotides by the phosphotriester approach. *Tetrahedron* 34:3143-3179.

Smith, M. (1980). Applications of synthetic oligodeoxyribonucleotides to problems in molecular biology. *Nucl. Acids. Res. Symp. Ser. No. 7* 387–395.

Wallace, R. B., Johnson, M. J., Hirose, T., Miyake, T., Kawashima, E. H., and Itakura, K. (1981). The use of synthetic oligonucleotides as hybridization probes. II. Hybridization of oligonucleotides of mixed sequence to rabbit β-globin DNA. *Nucl. Acids Res.* 9:879–894.

Wetzel, R., Heyneker, H., Goeddel, D., Jhurami, P., Shapiro, J., Crea, R., Low, T., McClure, J., Thurman, G., and Goldstein, A. (1980). Production of biologically active N^{α}-desacetylthymosin α_1 in *Escherichia coli* through expression of a chemically synthesized gene. *Biochem.* 19:6096–6104.

2

Synthetic Oligodeoxyribonucleotides as Probes for Nucleic Acids and as Primers in Sequence Determination

MICHAEL SMITH

INTRODUCTION

The hydrogen-bonded duplexes formed by complementary poly-nucleotide chains are among the most accessible to *in vitro* synthesis of the specific noncovalently bonded structures that are a striking and ubiquitous feature of biological systems. Given the importance of nucleic acids in biology, it therefore is not surprising that nucleic acid duplex structures have been the target of synthetic and physical chemical studies. As a consequence of such studies, it has been evident for some time (Astell & Smith, 1971) that synthetic oligo-deoxyribonucleotides have the potential to be powerful and unique tools for the isolation and sequence determination of individual genes, DNA fragments, small genomes (both DNA and RNA), and other RNAs (Gilham, 1962; Astell & Smith, 1972; Besmer et al., 1972; Wu, 1972; Astell et al., 1973; Sanger et al., 1973; Gillam, Waterman, & Smith, 1975; Wu, Bahl, & Narang, 1978). However, in the past the relative inaccessibility of synthetic oligodeoxyribo-nucleotides limited the extensive exploitation of the potential of these methods; the predominant use of synthetic oligodeoxyribo-nucleotides until recently has been in the synthesis of small genes (Wu, Bahl, & Narang, 1978; Khorana, 1979; Itakura & Riggs, 1980; Edge et al., 1981). Recent developments in methodologies (Gillam & Smith, 1980; Koster, 1980; Alvarado-Urbina et al., 1981; Caruthers, Chapter 1 of this volume) have now made synthetic oligodeoxyribonucleotides much more accessible. The commercial availability of completely automated machines that can synthesize oligodeoxyribonucleotides, together with the decreasing cost and increasing efficiency and reliability of custom synthesis, will make oligodeoxyribonucleotides of defined sequence available to all molecular biologists. This, together with the development of molecular cloning of DNA and of rapid methods for nucleic acid sequence determination, has reawakened interest in synthetic oligodeoxyribonucleotides as tools for nucleic acid isolation and sequence determination. Further stimuli are provided by the failure or unavailability of alternative methods of RNA and DNA purification such as genetic complementation, immunological purification of nascent proteins attached to polysomes, and monitoring of nucleic acid purification by *in vitro* translation (Agarwal, Brunstedt, & Noyes, 1981). The increasing sensitivity of protein sequence determination methods has greatly expanded the number of proteins for which at least partial sequence is available; the genetic code allows the prediction of the sequences of a family of oligodeoxyribonucleotides corresponding to a peptide. Instances where synthetic oligodeoxyribonucleotide-primed DNA synthesis, on an RNA or DNA template, provides the only, or the most convenient, method for sequence determination are increasing dramatically.

With this background, it is timely to review the general principles that have emerged from model studies concerning the stability and specificity of oligodeoxyribonucleotide duplexes with polyribonucleotides and with polydeoxyribonucleotides. This leads into a discussion of the strategy and logic of the use of synthetic oligodeoxyribonucleotides as probes in the isolation and characterization of DNA fragments and RNA together with their use in sequence determinations.

BACKGROUND

The objective of this section is to evaluate those features of oligonucleotide-polynucleotide duplex formation, stability, and specificity which are of practical importance to the use of synthetic oligodeoxyribonucleotides as specific probes and primers. With this in mind, the discussion of model studies on oligonucleotide duplex formation will be selective. More extensive information can be obtained, *inter alia*, from Steiner and Beers (1961), Magee, Gibbs, and Zimm (1963), Michelson and Monny (1967), and Pohl (1974).

Ideally, a synthetic oligodeoxyribonucleotide will form a stable and unique duplex structure with a complementary region of the genome under study or with one of the mixture of RNA transcripts derived from the genome. Most of the model studies that provide useful data for evaluating this possibility have involved simple synthetic oligonucleotides and polynucleotides. It is a basic premise that, with reasonable discretion in extrapolation, these model studies provide a workable set of ground rules that can be used to make empirical predictions about the behavior of specific synthetic oligodeoxyribonucleotide probes and primers.

OLIGONUCLEOTIDE LENGTH AND DUPLEX STABILITY

For polynucleotide duplexes longer than about 200 n.p. (nucleotide pairs) the duplex stability is independent of length (Steiner & Beers, 1961; McConaughy & McCarthy, 1967). Thus, a duplex of 50 n.p. can have a Tm (the temperature of the midpoint of the thermal denaturation transition; Marmur, Rownd, & Schildkraut, 1963) which is $10°$ lower than that of a duplex with the same base composition of longer than 200 n.p. This possibility is sometimes ignored in studies of the relationship of nucleotide composition and Tm, which may account, at least in part, for the different values derived for the incremental effect of dG,dC composition on Tm (Wada, Yabuki, & Husimi, 1980; Hillen, Goodman, & Wells, 1981). For polynucleotide duplexes longer than 200 n.p., the major structural, as opposed to environmental, determinants of duplex stability are base composition and nucleotide mismatches. A dG,dC duplex has a Tm about $40°$ higher than a dA,dT duplex (Marmur, Rownd, & Schildkraut, 1963; Wada, Yabuki, & Husimi, 1980). Base-pair mismatches destabilize duplexes; the effect depends on the type of mismatch (Lomant & Fresco, 1975; Topal & Fresco, 1976). In addition, the sugar residue (polyribonucleotide or polydeoxyribonucleotide) can have a dramatic effect, as can the substitution of U for T (Chamberlin, 1965). The sequence, in duplexes of the same dG,dC content, also can have a significant influence on the Tm (Wells & Wartell, 1974; Wells et al., 1980). In the interaction of oligonucleotides and complementary oligonucleotides or polynucleotides, an additional major factor is, as implied earlier, the length of the oligonucleotide (Steiner & Beers, 1961; Pohl, 1974). The general form of the relationship between oligonucleotide length and duplex stability is that shown in Fig. 2-1 for the interaction of oligoribonucleotides

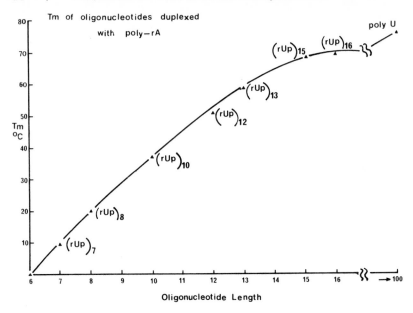

Figure 2-1. The relationship between oligouridylate length and the *Tm* of duplexes with polyadenylate in 0.02 M NaCl, 0.01 M sodium cacodylate, pH 7.0. Drawn from data of Michelson and Monny (1967).

of the series (rUp) with polyriboadenylic acid (data from Michelson & Monny, 1967). This same general type of effect is observed for oligoribonucleotides interacting with denatured DNA (Niyogi, 1969) and for oligodeoxyribonucleotides interacting with denatured DNA (McConaughy & McCarthy, 1967). Figure 2-1 demonstrates that the transition between dependence of *Tm* on oligonucleotide length and independence is a gradual one, although the incremental increase in *Tm* for an oligonucleotide length increase of 1 nucleotide becomes very small for lengths of greater than 15 nucleotides. A number of mathematical representations of the nonlinear relationship of oligo-nucleotide length and *Tm* have been formulated (Michelson & Monny, 1967). To a reasonable approximation the reciprocal of the transition temperature ($1/Tm$, °Kelvin) is a linear function of the reciprocal of the length of the oligonucleotide ($1/n$) for oligonucleo-tide-polynucleotide interactions where the oligonucleotide duplex is perfect (Magee, Gibbs, & Zimm, 1963). While this relationship is imprecise in that it does not predict the length independence of the *Tm* of long polynucleotide duplexes (Michelson & Monny, 1967), it is a valid and useful relationship for oligonucleotides of the lengths that are of interest to the types of experiments discussed in this chapter. It can be used, in limited extrapolation, to predict oligonucleotide duplex stabilities.

Because of their ready availability, oligoribonucleotides were used in the first studies of the interactions of characterized oligonucleotides and polynucleotides (Lipsett, Heppel, & Bradley, 1961). Because the available amounts of individual oligodeoxyribonucleotides were always small, studies on oligodeoxyribonucleotide duplexes of defined sequence were most readily carried out using thermal chromatography for the determination of duplex stability. In this method, the *Tm* values are determined from the temperatures at which oligonucleotides are eluted from their complements, which are irreversibly bound to an insoluble matrix. The method is more sensitive than measurement of *Tm* from thermochromic changes, allowing analysis of small samples which are radioactive (Gilham, 1962, 1964; Niyogi & Thomas, 1968). In particular, the development of oligonucleotide-celluloses (Gilham, 1962, 1964; Astell & Smith, 1971, 1972) allowed the possibility of a systematic study, with pure synthetic oligonucleotides, of the effect of oligonucleotide length, base composition, type of pentose, and base-pair mismatch on duplex stability (Astell & Smith, 1971, 1972; Astell et al., 1973; Gillam, Waterman, & Smith, 1975; Agarwal, Brunstedt, & Noyes, 1981). Figure 2-2, which is drawn from the data of Astell and Smith (1971, 1972) and Astell et al. (1973), shows that oligodeoxyribonucleotides of the series pdA_n, interacting with a complementary cellulose-pdT_9 column, behave in a qualitatively similar way to that previously observed for oligoribonucleotides interacting with polyribonucleotides in solution (cf. Fig. 2-1) or when the complementary polynucleotide is bound to hydroxyapatite (Niyogi & Thomas, 1968). In the case of the series pdA_6 to pdA_9, the incremental increase in duplex stability in interactions with cellulose-pdT_9 makes possible the complete resolution of a mixture of the oligodeoxyribonucleotides by thermal elution from cellulose-pdT_9 (Astell & Smith, 1972), despite the relatively wide breadth of the thermal transition for the melting of oligonucleotide duplexes (Steiner & Beers, 1961; Gilham & Naylor, 1966; Michelson & Monny, 1967). It is also possible to resolve a mixture of oligoribonucleotides by the same type of procedure (Gilham & Robinson, 1964). These data imply that, for short oligonucleotide duplexes, appropriate choice of temperature should allow completely specific discrimination in favor of a perfectly matched duplex relative to a duplex that is perfect but one base-pair shorter. It is not clear from presently available data where, with increasing oligonucleotide length, this discrimination is not possible.

Application of the relationship that $1/Tm$ is linearly proportional to $1/n$ to the data of Fig. 2-2 allows calculation of the approximate incremental increase in Tm for pairs of oligodeoxyribonucleotides differing in length by one nucleotide. For duplexes with complementary poly-dT, the increase in Tm going from pdA_6 to pdA_7 is

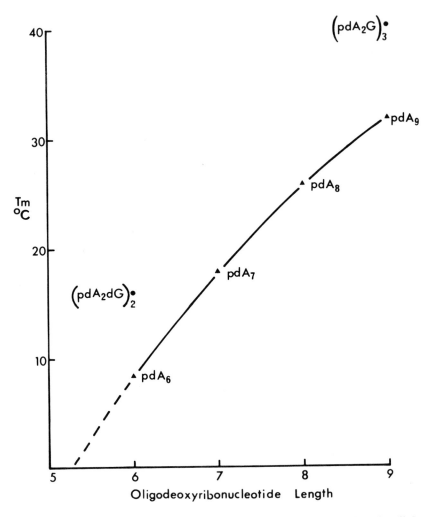

Figure 2-2. Tm values for oligodeoxyribonucleotides duplexed with cellulose-pdT_9 or cellulose-$pdCdT_2$. The Tm values were determined from the temperatures at which oligonucleotides were eluted from complementary oligonucleotide-cellulose columns in 1 M NaCl, 0.1 M sodium phosphate, pH 7.0. Drawn from data of Astell and Smith (1971, 1972) and Astell et al. (1973).

$9°$ ($4.5°$ per H bond); from pdA_9 to pdA_{10}, $5°$ ($2.5°$ per H bond); from pdA_{12} to pdA_{13}, $3°$ ($1.5°$ per H bond); and for pdA_{15} to pdA_{16}, $2.5°$ ($1.25°$ per H bond). This gives a reasonable idea of the duplex stabilities one might anticipate. A similar empirical relationship, $2°$ per dA,dT and $4°$ per dG,dC nucleotide pair, has been arrived at for oligodeoxyribonucleotide duplexes of 15–20 nucleotide pairs in other studies (Suggs et al., 1981).

Because of the uncertainty of either of these empirical formulas, a rapid annealing and radio-autographic procedure has been developed to allow scanning through a range of temperatures to establish the optimum for a specific oligodeoxyribonucleotide probe (Zoller & Smith, 1982). This whole procedure can be carried out in less than one day.

EFFECT OF BASE COMPOSITION ON OLIGONUCLEOTIDE DUPLEX STABILITY

The enhanced stability of GC-rich oligonucleotide duplexes has long been established (Lipsett, 1964; Michelson & Monny, 1967). Quantitative estimates of the effect of GC base pairs on oligoribonucleotide duplexes are provided by the studies of Uhlenbeck, Martin, and Doty (1971), for mixtures of oligoribonucleotides interacting with DNAs by Niyogi (1969), and for oligodeoxyribonucleic acid duplexes by Astell et al. (1973). Data from the latter experiments is included in Fig. 2-2. The incremental increase in *Tm* for a hexanucleotide duplex containing two dG,dC nucleotide pairs relative to an all dA,dT duplex is $7.5°$, and for a nonanucleotide containing three dG,dC nucleotide pairs relative to an all dA,dT duplex it is $8.5°$. Because of the limited amount of data available and the difficulty in comparing data for different oligodeoxyribonucleotide-cellulose columns, an average set of *Tm* values calculated from data obtained from several columns is shown in Table 2-1. This probably provides a more realistic prediction of duplex stabilities than the two experiments shown in Fig. 2-2. From Table 2-1, the increases in *Tm* resulting from replacing dA,dT base pairs by dG,dC base pairs are as follows: in a hexanucleotide duplex, two dG,dC pairs increase the *Tm* by $10.8°$; in an octanucleotide duplex, two dG,dC base pairs increase the *Tm* by $12.1°$; and for a nonanucleotide duplex, three dG,dC base pairs increase the *Tm* by $12°$. Clearly this data is inadequate for accurate prediction. However, it is not unreasonable to assume three

Table 2-1. Summary of Data on *Tm* Values, as Measured by the Temperature of Elution (in 1 M NaCl, 0.01 M Sodium Phosphate, pH 7.0) of Oligonucleotides Bound to Complementary Oligonucleotide-Cellulose Columns

Base pairs		Mean *Tm*
dA,dT	dG,dC	($^\circ$C)
6	0	5.3
4	2	16.1
8	0	21.4
6	2	33.5
9	0	28.5
6	3	40.5

Source: Data from Astell et al. (1973).

H bonds per dG,dC base pair and to apply the same incremental effects of added base pairs on duplex stability that were discussed earlier for dA,dT base pairs. The work of Niyogi (1969) and Uhlenbeck, Martin, and Doty (1971) suggests that the increment in *Tm* due to rG- or rC-containing base pairs may be slightly larger.

Studies on the interaction of mixtures of oligodeoxyribonucleotides of different length with denatured DNA (McConaughy & McCarthy, 1967) and on a few specific oligodeoxyribonucleotides interacting with DNA in solution (Besmer et al., 1972; Wu, 1972; Doel & Smith, 1973) are supportive of the idea that the interactions of complementary pairs of oligodeoxyribonucleotides and those of oligodeoxyribonucleotides and DNA obey similar rules.

While no extensive studies on the effect of sequence and concentration have been carried out with oligodeoxyribonucleotide-DNA duplexes in solution, such studies have been reported for oligoribonucleotide duplexes (Borer et al., 1974). Thus, a 10-fold decrease in the concentration of both oligodeoxyribonucleotides (100–10 μM) results in *Tm* decreases of 3–10° for different duplexes. Differences in sequence for oligoribonucleotide duplexes of the same base composition can result in *Tm* differing by up to 30° (Borer et al., 1974). It may be that oligodeoxyribonucleotide-polynucleotide duplexes do not have the same sensitivity to concentration and sequence. However, until more systematic data are available, the possibility should be borne in mind.

STABILITY OF OLIGODEOXYRIBONUCLEOTIDE-OLIGORIBONUCLEOTIDE DUPLEXES

The interactions of oligodeoxyribonucleotides of the series rA_n and rU_n with cellulose-pdT_9 and with cellulose-pdA_{10} have been determined (Astell et al., 1973). Comparison with the equivalent data for pdA_n and pdT_n reveals interesting differences in duplex stability. Thus, Fig. 2-3 shows that in interaction with cellulose-pdT_9 an oligoribonucleotide of the series rA_n forms a duplex that is 4-6° less stable than the duplex with pdA_n of the same length. Since a 5' phosphate destabilizes an oligonucleotide duplex by 1-2° (Lipsett, Heppel, & Bradley, 1961; Astell & Smith, 1971), the reduced stability of a completely equivalent prA_n duplex relative to a pdA_n duplex with cellulose-pdT_9 would be 5-8°. This increment is close to that for equivalent polymer duplexes; a poly-rA, poly-dT duplex is 4.5° less stable than a poly-dA, poly-dT duplex (Chamberlin, 1965). Thus, the incremental difference in duplex stability for the two series is essentially independent of polynucleotide length.

In interaction with cellulose-pdA_{10}, an oligoribonucleotide of the series rU_n is much less stable than the equivalent pdT_n (Fig. 2-4). When allowance is made for the effect of the 5' phosphate, the overall difference in stability of duplexes of the same length averages 26°. Again, the effect of length is minimal; a poly-rU, poly-dA duplex is 23° less stable than a poly-dT, poly-dA duplex (Chamberlin, 1965). From the practical point of view, a synthetic oligodeoxyribonucleotide probe interacting with rU-rich region of an RNA will have to be several nucleotides longer than the equivalent DNA probe to obtain a stable duplex. Although it is not germane to the theme of this chapter, it is also important to note that the reduced stability of a dA,rU duplex can result in cleavage by the single-strand specific nuclease, S1, in mRNA-DNA duplex mapping experiments (D. W. Leung & M. Smith, unpublished results). Misinterpretation of such an experiment could lead to the incorrect conclusion that an RNA is shorter than, in fact, it is or that there is an intervening sequence in the DNA.

No studies comparing the stabilities of dG_n,rC_n, dG_n,dC_n, and rG_n,dC_n oligonucleotide duplexes have been reported. Studies on polynucleotide duplexes (Chamberlin, 1965) show that a rG_n,dC_n duplex has a stability 25° greater than a dG_n,dC_n duplex. The Tm of a dG_n,rC_n duplex is 5° greater than that of dG_n,dC_n. If the principle that the incremental difference in duplex stabilities are independent of length is valid, then equivalent oligonucleotide duplexes will have similar differences in stability. For a long natural polyribonucleotide

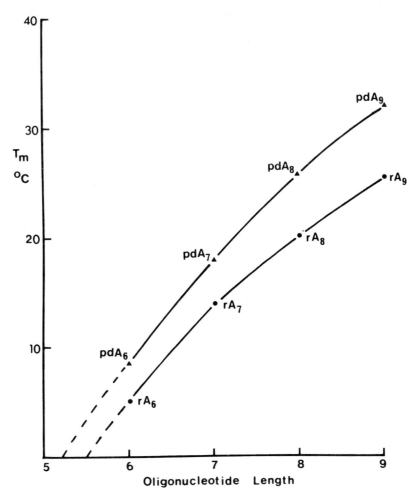

Figure 2-3. The *Tm* values of the pdA$_n$ and rA$_n$ series of oligonucleotides interacting with cellulose-pdT$_9$ in 1 M NaCl, 0.01 M sodium phosphate, pH 7.0. Drawn from data of Astell et al. (1973).

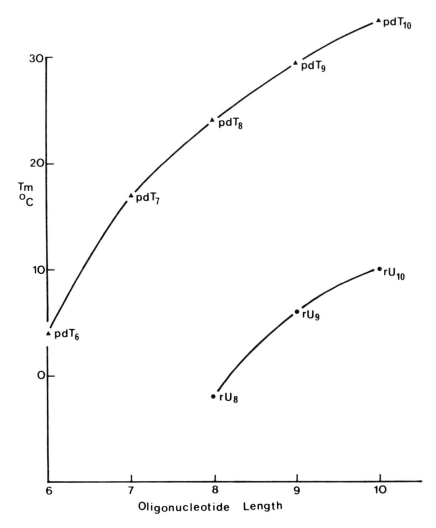

Figure 2-4. The *Tm* values of the pdT$_n$ and the rU$_n$ series of oligonucleotides interacting with cellulose-pdA$_{10}$ in 1 M NaCl, 0.01 M sodium phosphate, pH 7.0. Drawn from data of Astell et al. (1973).

these increases and decreases in stability will average out; it is important to remember that many oligodeoxyribonucleotide probes and primers will not be of average composition.

INFLUENCE OF BASE-PAIR MISMATCH ON
OLIGONUCLEOTIDE DUPLEX STABILITY

A considerable amount of data on the effect on duplex stability of mismatches is available for polyribonucleotides (Lomant & Fresco, 1975). However, there have been relatively few studies with oligodeoxyribonucleotide duplexes. One systematic study on the interaction of mismatched oligoribonucleotides and oligodeoxyribonucleotides with cellulose-pdT$_9$ provides some useful and surprising insights into the type of effect that can be anticipated (Gillam, Waterman, & Smith, 1975). Figure 2-5 shows that when, in the oligodeoxyribonucleotide series, a central nucleotide in pdA$_n$ is replaced by dT or dG there is a decrease of about 15° in the Tm of the duplex with cellulose-pdT$_9$. This destabilization has been quantitatively demonstrated for an oligodeoxyribonucleotide duplex in solution (Haasnoot et al., 1979). The position of a mismatch relative to the end of the duplex also has an effect; the duplex with a central mismatch is less stable (Table 2-2). Hence the data in Fig. 2-5 cannot be used for any simple extrapolations. However, it is clear that a centrally placed mismatch destabilizes an oligodeoxyribonucleotide-oligodeoxyribonucleotide duplex by an amount equivalent to removing at least two dA,dT pairs. With the particular duplex structures employed in this study it is not clear whether there is a base-pair mismatch or if one nucleotide is looped out (Lomant & Fresco, 1975).

It has often been assumed that dG,dT pairs are stable (Wu, 1972; Szostak et al., 1979; Hudson et al., 1981). Figure 2-5 clearly demonstrates that this is not so, the dG,dT pair being no more stable than a dT,dT pair. This is very much in contrast with the known stability of rG,rU mismatches (Lomant & Fresco, 1975). That lack of stability in a G,T(U) mismatch is unique to the dG,dT pair is shown by Fig. 2-6 (Gillam, Waterman, & Smith, 1975). In this experiment, stability of rA$_n$ duplexes with cellulose-pdT$_9$ is compared with the rA$_n$ series with an rU or an rG replacing a central nucleotide. The rG,dT pair is only slightly less stable than the rA,dT pair. More extended tracts of rG,dT pairs have also proved to be reasonably stable (Agarwal, Brunstedt, & Noyes, 1981). The rG,dT-

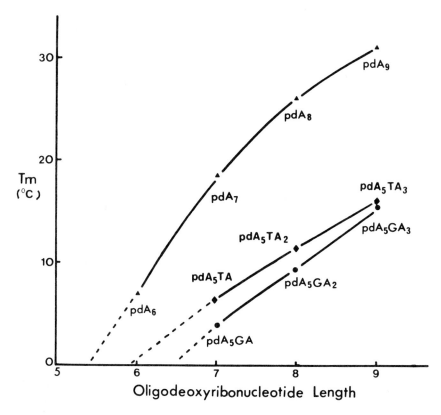

Figure 2-5. The *Tm* values of the duplexes with cellulose-pdT$_9$ of complementary oligodeoxyribonucleotides containing a single base-pair mismatch in 1 M NaCl, 0.01 M sodium phosphate, pH 7.0. Drawn from data of Gillam, Waterman, and Smith (1975).

Table 2-2. Effect of Position of Single Base-pair Mismatches on Oligodeoxyribonucleotide Duplex Stability using a Cellulose-pdT$_9$ Column as Complement (*Tm* values for elution from the column were determined in 1 M NaCl, 0.01 M sodium phosphate, pH 7.0)

	Tm		*Tm*		*Tm*		*Tm*
pdA$_9$	31°			pdA$_6$-dG-dA$_2$	23°	pdA$_5$-dG-dA$_3$	15.5°
pdA$_8$	26°	pdA$_7$-dG	22°	pdA$_6$-dG-dA	18°	pdA$_5$-dG-dA$_2$	9.5°

Source: Data taken from Gillam, Waterman, and Smith (1975).

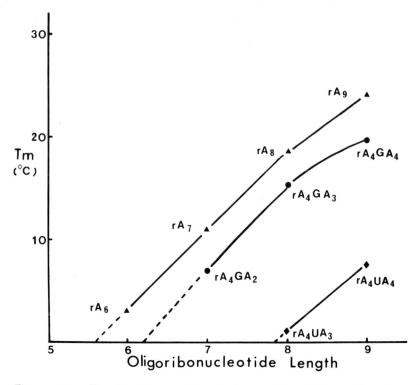

Figure 2-6. The *Tm* values of the duplexes with cellulose-pdT$_9$ of complementary oligoribonucleotides containing a single base-pair mismatch in 1 M
NaCl, 0.01 M sodium phosphate, pH 7.0. Drawn from data of Gillam, Waterman,
and Smith (1975).

containing duplexes are very much more stable than those containing
rU,dT mismatches (Fig. 2-6).

Specific rejection of a dG,dT pairing clearly is of major importance in maintaining the sequence of a replicating duplex. However,
in the context of the use of synthetic oligonucleotides as probes and
primers, this greater specificity of oligodeoxyribonucleotide-DNA
duplexes relative to oligoribonucleotide-DNA duplexes is a powerful
argument in favor of the use of oligodeoxyribonucleotides; ease of
synthesis and greater chemical stability are also very significant
factors.

Parenthetically, it is of interest that the demonstration of the
existence, at low temperature, of stable oligodeoxyribonucleotide
duplexes containing mismatches (Fig. 2-5, Table 2-2) was a major
stimulus to the development of methods that use such mismatched
duplexes as highly specific point mutagens (Smith & Gillam, 1981).

IONIC STRENGTH, POLYVALENT CATIONS, AND OLIGODEOXYRIBONUCLEOTIDE DUPLEX STABILITY

While temperature is a highly significant and useful factor in determining duplex stability, ionic strength and polyvalent cations also have very profound effects. The Tm of a nucleic acid duplex is directly proportional to the logarithm of the ionic strength; for polynucleotides there is an increase in Tm of 15–25° with an order of magnitude (0.1–1.0 M) increase in ionic strength (Inman & Baldwin, 1962; Marmur, Rownd, & Schildkraut, 1963; Oliver, Wartell, & Ratcliff, 1977; Wada, Yabuki, & Husimi, 1980; Hillen, Goodman, & Wells, 1981). Similar incremental increases in duplex stability are found for oligonucleotide duplexes (Lipsett, Heppel, & Bradley, 1961; Agarwal, Brunstedt, & Noyes, 1981). Thus, while Tm values for polynucleotide duplexes are often determined in 0.15 M NaCl, 0.015 M citrate, pH 7.0 (standard saline citrate, SSC), values for oligonucleotide duplexes are obtained at higher ionic strength — e.g., 1 M NaCl, 0.01 M Na phosphate, pH 7.0 (Astell & Smith, 1971, 1972) — to enhance duplex stability. Experiments using oligonucleotides as nucleic acid probes are carried out in high ionic strength buffers such as 4 × SSC (Montgomery et al., 1978; Noyes et al., 1979), thus increasing Tm values by 10–25°. A second potential benefit of the use of high ionic strength buffers is the possibility of better discrimination against mismatches and loopouts (Lomant & Fresco, 1975). This possibility has yet to be examined experimentally with oligonucleotide probes.

While high ionic strength and citrate buffer are entirely compatible with the use of oligonucleotides as probes, they are completely incompatible with the use of oligonucleotides as enzyme substrates, i.e., as primers for DNA polymerases. In this case, monovalent cation concentration of 0.1 M and Mg^{2+} concentrations of 0.01 M are required. An Mg^{2+} concentration of 0.1 M increases the stability of oligonucleotide duplexes by 10–30° (Lipsett, Heppel, & Bradley, 1961; Michelson & Monny, 1967). In effect, the Mg^{2+} almost exactly compensates for a decrease in monovalent cation concentration from 1 to 0.1 M. However, Mg^{2+} can have other effects; it promotes the formation of triple-stranded structures (Lipsett, Heppel, & Bradley, 1961). Also, in a study of the interaction of a synthetic dodecadeoxyribonucleotide, pTGGACTTTTTGT, with the strand of bacteriophage T4 DNA which contains its complement, stoichiometric binding was obtained in SSC (Doel & Smith, 1973) whereas in Mg^{2+}-containing buffers, many (10–100) oligodeoxyribonucleotide molecules were bound to each DNA molecule (M. T. Doel

& M. Smith, unpublished results) showing that Mg^{2+} can promote nonspecific binding of an oligonucleotide to DNA. Nonstoichiometric binding of oligodeoxyribonucleotides to DNA has also been observed in SSC (Besmer et al., 1972) although in these cases only one or two additional molecules were bound, presumably at sequences very similar to the primary targets.

In principle, when a synthetic oligodeoxyribonucleotide is used as a primer for a DNA polymerase, it does not have to form more than a transiently stable duplex to act as a primer. Thus an oligonucleotide, either perfectly or imperfectly matched, could function as a primer at a higher temperature than that at which a duplex could be detected in a probe experiment. For this reason, and because of the nonspecific binding in the presence of Mg^{2+}, it may be that an oligonucleotide is a less-specific primer than it is a probe. On the other hand, a mismatch near the 3' end of an oligonucleotide duplex may reduce its efficacy as a primer more than its efficacy as a probe.

Chaotrophic ions, organic solvents, and quaternary ammonium ions all have dramatic effects on the *Tm* of polynucleotide duplexes (Marmur, Rownd, & Schildkraut, 1963; Wada, Yabuki, and Husimi, 1980). Because these agents tend to reduce the *Tm* values of a duplex, so far they have not been used in molecular biology studies of the type discussed in this chapter.

OLIGONUCLEOTIDE DUPLEXES FORM VERY RAPIDLY

Duplexes involving simple sequences or high copy-number sequences are formed very rapidly. Duplexes involving single-copy sequences in complex genomes take much longer to form, and a period of annealing at some temperature below the *Tm* is required (Marmur, Rownd, & Schildkraut, 1963). Consequently, when using synthetic oligonucleotides as probes or primers, it is common practice to incubate the oligonucleotide and polynucleotide for several hours. In fact, this is not necessary. It has been shown that synthetic oligodeoxyribonucleotides form duplexes with complementary polynucleotides quite rapidly both in experiments where the oligonucleotide is used as a probe for a DNA in a Southern blot experiment (Montgomery et al., 1978) and where it is used as a primer for DNA synthesis and sequence determination of DNA and RNA (Smith et al., 1979; Kurjan et al., 1980; Sasavage et al., 1981).

LENGTH OF A GENETICALLY UNIQUE OLIGONUCLEOTIDE

Statistically, an oligonucleotide of n nucleotides occurs once in a polynucleotide target when 4^n is the number of nucleotides in the polynucleotide or when $4^n/2$ is the number of base pairs in a DNA duplex (Thomas, 1966; McConaughy & McCarthy, 1967; Astell & Smith, 1972). Table 2-3 lists some examples of the number of base pairs in haploid genomes of increasing complexity. For the smallest genomes (those of small phage and viruses) $n = 7$; for mammals, $n = 17$. These lengths are well within the compass of the newer methods of oligonucleotide synthesis, and, interestingly enough, correspond to the lengths for which stable duplexes have been demonstrated.

SYNTHETIC OLIGODEOXYRIBONUCLEOTIDE AS PROBE FOR GENOMIC DNA FRAGMENT PURIFICATION

In principle, a synthetic oligonucleotide can be attached to an insoluble matrix and used in this form to isolate a specific DNA fragment containing a complementary sequence. One significant application of this strategy uses cellulose-pdT$_n$ to isolate an ordered set of DNA fragments, all with poly-dA at one end, produced by

Table 2-3. Some Examples of the Number of Base Pairs in Haploid Genomes of Increasing Complexity

Organism	Length of DNA (μm)	Base pairs	n
Papovaviruses	1.7	5×10^3	7
Adenoviruses	12	35×10^3	9
E. coli	1.4×10^3	3×10^6	12
S. cerevisiae	5.0×10^3	15×10^6	13
D. melanogaster	56.0×10^3	165×10^6	15
H. sapiens	1.0×10^6	3×10^9	17

Note: An oligonucleotide to be a unique probe must contain n nucleotides, where $4^n > 2 \times$ number of base pairs in target genome.

partial restriction endonuclease digestion of a DNA containing poly-dA added enzymatically to the 3' end of one strand (Boseley, Moss, & Birnstiel, 1980; Moss, Boseley, & Birnstiel, 1980). Provided that appropriate precautions are taken to avoid nonspecific polynucleotide binding to cellulose (Bantle, Maxwell, & Hahn, 1976) the procedure works very well; elution of the bound, poly-dA tailed, DNA from the cellulose-pdT$_n$ is achieved by reducing the ionic strength from that of 0.4 M NaCl to that of 0.01 M tris-HCl rather than by increasing the temperature.

It is unlikely that the oligonucleotide-matrix approach to DNA fragment isolation will be generally applicable. For the less complex genomes, fractionation by gel electrophoresis is a convenient method for isolation of a specific DNA fragment. If a genome is so complex that individual DNA fragments are not resolved by gel electrophoresis, then the unique power of molecular cloning in plasmid or viral vectors allows isolation of specific fragments which can be less than 10^{-6} of the genome and at the same time permits amplification of the amount of the DNA. This powerful combination ensures that biological cloning is, and will continue to be, the preferred method of purification. Consequently, the role of a synthetic oligodeoxyribonucleotide will be as a probe to screen for the desired clone.

Prior to discussing the isolation of specific clones and the various strategies for gene isolation, it will be useful to review model studies that have been directed at using a synthetic oligodeoxyribonucleotide as a genomic probe. In particular, model studies where the experimental conditions are appropriate for fragment isolation and characterization will be discussed. The two most important types of experiments are the probing of electrophoretically fractionated fragments of DNA which have been transferred to a support such as nitrocellulose, and the probing of recombinant DNAs in colonies of transformed bacterial cells or in plaques produced by viral clones.

When single-stranded (denatured) DNA fragments are bound to supports such as nitrocellulose, the fragments are present in fmol or lesser amounts. With oligodeoxyribonucleotides as probes under such conditions, it can require 50–100-fold excess of the probe at concentrations of 1 pmol/ml to saturate the target site (Doel & Smith, 1973; Montgomery et al., 1978) although stoichiometric binding to the DNA target has been achieved with oligodeoxyribonucleotides in only 6-fold excess (Besmer et al., 1972). A number of factors must contribute to this order-of-magnitude difference in the optimum amount of probe. These include the concentrations of probe and DNA, the oligodeoxyribonucleotide length, base composition and sequence, the temperature of duplex formation, solvent composition, and the propensity of the target sequence of the DNA to form a hairpin structure which blocks access by the probe. In addition, the rapidity with which an oligonucleotide forms a duplex means that particular care must be taken to control the temperature at which washing of the hybrid product is carried out; clearly this can influence the efficiency and specificity of the probe (Besmer et al., 1972).

From this type of study it is clear that an oligodeoxyribonucleotide of 12 nucleotides and more can, in principle, clearly identify a DNA fragment containing a complementary sequence (Montgomery et al., 1978; Szostak et al., 1979; Wallace et al., 1979, 1981). As would be anticipated from the studies on oligodeoxyribonucleotide duplexes, it is possible to establish a temperature at which a perfect duplex is the only one formed between an oligodeoxyribonucleotide probe and a target DNA fragment bound to nitrocellulose (Wallace et al., 1979, 1981) although a mismatch near the end of the duplex is not always so different in stability that it can be discriminated against (Szostak et al., 1979). In addition, too large an excess of the probe can reduce its specificity (Besmer et al., 1972).

OLIGODEOXYRIBONUCLEOTIDES AS PROBES FOR MONITORING ISOLATION OF A GENOMIC DNA CLONE

The only genomic DNA fragment whose purification using this strategy has been reported is that encoding the iso-1-cytochrome *c* gene of *Saccharomyces cerevisiae* (Montgomery et al., 1978; Smith et al., 1979). Because of extensive studies on the *N*-terminal sequences of mutant iso-1-cytochrome *c* proteins, an unambiguous prediction of the corresponding encoding nucleic acid sequence is possible (Fig. 2-7) (Stewart & Sherman, 1974). The synthetic oligodeoxyribonucleotide pTTAGCAGAACCGG (Gillam et al., 1977) was used as the probe; 13 nucleotides should provide a unique sequence in a genome of the size of yeast (Table 2-3). When the oligonucleotide was used to identify the fragment containing the iso-1-cytochrome *c* gene in an *Eco*RI digest of *S. cerevisiae* DNA, after gel electrophoresis and Southern blotting of the denatured DNA fragments to nitrocellulose, about seven fragments gave a positive signal under optimum conditions (Montgomery et al., 1978). Optimum conditions included the use of 4 x SSC, 0.1% sodium dodecyl sulfate (SDS) as solvent, annealing for 30 min at 30–40° (and washing with the same solvent at the same temperature) with a concentration of probe of about 5 pmol/ml, labeled at a specific activity of about

WILD-TYPE:

Met—Thr—Glu—Phe—Lys—Ala—Gly—Ser—Ala—Lys—Lys—

ATG ACN GAR TTY AAR GCN GGN TCN GCN AAR AAR ...

N = A, G, C, T

R = A, G

Y = T, C

MUTANT:

Met—Thr—Glu—Phe—Lys—Pro—Val—Leu—Leu—Arg—Lys—

ATG ACN GAR TTY AAR CCN GTN CTN CTN AGR AAR ...

                         ↑   TTR TTR CGN ↑

           A or G Deleted   A or G inserted

CONCLUSION:

Sequence of wild-type DNA is

ATG ACN GAR TTY AAR GCC GGT TCT GCT AAF AAR ...
TAC TGN CTY AAR TTY CGG CCA AGA CGA TTC TTY ...

(Underlined sequence is 13-nucleotide synthetic probe)

Figure 2-7.   Prediction of part of yeast iso-1-cytochrome c gene from sequence of double-frameshift mutant. Data from Stewart and Sherman (1974).

2×10^6 cpm/pmol. The fragment containing the coding sequence for iso-1-cytochrome c was identified by comparison with the patterns of fragments identified in EcoRI digests of two mutant yeasts: one where the gene was deleted, and one where the EcoRI cleavage near the N-terminus of the gene was eliminated by a point mutation (Montgomery et al., 1978). Only one DNA fragment that was detected in wild-type DNA was absent from the mutants. This, therefore, contained the main body of the coding region for iso-1-cytochrome c. Once the fragment of wild-type yeast containing the coding sequence and also, of course, the larger fragment containing the gene from the mutant lacking the EcoRI site were identified, it was possible to screen a lambda phage recombinant pool made from a total EcoRI digest of yeast DNA. Hybridization of the probe to nitrocellulose replicas of phage plaques, again at a temperature of 30–40°, detected recombinants giving positive signals. A subset of these contained a yeast fragment of the correct size. To obtain a detectable signal in the plaque hybridization assay, it was important to ensure that there was an adequate amount of recombinant DNA in the plaque (Montgomery et al., 1978).

Given that only seven EcoRI fragments gave a positive signal to the probe, it would be possible to detect the gene by sequence determination of a clone of each fragment. However, it is desirable to have a more specific probe in the case where mutants that allow identification of the gene are not available. A 15-nucleotide probe, pAGCACCTTTCTTAGC, for the iso-1-cytochrome *c* gene is more specific in that it only detected two *Eco*RI fragments (Szostak et al., 1979), one of which encodes the iso-1-cytochrome *c* gene (Stiles et al., 1981). As a general principle, if two probes corresponding to different regions of a gene are constructed and each detects a small family of fragments, there is a high probability that only one fragment is common to the two families and this could provide a basis for unambiguous identification of a target DNA fragment.

In the case of both of the oligodeoxyribonucleotide probes discussed above, at lower than optimum temperature for duplex formation in Southern blot experiments, additional positive signals were detected. Above the optimal temperature no fewer positive signals were found; all signals disappeared as the temperature was progressively increased (Montgomery et al., 1978; Szostak et al., 1979). Given the variations in duplex stability which can occur, it seems likely that an optimal temperature will have to be established experimentally for any particular oligodeoxyribonucleotide probe. Under the type of conditions described above for the iso-1-cytochrome *c*, the temperature is likely to be in the range of 10–50° for probes of 11–17 nucleotides depending on the length, base composition, sequence, and concentration of the probe and the ionic strength of the buffer.

The iso-1-cytochrome *c* gene is a special case in that the sequence of part of its coding region is precisely predictable because of the availability of the amino acid sequences of a number of mutants, especially double-frameshift mutants (Stewart & Sherman, 1974). For most proteins only a single sequence or partial sequence is available. Only the codons for Met and Trp are unambiguous (Table 2-4). However, the codons of nine other amino acids (Asp, Asn, Cys, His, Phe, Tyr, Glu, Gln, and Lys) are minimally degenerate, with a purine or with a pyrimidine deoxyribonucleotide in the third position. In addition only three of the remaining nine amino acids (Arg, Leu, and Ser) are degenerate in the first two nucleotides of their codons. Thus a segment of a protein or peptide sequence which is rich in the amino acids Met, Trp and the nine amino acids with the least codon degeneracy can minimize the size of the family of oligodeoxyribonucleotides required to provide for all possibilities for the coding sequence corresponding to the peptide. If a DNA is the target of the oligodeoxyribonucleotide probe(s), all possible oligomers must be synthesized, rather than making the incorrect assumption that dG,dT pairs are stable. One simplifying principle is to use codons that are used preferentially in other genes in the same or related organisms

(Grantham, Gautier, & Gouy, 1980). However, there are occasions when a protein does not follow the normal pattern of codon usage (Astell et al., 1981).

A useful set of model studies that define conditions under which optimal specificity of oligodeoxyribonucleotide-DNA duplexes is obtained have been described by Wallace et al. (1979, 1981a). This information is particularly useful when a family of oligodeoxyribo-

Table 2-4. The Genetic Code (64 Triplet Codons for 20 Amino Acids and 3 Stop Signals)

Alanine	GCT	Glycine	GGT	Proline	CCT
	GCC		GGC		CCC
	GCA		GGA		CCA
	GCG		GGG		CCG
Arginine	CGT	Histidine	CAT	Serine	TCT
	CGC		CAC		TCC
	CGA				TCA
	CGG	Isoleucine	ATT		TCG
	AGA		ATC		AGT
	AGG		ATA		AGC
Asparagine	AAT	Leucine	TTA	Threonine	ACT
	AAC		TTG		ACC
			CTT		ACA
Aspartic	GAT		CTC		ACG
acid	GAC		CTA		
			CTG	Tryptophan	TGG
Cysteine	TGT				
	TGC	Lysine	AAA	Tyrosine	TAT
			AAG		TAC
Glutamine	CAA				
	CAG	Methionine	ATG	Valine	GTT
					GTC
Glutamic	GAA	Phenyl	TTT		GTA
acid	GAG	alanine	TTC		GTG
		STOP	TAA	TAG	TGA

nucleotide probes is being used on the target DNA, in order to minimize false positive signals. Although the *Tm* values of the oligo-deoxyribonucleotide-DNA duplexes are lower by about $20°$ than would be predicted by the earlier oligodeoxyribonucleotide cellulose-pT$_9$ studies (Astell & Smith, 1971, 1972; Astell et al., 1973), this is not unexpected given the marked effects of concentration and sequence differences on duplex stability (Borer et al., 1974). The incremental effects of sequence length and also the destabilizing effects of mismatches are very similar to those discussed earlier (Gillam, Waterman, & Smith, 1975). The overall conclusion from these studies is that duplex formation, or washing, at a temperature about $10° ± 5°$ below the *Tm* of a perfect oligodeoxyribonucleotide-DNA duplex will clearly reject mismatches of one base pair. For the oligodeoxyribonucleotides tested, the *Tm* values were 11-mer (6dA,dT), $33°$; 14-mer (7dA,dT), $40.5°$; 17-mer (7dA,dT), $55°$. Under similar conditions, a 13-mer (6dA,dT) of different sequence could form a duplex with globin DNA at $37°$ where another 13-mer with a single mismatch did not form a detectable duplex (Wallace et al., 1981a). The probes in these latter experiments were tested with Southern blots onto nitrocellulose of gel electrophoretically separated DNA fragments and also on colonies of *E. coli* transformed with a recombinant plasmid DNA and with the DNA fixed to Whatman 540 paper. In both types of experiments, the experimental protocols gave low backgrounds; high backgrounds have been a problem in other studies with oligodeoxyribonucleotide probes (Montgomery et al., 1978; Szostak et al., 1979) in part because of the relatively large excess of an oligonucleotide (100–1000-fold) which can be needed to saturate the target DNA (Doel & Smith, 1973), and also as a consequence of the high salt concentrations that are used in the wash buffer.

The direct extrapolation of these model studies to the probing of genomic DNA fragments is of interest. The model study demon-strated this successfully with fragments from a small genome—that of ϕX174 (Wallace et al., 1979). However, the genetic complexity of a mammalian genome for which the unique oligonucleotide length is 17 creates a rather complex problem. A probe 17 nucleotides in length predicted from the codons of five adjacent minimally degener-ate amino acids (excluding Met and Trp) and the first two nucleotides of the next (carboxyl end) amino acid which is not Arg, Leu, or Ser requires a family of 32 oligodeoxyribonucleotides. Synthesis of com-plex families of oligodeoxyribonucleotides is possible; families of eight 13 nucleotide probes have been synthesized (Wallace et al., 1981a), as have families of twelve 11-mer and twenty-four 12-mer probes (Goeddel, Shepard, et al., 1980; Goeddel, Yelverton, et al., 1980). A family of one hundred and twenty-eight 17-mers has been synthesized in a single experiment (A. F. Markham, personal communication). However, if such a family of oligodeoxyribonucleotides, 17 long, is used as a probe for genomic DNA fragments from a higher eukaryote, it is possible that each member of the family will elicit a positive signal, these being false positives. Thus some sort of secondary screening would be needed either by using a second family of probes for the same gene or by DNA sequence determination.

An alternative way of solving this problem has proved to be more attractive to date; in this procedure, a cDNA pool of clones obtained from the mRNA of a cell line or tissue which produces the protein of interest is screened using the synthetic oligodeoxyribonucleotide family of probes. The genome of a higher eukaryote contains the coding capacity for a few million genes, whereas the mRNA pool of a tissue probably contains significant amounts of the transcripts of only a few thousand genes. This decrease of about 1,000-fold in genetic complexity reduces the size of a unique probe to 12–13 nucleotides, i.e., the coding sequence of about four amino acids. If a Trp or Met is present in the peptide sequence, this requires a relatively small family of eight probes. Before discussing the probing of cDNA pools derived from mammalian mRNAs, it should be emphasized that direct probing of genomic fragments of genomes of intermediate complexity such as that of bacteria and yeasts is still an attractive approach to gene isolation, even with a mixture of probes.

OLIGODEOXYRIBONUCLEOTIDES AS PROBES FOR MONITORING ISOLATION OF A cDNA CLONE

A good example of the use of families of synthetic oligodeoxyribonucleotide probes for the isolation of a cDNA clone is that of human leukocyte interferon cDNAs (Goeddel et al., 1980b). Peptide sequence analysis provided sequences of two regions of leukocyte interferon, Glu—Ile—Met—Arg and His—Glu—Met—Ile. A family of twelve 11-mers was required to cover all possibilities for the first peptide (omitting the last nucleotide of the Arg codons) and four 11-mers for the second peptide (omitting the last nucleotide of the Ile codon). A subset of cDNA-containing plasmids obtained from leukocyte mRNA was individually purified and the DNAs were linearized, denatured, spotted, and fixed to nitrocellulose. The protocol of Wallace et al. (1979b) was used to hybridize the ^{32}P-labeled probes with the DNA at 15°. A member of each family of probes formed duplexes with a number of the cDNA clones, but only eight of the cDNAs were recognized by probes corresponding to both peptides. One of these yielded an interferon coding sequence (Goeddel et al., 1980b). Apart from the positive identification, which this strategy of using two independent short probes provided, a striking observation of the experiment was the high specificity of the duplex formation. Only the perfectly matched duplexes gave a signal. This is an impressive confirmation of the marked difference in stability of perfect oligodeoxyribonucleotide duplexes and those with any type of mismatch (Gillam, Waterman, & Smith, 1975).

Another example of the application of families of synthetic oligodeoxyribonucleotides as probes to isolate a specific cDNA clone from a pool of cDNA clones is the isolation of cDNA for human β_2-microglobulin from a pool of cDNAs prepared from the mRNA of a human lymphoblastoid cell line (Suggs et al., 1981). The family of 24 oligonucleotides corresponding to the amino acid sequence Trp—Asp—Arg—Asp—Met and a family of 8 undecanucleotides corresponding to the sequence Lys—Asp—Glu—Tyr (excepting the last nucleotide of the Tyr codons) were used as probes. A subset of the family of 24 pentadecanucleotides gave a single positive signal with one out of 535 colonies of transformed bacteria tested. One observation made in these experiments is worth particular attention. Although the family of 8 undecanucleotides contained one member corresponding to the cloned cDNA sequence, it did not give a positive signal during the screening of bacterial colonies, although it did give a positive signal in a Southern blot analysis. There are a number of possible reasons for this discrepancy (Suggs et al., 1981), but the take-home message is fairly clear; to be sure of obtaining a positive signal, use oligonucleotides with more than 11 nucleotides as probes.

A cDNA sequence, possibly derived from an mRNA which is a product of missplicing, corresponding to part of the murine transplantation antigen, H—2K^b, amino acid sequence has also been isolated using a family of eight hexadecanucleotides corresponding to part of the amino acid sequence Trp—Met—Glu—Gln—Glu—Gly. One positive signal was obtained from 30,000 colonies of *E. coli* transformed with recombinant plasmids derived from cDNA produced from mRNA isolated from a mouse cell line (Reyes et al., 1980). The cDNA of bovine preproenkephalin mRNA has been isolated using a mixture of two tetradecanucleotides corresponding to the sequence Glu—Trp—Trp—Met—Asp [omitting the last nucleotide of the Asp codon (Noda et al., 1982)]. A total of 190,000 transformants were screened to yield 19 positive clones, of which at least 6 had the anticipated restriction endonuclease digestion pattern.

Bovine proenkephalin mRNA cDNA has also been detected by using a hexadecanucleotide probe for the coding sequence Trp—Trp—Met—Asp—Tyr—Glu (omitting the last two nucleotides of the Glu codon). This is a very favorable case, in that a family of four oligonucleotides covers all possibilities. In fact, only one oligonucleotide was synthesized assuming that G was the third nucleotide of the Asp and Tyr codons (Gubler et al., 1981, 1982). A mixture of four pentadecanucleotides was used at 30° in 4 x SSC to probe for human proenkephalin cDNA (Comb et al., 1982). It is of interest that this probe was successful even though there is a nucleotide mismatch with the target.

Two striking recent applications of the use of two families of oligonucleotide sequences corresponding to two regions of coding sequence in the same protein as probes for cDNA clones are the isolation of cDNA clones for human complement protein factor B (Woods et al., 1982) and for *Torpedo* acetylcholine receptor alpha-subunit (Noda et al., 1982b).

Another useful application of an oligodeoxyribonucleotide probe for mRNA is an investigation of the precision of RNA splicing in yeast mitochondrial RNA using a probe that bridges the splice points (Tabak et al., 1981).

OLIGODEOXYRIBONUCLEOTIDES AS PRIMERS FOR REVERSE TRANSCRIPTASE TO PRODUCE SPECIFIC cDNA PROBES

Synthetic oligodeoxyribonucleotides can provide specific primers for reverse transcriptase on an mRNA template. In this way a specific radioactive probe can be obtained from a mixture of mRNAs. For example, the hexadeoxyribonucleotide TTGGGT (Gillam, Jahnke, & Smith, 1978) when used as a primer with, as template, impure immunoglobulin L-chain mRNA from mouse T-cells yielded a pure cDNA (Rabbitts et al., 1977). Likewise a decadeoxyribonucleotide, pCCTCCACCAG, corresponding to a sequence close to the *N*-terminus of rat insulins I and II was used as a primer on mRNA from an insulin-producing tumor (Chan et al., 1979). The principle of using a primer corresponding to a sequence near the *N*-terminus of the gene is important for two reasons. First, most of the reverse transcripts will terminate at the 5' end(s) of the target mRNA. This will result in a distinct fragment readily identified after gel electrophoresis, even if there is a high background. The fragment is usually 50–150 nucleotides long since eukaryote mRNAs are monocistronic and they do not have long 5'-untranslated regions. Second, when the reverse transcript is used as a probe with a cDNA pool, it will only identify cDNAs that are full length or close to full length, because the probe is specific to the 5' end of the mRNA.

In some cases, a single primer may produce more than one distinct reverse transcript. This can be a consequence of the mRNA having heterogeneous 5' ends, as is found with yeast iso-1-cytochrome *c* mRNA (Faye et al., 1981), or because it targets different sites. When there is only a small number of reverse transcripts, they can be separated electrophoretically and the correct one identified by DNA sequence determination. This strategy has been applied in the case of rat insulin mRNA (Chan et al., 1979) and hog gastrin mRNA (Noyes et al., 1979). In these cases, the priming site on the mRNA was downstream of a region coding for the established protein sequence. This allows unambiguous identification of the desired reverse transcript.

It is possible to use the total product of reverse transcription as a probe without further purification. Examples where this was successful are for cDNAs of human leukocyte interferon mRNA (Goeddel, Yelverton, et al., 1980) and rat relaxin mRNA (Hudson et al., 1981). With relaxin mRNA, a specific 150-nucleotide fragment produced with the oligodeoxyribonucleotide primer was purified and used as a secondary probe.

In some cases the oligodeoxyribonucleotide primer will be complementary to the sequence in the mRNA encoding the *N*-terminus of the established sequence. In this case the reverse transcript sequence cannot be used to identify the desired product. In such an instance, with a primer for human fibroblast interferon, a specific reverse transcript was observed which was only produced by mRNA isolated from induced fibroblasts and not with mRNA from noninduced fibroblasts (Houghton et al., 1980). This was used to identify a cDNA clone from a pool produced from the mRNA of interferon-producing fibroblasts. The sequence of the cDNA confirmed that it encoded the interferon.

Use of an oligodeoxyribonucleotide primer for production of a specific cDNA probe for a bacterial mRNA is nicely demonstrated by the isolation of the bacteriorhodopsin gene from *Halobacterium halobium* (Dunn et al., 1981).

Strategies have been developed to deal with the case where a specific reverse transcript cannot be identified above the background. The first of these was used in another isolation of a human fibroblast interferon cDNA (Goeddel et al., 1980a). A family of twenty-four 12-mer primers (in six groups of four) was used; the primers correspond to the N-terminal peptide Met—Ser—Tyr—Asn. The total reverse transcript from each of the set of primers was isolated and duplexed with mRNA from fibroblasts that were not producing interferon. The desired interferon cDNA probe, which is single-stranded, was isolated by hydroxyapatite chromatography. In the case of the crude cDNA probe for human leukocyte interferon, which was described earlier, a second probe using a primer for another region of the coding sequence of the mRNA was constructed and used to identify cDNA clones that responded to both of the impure probes (Goeddel, Yelverton, et al., 1980). A third strategy uses reverse transcription from a specific oligodeoxyribonucleotide probe carried out with three 2'-deoxyribonucleoside-5'-triphosphates together with one 2',3'-dideoxyribonucleoside-5'-triphosphate which terminates DNA synthesis whenever it is incorporated (Sood, Pereira, & Weissman, 1981). In this experiment the primer was an 11-mer with mRNA from cells producing human histocompatibility antigen HLA-B. The experiment was carried out four times using different combinations of triphosphates. The experiments with ddTTP gave a 30-nucleotide product, *inter alia*, and that with ddGTP gave a 16-nucleotide product. The 30-nucleotide product proved to be complementary to the sequence of HLA-B mRNA. The important point about this strategy is that it limits the number of significant products, and these are easily characterized by sequence determination. The same effect can be achieved by carrying out the reverse transcription in the presence of only three of the 2'-deoxyribonucleoside-5'-triphosphates. The strategy requires that the coding sequence of interest should not contain one of the four nucleotides for at least two codons adjacent to the priming site, to allow a unique product to emerge from the background of products resulting from more limited extension at other priming sites. The same strategy has been used to isolate a cDNA clone for human HLA-Dr antigen α-chain (Stetler et al., 1982).

Some general comments about the use of synthetic oligodeoxyribonucleotide primers are appropriate at this time. First, it should be noted that the reverse transcription is carried out under rather different conditions from those used for Southern blot experiments with genomic DNA. A typical reverse transcription mix contains, in 10 μl of Mg^{2+}-containing buffer, about 20 fmol of the target mRNA (assuming it is about 0.1% of the total mRNA and contains about 1,500 nucleotides) together with about 1 pmol of oligodeoxyribonucleotide primer. Under these conditions, very short oligodeoxyribonucleotides can prime DNA synthesis very well even at $37°$, as in the case of TTGGGT mentioned above (Rabbitts et al., 1977). Thus it is possible that a longer oligodeoxyribonucleotide which is a completely specific probe in a duplex experiment is a less specific primer by virtue of mismatching near its $5'$ end. This reduced specificity (although the nature of the mismatch is not known) has been observed on studies with gastrin mRNA (Mevarech, Noyes, & Agarwal, 1979; Noyes et al., 1979). It seems likely that a practice of using an exceedingly large excess of primer should not be followed blindly; rather, a systematic definition of the optimum amount of primer should be carried out (Houghton et al., 1980). It is of interest that, when the sequence of the target region of the mRNA is known for primers in the size range most studied (11 and 12 nucleotides), a perfectly matched oligodeoxyribonucleotide is the only effective primer (Goeddel, Shepard, et al., 1980; Dunn et al., 1981). This was even the case with a pair of 15-nucleotide primers; the one with a single mismatch did not work (Houghton et al., 1980).

In a number of instances, preannealing of the primer and mRNA in high ionic strength buffer has been carried out (Houghton et al., 1980; Sood, Pereira, & Weissman, 1981). The rapid association and dissociation of oligodeoxyribonucleotide duplexes indicates that this step can safely be omitted (Sasavage et al., 1980, 1982). One pretreatment of the mRNA template that clearly is useful is the dissociation of secondary structure which blocks interaction with the primer. This can be achieved by heat denaturation in the presence of EDTA or by limited alkaline hydrolysis (Agarwal, Brunstedt, & Noyes, 1981).

In the experiments discussed in this section, there are examples where only a limited number of primer oligodeoxyribonucleotides were prepared and others where a complete family was made. One interesting comparison is provided by the two isolations of a cDNA for human fibroblast interferon. The sequence of the *N*-terminus is Met—Ser—Tyr—Asn—Leu. The first four amino acids require a family of 24 dodecanucleotides; these have been synthesized (Goeddel, Shepard, et al., 1980a), and one of the six pools of four oligodeoxyribonucleotides generated a specific probe. In the second set of experiments only two of the one hundred and twenty-four possible 15-mers corresponding to the *N*-terminal pentapeptide were synthesized, and one of them was the correct primer (Houghton et al., 1980). The two oligodeoxyribonucleotides were chosen by comparison of the peptide sequence of the human interferon and the corresponding sequence in mouse interferon A and B together with taking into account codon preferences in human genes. When all the successful experiments using the oligodeoxyribonucleotide primers are examined, it is clear that only the perfectly matched oligodeoxyribonucleotides generated the desired reverse transcripts. Thus, even mismatches of the rG,dT or rU,dG type which would be relatively stable (Gillam, Waterman, & Smith, 1975) are poorly tolerated for oligodeoxyribonucleotides of lengths from 11 to 15 nucleotides under the standard conditions for using reverse transcriptase. This means that it is not possible to use these base pairings to reduce the numbers of oligodeoxyribonucleotides that have to be synthesized as has been supposed (Smith, 1980b; Agarwal, Brunstedt, & Noyes, 1981). Clearly the simplifying principles used by Houghton et al. (1980) have been more fruitful.

USE OF SYNTHETIC OLIGODEOXYRIBONUCLEOTIDES TO PREPARE DOUBLE-STRAND cDNAs

The use of pdT_n to prime the synthesis of cDNAs on poly-rA containing mRNA has been reviewed recently (Efstratiadis & Villa-Komaroff, 1979) and will not be discussed further here, nor will the production of double-stranded cDNA using as a priming site a hairpin duplex at the 3' end of the initial single-stranded cDNA product. The objective of this section is to describe the use of synthetic oligodeoxyribonucleotides in alternative and rather more specific methods for synthesizing double-stranded cDNAs.

The standard primer pT_n ($n = 12$–18) is nonspecific both in the position with which it interacts with the poly-rA tract on the mRNA and with regard to the mRNA with which it interacts. The series of oligodeoxyribonucleotides pdT_8dNdN' ($N =$ A, C, or G; $N' =$ A, C, G, or T) has been synthesized to provide specific phasing on a template mRNA and also to select for a subset (hopefully of one mRNA) in a partially purified mRNA preparation (Gillam & Smith, 1980). While this set of oligodeoxyribonucleotides has been used most extensively in determination of the sequence of the 3' end of mRNAs (see below), there has been one report of the use of $pdT_{10}dGdC$ in the preparation of a cDNA, that of rat preproinsulin (Villa-Komaroff et al., 1978). It is surprising that this approach has not been used more extensively to generate clones of cDNAs; as well as having the potential to reduce by an order of magnitude the number of clones that have to be screened, it could be used very fruitfully in concert with hybridization-arrested translation (Paterson, Roberts, & Kuff, 1977).

The "hairpin priming" method of generating double-stranded cDNAs for cloning suffers from two disadvantages. First, not all full-length single-stranded cDNAs form a hairpin at their 3' ends. Second, S1 nuclease treatment of a hairpin which is required to generate a ligatable molecule is accompanied by exonucleolytic degradation of the cDNA (Volckaert et al., 1981). Two useful strategies have been developed to circumvent these problems. In the first of these, a tract of dC residues is added to the 3' end of the single-stranded cDNA and oligo-dG is used to prime double-stranded DNA synthesis (Land et al., 1981) of a specific oligodeoxyribonucleotide whose sequence is identical to part of the 5' end of the mRNA. This method is used in concert with a synthetic oligodeoxyribonucleotide probe that is complementary to the 5' end of the mRNA as described earlier. The synthesis of double-stranded cDNA of human leukocyte interferon by this method has been described by Houghton et al. (1980). In this case, the oligodeoxyribonucleotide was chosen to produce a cDNA starting eight nucleotides upstream of the first codon of the mature interferon. A variant of this approach was also used with human fibroblast interferon cDNA; in this case the product double-stranded cDNA started at, or one nucleotide before, the first codon of the mature protein (Goeddel, Yelverton, et al., 1980). This provided a DNA suitable for cloning in an *E. coli* expression vector. A special example of use of synthetic oligodeoxyribonucleotides for producing cDNAs is that of influenza virus RNAs. The virus has a segmented genome of eight fragments (genes) each with identical 5' ends of 13 nucleotides and 3' ends of 12 nucleotides. Thus a synthetic dodecadeoxyribonucleotide complementary to the 3' ends and a tridecadeoxyribonucleotide corresponding to the 5' ends allows construction of double-stranded cDNA for all the genes (Lai et al., 1980; Winter et al., 1981).

The second strategy, which is very efficient in producing clones, involves attaching the two oligonucleotide primers to the vector DNA prior to cDNA synthesis and its conversion to double-stranded DNA (Okayama & Berg, 1982).

OLIGODEOXYRIBONUCLEOTIDES AS PRIMERS FOR DNA SEQUENCE DETERMINATION

An early application of synthetic oligodeoxyribonucleotides was as primers for limited DNA synthesis in sequence determinations

(Besmer et al., 1972; Sanger et al., 1973) and also in development of the earliest of the rapid methods of DNA sequence determination (Sanger & Coulson, 1975). The advent of restriction endonucleases made specific DNA fragments readily available, and consequently such DNA fragments have been the main avenue for sequence determination in recent years (Sanger, Nicklen, & Coulson, 1977; Smith, 1980a; Maxam & Gilbert, 1977, 1980). However, there are a number of instances where synthetic oligodeoxyribonucleotide primers provide considerable advantages for sequence by the rapid ladder methods, and with the growing accessibility of synthetic oligodeoxyribonucleotides it is likely that these methods will be used much more extensively.

The most obvious application is when the same primer is used repeatedly. A major application of this is in sequence determination by the enzymatic terminator method (Sanger, Nicklen, & Coulson, 1977) after cloning of the target template DNA in a M13 phage-derived single-stranded DNA vector (Messing, Crea, & Seeburg, 1981).

The low Tm values of oligodeoxyribonucleotide duplexes and their rapid rate of formation make it possible to use an oligodeoxyribonucleotide as a primer, using denatured double-stranded DNA as template, to sequence an adjacent region of the DNA (Smith et al., 1979). A particular application of this is in the determination of the sequence of mutants which map at a defined locus. For example, a large number of mutants at the SUP4-0 locus of *S. cerevisae* have been prepared (Kurjan et al., 1980). The mutant loci were cloned into the double-stranded *E. coli* vector pBR322. An oligodeoxyribonucleotide pAAAAACAAA which is complementary to the transcription termination sequence of the SUP4-*o* gene (encoding a tyrosine-inserting ochre suppressor tRNA) acts as an efficient primer for sequence determination by the terminator method when the denatured recombinant clone is used as template (Kurjan et al., 1980; Koski et al., 1980). Because of the simplicity and speed of this method of sequence determination, the characterization of a large number of mutants is made quite trivial. This method has recently been applied to a mutant in the iso-1-cytochrome *c* gene (Ernst, Stewart, & Sherman, 1981).

The same approach to sequencing of a particular region of many clones of DNA in a single-stranded DNA is also useful. A particular application lies in site-directed mutagenesis where the sequencing of several clones is sometimes required to identify a desired mutant genotype (Gillam, Astell, & Smith, 1980). A set of oligodeoxyribonucleotide primers for each side of the major restriction endonuclease cleavage sites in pBR322 has been synthesized (Wallace et al., 1981b).

A second application of synthetic oligodeoxyribonucleotides in DNA sequencing is in a systematic approach to sequence determination. This is sequence determination by a strategy analogous to "walking" along a genome to construct an extended DNA fragment map. It is usually fairly straightforward to establish the sequence of the ends of a DNA fragment of interest, often by the chemical ladder method (Maxam & Gilbert, 1980). The two most-favored approaches for obtaining the rest of the DNA sequence are sequencing of subfragments by the chemical method or by cloning in an M13 vector and sequence determination by the enzymatic method. In each case it is necessary to carry out the sequence determination with sets of fragments produced by several different restriction endonucleases in order to obtain all the sequence and to determine the orientations and order of all the fragments. In essence, these shotgun approaches involve obtaining a large fraction of the sequence several times over, and often there still are missing segments. The use of synthetic oligodeoxyribonucleotides as primers allows the sequence to be obtained stepwise, systematically, and completely. In addition, the substrate is the primary clone in pBR322 and there is no possibility of loss of subfragments either during fractionation or cloning in an M13 vector. This method was applied extensively in the determination of the sequence of the gene for iso-1-cytochrome *c* (Smith et al., 1979). The greater availability of synthetic oligodeoxyribonucleotides makes it an increasingly attractive strategy.

In general, oligodeoxyribonucleotide of 7 to 15 nucleotides work well in this type of experiment, the primary determinant of oligodeoxyribonucleotide length being that required to recognize a unique site on the target DNA. The primer is normally used in relatively small excess, 2- to 10-fold. On occasion a larger excess, up to 50-fold, is required when the target DNA contains a moderately stable hairpin structure at the priming site. It seems likely that very stable hairpins with 10 or more base pairs will be inaccessible to short oligodeoxyribonucleotide primers.

USE OF SYNTHETIC OLIGODEOXYRIBONUCLEOTIDES AS PRIMERS FOR RNA SEQUENCE DETERMINATION

The basic principle of this method is to use a specific oligodeoxyribonucleotide as a primer for DNA synthesis on an RNA template. Because of the specificity of the duplex structure, the method can be applied to impure RNA preparations. The first reported application involved the use of pT_{10}—G—C as a primer at the 3' end of rabbit β-globin mRNA (Cheng et al., 1976) using *E. coli* DNA polymerase. However, the most useful system involves the use of reverse transcriptase to produce the cDNA. The sequence strategy can use a $(5'-{}^{32}P)$-labeled oligodeoxyribonucleotide to make a cDNA whose sequence is determined by the method of Maxam and Gilbert; this was first applied to determination of the sequence at the 5' end of the mRNA of iso-1-cytochrome *c* of yeast which represented 0.1– 0.5% of the RNA in the total yeast mRNA used as substrate (Szostak et al., 1977). This approach is particularly useful if the oligodeoxyribonucleotide priming results in the synthesis of several fragments; the mixture of transcripts can be separated electrophoretically and then sequenced individually (Szostak et al., 1977; Chan et al., 1979; Noyes et al., 1979; Chang et al., 1981).

Although synthetic oligodeoxyribonucleotides have been used as primers for DNA synthesis on RNA templates and the resultant DNA sequenced by the wandering-spot method (Cheng et al., 1976; Schwartz, Zamecnik, & Weith, 1977) or the plus-minus method (Baralle, 1977; Bernard et al., 1977; Brownlee & Cartwright, 1977; Hamlyn et al., 1977; Proudfoot et al., 1977), the best direct method is the Sanger terminator method modified for use with reverse transcriptase (Zimmern & Kaesberg, 1978; McGeoch & Turnbull, 1978). Using relatively pure mRNAs for specific proteins, this strategy has been used to obtain the sequence of a mouse immunoglobulin in light chain mRNA (Hamlyn, Gait, & Milstein, 1981) and an α-interferon (Houghton et al., 1980). The shortest effective primer reported to date to give a single reverse transcript is the hexanucleotide TTGGGT (Gillam, Jahnke, & Smith, 1978) used for a segment of the immunoglobulin light chain mRNA (Hamlyn, Gait, & Milstein, 1981) and also to obtain a specific light chain cDNA probe (Rabbitts et al., 1977). More recently, the 3'-hexanucleotide segment of a heptanucleotide has also been used as a primer for cDNA production with rat prolactin mRNA; several cDNAs were produced, but one corresponded to the desired priming site complementary to the codons for two adjacent Met residues (Taylor et al., 1981).

A novel and very effective approach has been used in the determination of the sequence of tobacco mosaic virus RNA (Goelet et al., 1982)—a mixture of very short oligonucleotides which primed reverse transcriptase at random sites on the viral RNA templates. Double-stranded DNAs derived from the initial cDNA were then sequenced by the Sanger method after cloning in an M13 vector.

The availability of a set of the 12 oligodeoxyribonucleotides of the series $pTTTTTTTTNN'$ (N = C, A, or G; N' = T, C, A, or G) makes it possible to define the sequence of the 3′ end of a purified or enriched polyadenylated mRNA (Gillam & Smith, 1980). The most convenient strategy is to use each oligodeoxyribonucleotide in turn as a primer in the terminator sequencing method for just one dedeoxyribonucleotide (preferably T, because it allows the ready identification of the sequence AAUAAA near the 3′ end of the template mRNA; Sasavage et al., 1980). The method has to be used to define the 3′ ends of the mRNAs of bovine growth hormone (Sasavage et al., 1980) and yeast iso-1-cytochrome *c* (Boss et al., 1981). A particularly striking result obtained with this method is the demonstration of microheterogeneity at the 3′ end of bovine prolactin mRNA (Sasavage et al., 1982).

The 3′-end sequence of an RNA that does not normally have poly-A attached can be determined after *in vitro* addition of poly-A (Hagenbüchle et al., 1978).

CONCLUSION

This chapter has described the general principles involved in the use of synthetic oligodeoxyribonucleotides as probes for nucleic acids and as primers in sequence determinations. The unique potential of these applications together with the growing availability of synthetic oligodeoxyribonucleotides is certain to ensure the continued use of the strategies described in this chapter as major elements of modern molecular genetics.

REFERENCES

Agarwal, K. L., Brunstedt, J., and Noyes, B. E. (1981). A general method for detection and characterization of an mRNA using an oligonucleotide probe. *J. Biol. Chem.* 256:1023–1028.

Alvarado-Urbina, G., Sathe, G. M., Liu, W.-C., Gillen, M. F., Duck, P. D., Bender, R., and Ogilvie, K. K. (1981). Automated synthesis of gene fragments. *Science* 214:270–274.

Astell, C. R., and Smith, M. (1972). Synthesis and properties of oligonucleotide-cellulose columns. *Biochem.* 11:4114–4120.

Astell, C. R., and Smith, M. (1971). Thermal elution of complementary sequences of nucleic acids from cellulose columns with covalently attached oligonucleotides of known length and sequence. *J. Biol. Chem.* 246:1944-1946.

Astell, C. R., Ahlstrom-Jonasson, L., Smith, M., Tatchell, K., Nasmyth, K. A., and Hall, B. D. (1981). The sequence of the DNAs coding for the mating-type loci of *Saccharomyces cerevisiae. Cell* 27:15-23.

Astell, C. R., Doel, M. T., Jahnke, P. A., and Smith, M. (1973). Further studies on the properties of oligonucleotide cellulose columns. *Biochem.* 12:5068-5074.

Aviv, H., and Leder, P. (1972). Purification of biologically active globin messenger RNA by chromatography on oligothymidylic acid-cellulose. *Proc. Natl. Acad. Sci. USA* 69:1408-1412.

Bantle, J. A., Maxwell, I. H., and Hahn, W. E. (1976). Specificity of oligo(dT)-cellulose chromatography in the isolation of polyadenylated RNA. *Anal. Biochem.* 72:413-427.

Baralle, F. E. (1977). Complete nucleotide sequence of the 5' noncoding region of rabbit β-globin mRNA. *Cell* 10:549-558.

Bernard, O. D., Jackson, J., Cory, S., and Adams, J. M. (1977). Noncoding nucleotide sequence in the 3'-terminal region of a mouse immunoglobin kappa chain messenger RNA determined by analysis of complementary DNA. *Biochem.* 16:4117-4125.

Besmer, P., Miller, R. C., Jr., Caruthers, M. H., Kumar, A., Minamoto, K., Van de Sande, J. H., Sidarova, N., and Khorana, H. G. (1972). Studies on polynucleotides. CXVII. Hybridization of polydeoxynucleotides with tyrosine transfer RNA sequences to the r-strand of $80psu_{III}^{+}$ DNA. *J. Mol. Biol.* 72:503-522.

Borer, P. N., Dengler, B., Tinoco, I., Jr., and Uhlenbeck, O. C. (1974). Stability of ribonucleic acid double-stranded heleces. *J. Mol. Biol.* 86:843-853.

Boseley, P. G., Moss, T., and Birnstiel, M. L. (1980). 5'-labeling and poly(dA) tailing. In *Methods in Enzymology*, Vol. 65, eds. L. Grossman and K. Moldave (New York: Academic Press), pp. 478-494.

Boss, J. M., Gillam, S., Zitomer, R. S., and Smith, M. (1981). Sequence of the yeast iso-1-cytochrome *c* mRNA. *J. Biol. Chem.* 256:12958-12961.

Brownlee, G. G., and Cartwright, E. M. (1977). Rapid gel sequencing of RNA by primed synthesis with reverse transcriptase. *J. Mol. Biol.* 114:93-117.

Chamberlin, M. J. (1965). Cooperative properties of DNA, RNA and hybrid homopolymer pairs. *Fed. Proc.* 24:1446-1457.

Chan, S. J., Noyes, B. E., Agarwal, K. L., and Steiner, D. F. (1979). Construction and selection of recombinant plasmids containing full-length complementary DNAs corresponding to rat insulins I and II. *Proc. Natl. Acad. Sci. USA* 76:5036-5040.

Chang, S. H., Majumdar, A., Dunn, R., Makabe, O., RajBhandary, U. L., Khorana, H. G., Ohtsuka, E., Tanaka, T., Taniyama, Y. O., and Ikehara, M. (1981). Bacteriorhodopsin: Partial sequence of mRNA provides amino acid sequence in the precursor region. *Proc. Natl. Acad. Sci. USA* 78:3398-3402.

Cheng, C. C., Brownlee, G. G., Carey, N. H., Doel, M. T., Gillam, S., and Smith, M. (1976). The 3' terminal sequence of chicken ovalbumin messenger RNA and its comparison with other messenger RNA molecules. *J. Mol. Biol.* 107:527-547.

Comb, M., Seeburg, P. H., Adelman, J., Eiden, L., and Herbert, E. (1982). Primary structure of the human Met- and Leu- enkephalin precursor and its mRNA. *Nature* 295:663-666.

Doel, M. T., and Smith, M., 1973. The chemical synthesis of deoxyribooligonucleotides complementary to a portion of the lysozyme gene of phage T4 and their hybridization to phage specific RNA and phage DNA. *FEBS Lett.* 34:99-102.

Dunn, R., McCoy, J., Simsek, M., Majumdar, A., Chang, S. H., RajBhandary, U. L., and Khorana, H. G. (1981). The bacteriorhodopsin gene. *Proc. Natl. Acad. Sci. USA* 78:6744-6748.

Edge, M. D., Greene, A. R., Heathcliffe, G. R., Meacock, P. A., Schnuch, W., Scanlon, D. B., Atkinson, T. C., Newton, C. R., and Markham, A. F. (1981). Total synthesis of a human leukocyte interferon gene. *Nature* 292:756-762.

Efstratiadis, A., and Villa-Komaroff, N. (1979). Cloning of double-stranded cDNA. In *Genetic Engineering, Principles and Methods*, Vol. 1, eds. J. K. Setlow and A. Hollaender (New York: Academic Press), pp. 15-36.

Ernst, J. F., Stewart, J. W., and Sherman, F. (1981). The cycl-11 mutation in yeast reverts by recombination with a non-allelic gene: Composite genes determining the iso-cytochromes *c*. *Proc. Natl. Acad. Sci. USA* 78:6334-6338.

Faye, G., Leung, D. W., Tatchell, K., Hall, B. D., and Smith, M. (1981). Deletion mapping of sequences essential for *in vivo* transcription of the iso-1-cytochrome *c* gene of yeast. *Proc. Natl. Acad. Sci. USA* 78:2258–2262.

Gilham, P. T. (1964). The synthesis of polynucleotide celluloses and their use in the fractionation of polynucleotides. *J. Amer. Chem. Soc.* 86:4982–4985.

Gilham, P. T. (1962). Complex formation in oligonucleotides and its application to the separation of polynucleotides. *J. Amer. Chem. Soc.* 84:1311–1312.

Gilham, P. T., and Naylor, R. (1964). Studies on some interactions and reactions of oligonucleotides in aqueous solution. *Biochem.* 5:2722–2728.

Gilham, P. T., and Robinson, W. E. (1964). The use of polynucleotide-celluloses in sequence studies of nucleic acids. *J. Amer. Chem. Soc.* 86:4985–4989.

Gillam, S., and Smith, M. (1980). Use of *E. coli* polynucleotide phosphorylase for the synthesis of oligodeoxyribonucleotides of defined sequence. In *Methods in Enzymology*, Vol. 65, eds. L. Gossman and K. Moldave (New York: Academic Press), pp. 687–701.

Gillam, S., Astell, C. R., and Smith, M. (1980). Site-specific mutagenesis using oligodeoxyribonucleotides: Isolation of a phenotypically silent X174 mutant, with a specific nucleotide deletion at very high efficiency. *Gene* 12:129–137.

Gillam, S., Jahnke, P., and Smith, M. (1978). Enzymatic synthesis of oligodeoxyribonucleotides of defined sequence. *J. Biol. Chem.* 253:2532–2539.

Gillam, S., Waterman, K., and Smith, M. (1975). The base-pairing specificity o-cellulose-pdT$_9$. *Nucl. Acids Res.* 2:625–634.

Gillam, S., Rottman, F., Jahnke, P., and Smith, M. (1977). Enzymatic synthesis of oligonucleotides of defined sequence: Synthesis of a segment of yeast iso-1-cytochrome *c* gene. *Proc. Natl. Acad. Sci. USA* 74:96–100.

Goeddel, D. V., Shepard, H. M., Yelverton, E., Leung, D., and Crea, R. (1980). Synthesis of human fibroblast interferon by *E. coli*. *Nucl. Acids Res.* 8:4057–4074.

Goeddel, D. V., Yelverton, E., Ullrich, A., Heyneker, H. L., Miozzari, G., Holmes, W., Seeburg, P. H., Dull, T., May, L., Stebbing, N., Crea, R., Maeda, S., McCandliss, R., Sloma, A., Tabor, J. M., Gross, M., Amilletti, P. C., and Pestka, S. (1980). Human leukocyte interferon produced by *E. coli* is biologically active. *Nature* 287:411–416.

Grantham, R., Gautier, C., and Gouy, M., (1980). Codon frequencies in 119 individual genes confirm consistent choices of degenerate bases according to genome type. *Nucl. Acids Res.* 8:1893–1912.

Goelet, P., Lomonossoff, G. P., Bulter, P. J. G., Akam, M. E., Gait, M. J., and Karn, J. (1982). Nucleotide sequence of tobacco mosaic virus RNA. *Proc. Natl. Acad. Sci. USA* 79:5818–5822.

Gubler, U., Kilpatrick, D. L., Seeburg, P. H., Gage, L. P., and Udenfriend, S. (1981). Detection and partial characterization of proenkephalin mRNA. *Proc. Natl. Acad. Sci. USA* 78:5484–5487.

Gubler, U., Seeburg, P., Hoffman, B. J., Gage, L. P., and Udenfriend, S. (1982). Molecular cloning establishes proenkephalin as precursor of eukephalin-containing peptides. *Nature* 295:206–208.

Haasnoot, C. A. G., den Hartog, J. H. J., de Rooij, J. F. M., van Boom, J. H., and Altona, C. (1979). Local destabilisation of a DNA double helix by a T-T wobble pair. *Nature* 281:235–236.

Hagenbüchle, O., Santer, M., Steitz, J. A., and Mans, R. J. (1978). Conservation of the primary structure at the 3' end of 18S rRNA from eukaryote cells. *Cell* 13:551–563.

Hamlyn, P. H., Gait, M. J., and Milstein, C. (1981). Complete sequence of an immunoglobin mRNA using specific priming and the dideoxynucleotide method of RNA sequencing. *Nucl. Acids Res.* 9:4485–4494.

Hamlyn, P. H., Gillam, S., Smith, M., and Milstein, C. (1977). Sequence analysis of the 3' non-coding region of mouse immunoglobin light chain messenger RNA. *Nucl. Acids Res.* 4:1123–1134.

Hillen, W., Goodman, T. C., and Wells, R. D. (1981). Salt dependence and thermodynamic interpretation of the thermal denaturation of small DNA restriction fragments. *Nucl. Acids Res.* 9:415–436.

Houghton, M., Stewart, A. G., Doel, S. M., Emtage, J. S., Eaton, M. A. W., Smith, J. C., Patel, T. P., Lewis, H. M., Porter, A. G., Birch, J. R., Cartwright, T., and Carey, W. H. (1980). The amino acid sequence of human fibroblast interferon as deduced from reverse transcripts obtained using synthetic oligonucleotide primers. *Nucl. Acids Res.* 8:1913–1931.

Hudson, P., Haley, J., Cronk, M., Shine, J., and Niall, H. (1981). Molecular cloning and characterization of cDNA sequences coding for rat relaxin. *Nature* 291:127–131.

Inman, R. B., and Baldwin, R. L. (1982). Helix-random coil transitions in synthetic DNAs of alternating sequence. *J. Mol. Biol.* 5:172–184.

Itakura, K., and Riggs, A. D. (1980). Chemical DNA synthesis and recombinant DNA studies. *Science* 209:1401–1405.

Khorana, H. G. (1979). Total synthesis of a gene. *Science* 203:614–625.

Koski, R. A., Clarkson, S. G., Kurjan, J., Hall, B. D., and Smith, M. (1980). Mutations of the yeast SUP4 tRNA^{Tyr} locus: Transcription of the mutant genes *in vitro*. *Cell* 22:415–425.

Koster, H. (1980). Nucleic acids synthesis: Applications to molecular biology and genetic engineering. *Nucl. Acids Symp. Ser. No. 7* 1–395.

Kurjan, J., Hall, B. D., Gillam, S., and Smith, M. (1980). Mutations at the yeast SUP4 tRNA^{Tyr} locus: DNA sequence changes in mutants lacking suppressor activity. *Cell* 20:701–709.

Lai, C. J., Markoff, L. J., Zimmerman, S., Cohen, B., Berndt, J. A., and Chanock, R. M. (1980). Cloning DNA sequences from influenza viral RNA segments. *Proc. Natl. Acad. Sci. USA* 77:210–214.

Land, H., Grez, M., Hausser, H., Lindenmaier, W., and Shutz, G. (1981). 5'-Terminal sequences of eukaryote mRNA can be cloned with high efficiency. *Nucl. Acids Res.* 9:2251–2266.

Lipsett, M. N. (1964). Complex formation between polycytidylic acid and guanine oligonucleotides. *J. Biol. Chem.* 239:1256–1260.

Lipsett, M. N., Heppel, L. A., and Bradley, D. F. (1961). Complex formation between oligonucleotides and polymers. *J. Biol. Chem.* 236:857–863.

Lomant, A. J., and Fresco, J. R. (1975). Structure and energetic consequences of non-complementary base oppositions in nucleic acid helices. In *Progress in Nucleic Acid Research and Molecular Biology*, Vol. 15, ed. W. E. Cohn (New York: Academic Press), pp. 185–218.

McConaughy, B. L., and McCarthy, B. J. (1967). The interaction of oligodeoxyribonucleotides with denatured DNA. *Biochim. Biophys. Acta* 149:180–189.

McGeoch, D. J., and Turnbull, N. T. (1978). Analysis of the 3'-terminal nucleotide sequence of vesicular stomatitis virus N protein mRNA. *Nucl. Acids Res.* 5:4007–4024.

Magee, W. S., Jr., Gibbs, J. H., and Zimm, B. H. (1963). Theory of helix-coil transitions involving complementary poly and oligo-nucleotides. I. The complete binding case. *Biopolymers* 1:133–143.

Marmur, J., Rownd, R., and Schildkraut, C. L. (1963). Renaturation and denaturation of deoxyribonucleic acid. In *Progress in Nucleic Acid Research*, Vol. 1, eds. J. N. Davidson and W. E. Cohn (New York: Academic Press), pp. 231–300.

Maxam, A. M., and Gilbert, W. (1980). Sequencing end-labeled DNA with base-specific chemical cleavages. In *Methods in Enzymology*, Vol. 65, eds. L. Grossman and K. Moldave (New York: Academic Press), pp. 499–560.

Maxam, A. M., and Gilbert, W. (1977). A new method for sequencing DNA. *Proc. Natl. Acad. Sci. U.S.A.* 74:560–564.

Messing, J., Crea, R., and Seeburg, P. H. (1981). A system for shotgun DNA sequencing. *Nucl. Acids Res.* 9:309–322.

Mevarech, M., Noyes, B. E., and Agarwal, K. L. (1979). Detection of gastrin-specific mRNA using oligodeoxyribonucleotide probes of defined sequence. *J. Biol. Chem.* 254:7472–7475.

Michelson, A. M., and Monny, C. (1967). Polynucleotides. X. Oligonucleotides and their association with polynucleotides. *Biochim. Biophys. Acta* 149:107–126.

Montgomery, D. L., Hall, B. D., Gillam, S., and Smith, M. (1978). Identification and isolation of the yeast cytochrome *c* gene. *Cell* 14:673–680.

Moss, T., Boseley, P. G., and Birnstiel, M. L. (1980). More ribosomal spacer sequences from *Xenopus laevis*. *Nucl. Acids Res.* 8:467–485.

Niyogi, S. K. (1969). The influence of chain length and base composition on the specific association of oligoribonucleotides with denatured deoxyribonucleic acid. *J. Biol. Chem.* 244:1576–1581.

Niyogi, S. K., and Thomas, C. A., Jr. (1968). The stability of oligoadenylate-polyuridylate complexes as measured by thermal chromatography. *J. Biol. Chem.* 243:1220–1223.

Noda, M., Furutani, Y., Takahashi, H., Toyosato, M., Hirose, T., Inayama, S., Nakanishi, S., and Numa, S. (1982a). Cloning and sequence analysis of cDNA for bovine adrenal preproenkaphalin. *Nature* 295:202–206.

Noda, M., Takahashi, H., Tanabe, T., Toyosato, M., Furutani, Y., Hirose, T., Asai, M., Inayama, S., Miyata, T., and Numa, S. (1982b). Primary structure of alpha-subunit precursor of *Torpedo california* acetylcholine receptor deduced from cDNA sequence. *Nature* 299:793–797.

Noyes, B. E., Mevarech, M., Stein, R., and Agarwal, K. L. (1979). Detection and partial sequence analysis of gastrin mRNA by using an oligodeoxyribonucleotide probe. *Proc. Natl. Acad. Sci. USA* 76:1770–1774.

Okayama, H., and Berg, P. (1982). High efficiency cloning of full-length cDNA. *Mol. Cell. Biology* 2:161–170.

Oliver, A. L., Wartell, R. M., and Ratliff, R. L. (1977). Helix coil transitions of $d(A)_n \cdot d(T)_n$, $d(A-T)_n \cdot d(A-T)_n$ and $d(A-A-T)_n \cdot d(A-T-T)_n$; evolution of parameters governing DNA stability. *Biopolymers* 16:1115–1137.

Paterson, B. M., Roberts, B. F., and Kuff, E. L. (1977). Structural gene identification and mapping of DNA mRNA hybrid-arrest cell-free translation. *Proc. Natl. Acad. Sci. USA* 74:4370–4374.

Pohl, F. M. (1974). The conformation and physical properties of polypeptide and polynucleotide chains in solution. In *MTP International Review of Science. Biochemistry, Series One, Vol. I, Chemistry of Macromolecules*, ed. H. Gutfreund (Baltimore: University Park Press), pp. 109–147.

Proudfoot, N. J., Gillam, S., Smith, M., and Langley, J. I. (1977). Nucleotide sequence of the 3′ terminal third of rabbit α-globin messenger RNA: comparison with human α-globin messenger RNA. *Cell* 11:807–818.

Rabbits, T. H., Forster, A., Smith, M., and Gillam, S. (1977). Immunoglobulin-like messenger RNA in a mouse T cell lymphoma. *Eur. J. Immunol.* 7:43–48.

Reyes, A. A., Johnson, M. J., Schold, M., Ito, H., Ike, Y., Morin, C., Itakura, K., and Wallace, R. B. (1980). Identification of a H-2K^b related molecule by molecular cloning. *Immunogenetics*, 14:383–392.

Sanger, F., and Coulson, A. R. (1975). A rapid method for determining sequences in DNA by primed synthesis with DNA polymerase. *J. Mol. Biol.* 94:441–448.

Sanger, F., Nicklen, S., and Coulson, A. R. (1977). DNA sequencing with chain-terminating inhibitors. *Proc. Natl. Acad. Sci. USA* 74:5463–5467.

Sanger, F., Donelson, J. E., Coulson, A. R., Kössel, H., and Fischer, D. (1973). Use of DNA polymerase I primed by a synthetic oligonucleotide to determine a nucleotide sequence in bacteriophage f1 DNA. *Proc. Natl. Acad. Sci. USA* 70:1209–1213.

Sasavage, N. L., Smith, M., Gillam, S., Astell, C., Nilson, J. H., and Rottman, F. (1980). Use of oligodeoxyribonucleotide primers to determine poly (adenylic acid) adjacent sequences in messenger ribonucleic acid. 3′-

Terminal noncoding sequencing of bovine growth hormone messenger ribonucleic acid. *Biochem.* 19:1737-1743.

Sasavage, N. L., Smith, M., Gillam, S., Woyehik, R. P., and Rottman, F. M. (1982). Variation in the polyadenylation site of bovine prolactin messenger RNA. *Proc. Natl. Acad. Sci. U.S.A.* 79:223-227.

Schwartz, D. E., Zamecnik, P. C., and Weith, H. L. (1977). Rous sarcoma virus is terminally redundant: The 3' sequence. *Proc. Natl. Acad. Sci. USA* 74:994-998.

Smith, A. J. H. (1980a). DNA sequence analysis by primed synthesis. In *Methods in Enzymology*, Vol. 65, eds. L. Grossman and K. Moldave (New York: Academic Press), pp. 560-580.

Smith, M. (1980b). New strategies for DNA sequence determination. In *Gene Structure and Expression*, eds. D. H. Dean, L. F. Johnson, P. C. Kimball, and P. S. Perlman (Columbus, Ohio: Ohio State University Press), pp. 81-106.

Smith, M. and Gillam, S. (1981). Constructed mutants using synthetic oligodeoxyribonucleotides as site-specific mutagens. In *Genetic Engineering Principles and Methods*, Vol. 3, eds. J. K. Setlow and A. Hollaender (New York: Plenum Press), pp. 1-32.

Smith, M., Leung, D. W., Gillam, S., Astell, C. R., Montgomery, D. L., and Hall, B. D. (1979). Sequence of the gene for iso-1-cytochrome *c* in *Saccharomyces cerevisiae*. *Cell* 16:753-761.

Sood, A. K., Pereira, D. and Weissman, S. M. (1981). Isolation and partial nucleotide sequence of a cDNA clone for human histocompatibility antigen HLA-B by use of an oligodeoxyribonucleotide primer. *Proc. Natl. Acad. Sci. USA* 78:616-626.

Steiner, R. F., and Beers, R. J., Jr. (1961). *Polynucleotides* (Amsterdam: Elsevier).

Stetler, D., Das, H., Nunberg, J. H., Saiki, R., Sheng-Dong, R., Mullis, K. B., Weissman, S. M., and Erlich, H. A. (1982). Isolation of a cDNA clone for the human HLA-DR antigen alpha-chain by using a synthetic oligonucleotide as a hybridization probe. *Proc. Natl. Acad. Sci. USA* 79: 5966-5970.

Stewart, J. W., and Sherman, F. (1974). Yeast frameshift mutations identified by sequence changes in iso-1-cytochrome *c*. In *Molecular and Environmental Aspects of Mutagenesis*, eds. L. Prakash, F. Sherman, M. W. Miller, C. W. Lawrence, and H. W. Taber (Springfield, Ill.: Charles C Thomas), pp. 102-107.

Stiles, J. A., Szostak, J. W., Young, A. T., Wu, R., Consaul, S., and Sherman, F. (1981). DNA sequence of a mutation in the leader region of the yeast iso-1-cytochrome *c* mRNA. *Cell* 25:277–284.

Suggs, S. V., Wallace, R. B., Hirose, T., Kawashima, E., and Itakura, K. (1981). Use of synthetic oligonucleotides as hybridization probes: isolation of cloned mRNA sequences for human β2-microglobulin. *Proc. Natl. Acad. Sci. USA* 78:6613–6617.

Suggs, S. V., Hirose, T., Miyake, T., Kawashima, E. H., Johnson, M. J., Itakura, K., and Wallace, R. B. (1981). Use of synthetic oligodeoxyribonucleotides for the isolation of specific cloned DNA sequences. In *Developmental Biology Using Purified Genes. ICN-UCLA Symposia on Molecular and Cellular Biology.* Vol. 23, eds. D. D. Brown and D. F. Fox (New York: Academic Press), pp. 683–693.

Szostak, J. W., Stiles, J. I., Bahl, C. P., and Wu, R. (1977). Specific binding of a synthetic oligodeoxyribonucleotide to yeast cytochrome *c* mRNA. *Nature New Biol.* 265:61–63.

Szostak, J. W., Stiles, J. I., Tye, B.-K., Chiu, P., Sherman, F., and Wu, R. (1979). Hybridization with synthetic oligonucleotides. In *Methods in Enzymology*, Vol. 68, ed. R. Wu (New York: Academic Press), pp. 419–428.

Tabak, H. F., van der Laan, J., Osinga, K. A., Schouten, J. P., van Boom, J. H., and Veeneman, G. H. (1981). Use of a synthetic oligonucleotide to probe the precision of RNA splicing in a yeast mitochondrial petite mutant. *Nucl. Acids Res.* 9:4475–4483.

Taylor, W. L., Collier, K. J., Weith, H. L., and Dixon, J. E. (1981). The use of a heptadeoxyribonucleotide as a specific primer for prolactin mRNA: A prediction of ambiguous RNA splicing. *Biochem. Biophys. Res. Commun.* 102:1071–1077.

Thomas, C. A., Jr. (1966). Recombination of DNA molecules. In *Progress in Nucleic Acid Research and Molecular Biology*, Vol. 5, eds. J. N. Davidson and W. E. Cohn (New York: Academic Press), pp. 315–337.

Topal, M. D., and Fresco, J. R. (1976). Complementary base pairing and the origin of substitution mutations. *Nature* 263:285–289.

Uhlenbeck, O. C., Martin, F. H., and Doty, P. 1971. Self-complementary oligoribonucleotides: Effects of helix defects and guanylic acid-cytidylic acid base pairs. *J. Mol. Biol.* 57:217–229.

Villa-Komaroff, L., Efstratiadis, A., Broome, S., Lomedico, P., Tizard, R., Naber, S. P., Chick, W. L., and Gilbert, W. (1978). A bacterial clone synthesizing proinsulin. *Proc. Natl. Acad. Sci. USA* 75:3727–3731.

Volckaert, G., Tavernier, J., Derynck, R., Devos, R., and Fiers, W. (1981). Molecular mechanisms of nucleotide-sequence rearrangements in cDNA clones of human fibroblast interferon mRNA. *Gene* 15:215-223.

Wada, A., Yabuki, S., and Husimi, Y. (1980). Fine structure in the thermal denaturation of DNA: High temperature-resolution spectrophotometric studies. In *CRC Critical Reviews of Biochemistry*, Vol. 9, ed. G. D. Fasman (Boca Raton: CRC Press), pp. 87-144.

Wallace, R. B., Johnson, M. J., Hirose, T., Miyake, T., Kawashima, E. H., and Itakura, K. (1981a). The use of synthetic oligonucleotides as hybridization probes. II. Hybridization of oligonucleotides of mixed sequence to rabbit β-globin DNA. *Nucl. Acids Res.* 9:3647-3656.

Wallace, R. B., Johnson, M. J., Suggs, S. V., Miyoshi, K., Bhatt, R., and Itakura, K. (1981b). A set of synthetic oligodeoxyribonucleotide primers for DNA sequencing in the plasmid vector pBR322. *Gene* 16:21-26.

Wallace, R. B., Shaffer, J., Murphy, R. F., Bonner, J., Hirose, T., and Itakura, K. (1979). Hybridization of synthetic oligodeoxyribonucleotides of φX174 DNA: The effect of single base-pair mismatch. *Nucl. Acids Res.* 6:3543-3557.

Wells, R. D., and Wartell, R. M. (1974). The influence of nucleotide sequence on DNA properties. In *MTP International Review of Science. Biochemistry, Series One*, Vol. 6, ed. K. Burton (Baltimore: University Park Press), pp. 41-64.

Wells, R. D., Goodman, T. C., Hillen, W., Horn, G. T., Klein, R. D., Larson, J. E., Müller, U. R., Neuendorf, S. K., Panayotatos, N., and Stirdivant, S. M. (1980). DNA structure and gene regulation. In *Progress in Nucleic Acid Research and Molecular Biology*, Vol. 24, ed. W. E. Cohn (New York: Academic Press), pp. 168-267.

Winter, G., Fields, S., Gait, M. J., and Brownlee, G. G. (1981). The use of synthetic oligodeoxyribonucleotide primers in cloning and sequencing segment 8 of influenza virus (A/PR/8/34). *Nucl. Acids Res.* 9:237-245.

Woods, D. E., Markham, A. F., Ricker, A. T., Goldberger, G., and Colten, H. R. (1982). Isolation of cDNA clones for the human complement protein factor B, a class III major histocompatibility complex gene product. *Proc. Natl. Acad. Sci. USA* 79:5661-5665.

Wu, R. (1972). Nucleotide sequence analysis of DNA. *Nature New Biol.* 236:198-200.

Wu, R., Bahl, C. P., and Narang, S. A. (1978). Synthetic oligodeoxyribonucleotides for analysis of DNA structure and function. In *Progress in Nucleic Acid Research and Molecular Biology*, Vol. 21, ed. W. E. Cohn (New York: Academic Press), pp. 101–141.

Zimmern, D., and Kaesberg, P. (1978). 3'-Terminal nucleotide sequence of encephalomyocarditis virus RNA determined by reverse transcriptase and chain terminating inhibitors. *Proc. Natl. Acad. Sci. USA* 75:4252–4261.

Zoller, M. J., and Smith, M. (1982). Oligonucleotide-directed mutagenesis using M13-derived vectors: An efficient and general procedure for the production of point mutations in any fragment of DNA. *Nucl. Acids Res.* 10:6487–6500.

3

Sequencing DNA by the Sanger Chain Termination Method

G. NIGEL GODSON

INTRODUCTION

The first polymeric biological molecule to be sequenced was the protein insulin (Sanger & Thompson, 1963), and the principles that Sanger devised to order the 51 amino acids of the two insulin poly-peptide chains became the basis of the methodology developed to sequence nuclei acids, first RNA in the mid-1960s and then DNA in the early 1970s. The basic principle of sequencing is to specifically cleave a macromolecule into fragments small enough to be totally analyzed by standard chemical methods and then to assemble the complete linear subunit sequence of the macromolecule by comparing the sequence of overlapping fragments. In the case of proteins, the polypeptide chain is cleaved into specific peptides by partial hydrolysis with proteolytic enzymes and the order of amino acids in each peptide is determined by partial and complete hydrolysis, both chemical and enzymatic. In the case of RNA, the nucleotide chain is specifically cleaved chemically or enzymatically and the order of nucleotides in each oligonucleotide fragment is determined chromato-graphically (Sanger, Brownlee, & Barrell, 1965; Brownlee & Sanger, 1969; Barrell, 1971). Such direct methods, however, do not work well for sequencing DNA because of the large size of DNA molecules and because their double-stranded nature gives rise to two comple-mentary sets of sequence data.

In spite of this the first DNA sequences were obtained by direct methods using single-stranded DNA and techniques analogous to those used to sequence RNA (Robertson et al., 1973; Ziff, Sedat, & Galibert, 1973), but the breakthrough that made extensive DNA sequencing possible was the introduction of indirect copying methods. In these methods, a DNA polymerase is used to copy a small region of a large DNA molecule into a radioactive complement, and the sequence of this small single-stranded piece of DNA is determined. The first successful method of primed synthesis copy sequencing was the so-called plus and minus method (Sanger & Coulson, 1975), which was a two-step procedure; the first step involved the generation of random chain extensions, and the second step the identification of the terminal nucleotide of each chain. The complete 5,386 nucleotide sequence of the ϕX174 DNA molecule was determined by this method (Sanger et al., 1978). The dideoxy chain termination method (Sanger, Nicklen, & Coulson, 1977) is a refinement of the plus and minus method in that the two-step procedure is replaced by a single-step procedure in which the chain extension and terminal nucleotide identification are performed in a single step. This method has now superseded all preceding methods and is the only method that will be described in this chapter.

PRINCIPLE OF PRIMED SYNTHESIS SEQUENCING AND THE CHAIN TERMINATION METHOD

The principle of primed synthesis DNA sequencing is illustrated in Fig. 3-1. A partially single-stranded primed template molecule is established, consisting of a primer strand (with a free 3' hydroxyl group) base-paired to a single-stranded template. (The terms primer and template are defined in Kornberg's early DNA polymerase paper; see Kornberg, 1969.) DNA polymerase is then used to copy the template strand in a chain extension reaction primed from the 3' hydroxyl, using all four deoxyribonucleotide triphosphates (dATP, dCTP, dGTP, and dTTP), one of them labeled α^{32}P. The newly synthesized ^{32}P-labeled DNA therefore represents information of only a small portion of the DNA and of only one of the DNA strands. (This strategy immediately gets around the difficulty of the size and double-stranded nature of DNA.) To produce a unique nucleotide sequence, the primer must be unique and prime at only one place on the template. In the early studies, the primers were synthetic oligonucleotides (Sanger et al., 1973), but in later studies (Sanger, Nicklen,

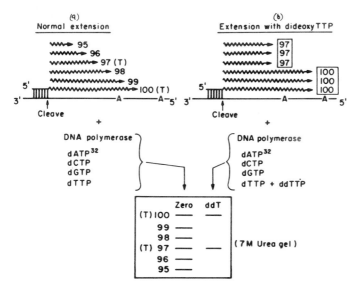

Figure 3-1. Principle of chain termination DNA sequencing. (*a*) Under normal conditions of primed synthesis, DNA polymerase extends the 3'-OH terminus of a primer in a continuous manner, inserting the complementary nucleotide opposite each base in the template DNA strand. Under conditions of low nucleotide triphosphate concentration (2μM), the extensions are incomplete, and a family of extensions are generated ending at every base in the sequence. This effect was utilized in the "plus and minus" method of primed synthesis sequencing (Sanger & Coulson, 1975) and gives rise to a ladder of radioactive bands as illustrated above. (*b*) By introducing into the reaction a specific nucleotide analog which lacks a 3'-OH terminus, some of the chain extensions terminate at each base in the template that is complementary to the analog. In the above illustration, ddT causes termination of some of the chains at the A in position 97 and 100 of the template. By using ddA, ddC, and ddG analog in parallel reactions, the position of all of the bases in the template can be determined from the length of the chain extension.

& Coulson, 1977; Sanger et al., 1977) restriction endonuclease DNA fragments were used. However, primers and templates can be either RNA or DNA in any combination.

In the chain termination method of DNA sequencing, nucleotide-specific chain terminators are used to generate a family of nucleotide-specific staggered chain extensions. The chain terminators used are either 2',3'-dideoxynucleotide triphosphates that lack a 3' hydroxyl group (see Fig. 3-2) or arabinonucleoside 5'-triphosphates (β-D-arabinofuranoside 5'-triphosphates) whose 3' hydroxyl is sterically unavailable for use. In both cases the 5' terminus is normal, and therefore these compounds can be incorporated into a growing DNA chain; but, once incorporated, because they lack the necessary 3' hydroxyl they

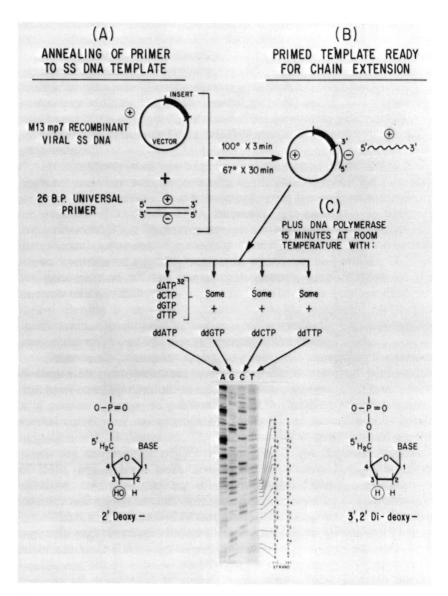

Figure 3-2. Chain termination system of sequencing using the filamentous phage vectors. This system uses the filamentous phages (Table 3-1) as cloning vectors, so that the recombinant DNA can easily be obtained as a single-stranded molecule. By annealing such a recombinant viral single-stranded DNA to an appropriate primer (in the above illustration, a "universal" primer is used, which is complementary to a sequence in the viral DNA close to the cloning site), a primed template is formed which is then divided four ways and used for chain termination sequencing as described in Fig. 3-1.

72

cannot be used for further chain extension. If, for example, dideoxy TTP (ddT) is added to a normal mixture of all four dNTPs (dATP, dCTP, dGTP, and dTTP), every time the polymerase needs to insert a T in a normal chain extension (i.e., opposite an A in the template), it will have a choice either of inserting a ddT and stopping further chain extension or of inserting a deoxy T (dT) and continuing chain extension until it reaches the next T position, when it will make the same choice again. In this way, the presence of a ddT induces a staggered chain extension, with some of the chains of the population stopping at every position of T in the nucleotide sequence. This is illustrated in Fig. 2-1. The other nucleotide-specific chain termina-tors—ddA, ddC, ddG, araA, and araC—act in the same way. Conse-quently, if a specifically primed template is divided four ways and each portion is incubated with all four dNTPs (one containing α-^{32}P), plus one of each of the four nucleotide-specific chain termi-nators, each incubation will give a different nucleotide-specific stagger and, by separating the chains on a polyacrylamide gel, the sequence of the template strand can be inferred by ordering the specific chain terminations by nucleotide length. This is illustrated in Figs. 3-1 and 3-2.

If the primers are long—i.e., greater than 100 nucleotides—it is necessary to cleave off the newly extended chain in order to keep their size within the resolving power of the polyacrylamide gel (up to approximately 400 nucleotides). It is normal, but not necessary, to do this with the same restriction enzyme that was used to generate the primer in the first place. The chains, however, will still have a common 5' end. Recleavage opens strategies of cleaving "backward" into the primer or "forward" into the chain extension, in order to gain access to different regions of the sequence (see below).

USE OF THE FILAMENTOUS PHAGE CLONING VECTORS FOR CHAIN TERMINATION SEQUENCING

The Sanger primed synthesis DNA sequencing methods were origi-nally developed using the single-stranded DNA phage ϕX174 (see Introduction), and although various adaptations of the method have been developed for sequencing double-stranded DNA (see Appendix E), the methods still work best on single-stranded phage DNA templates. The introduction of the filamentous phages as general DNA cloning vectors (Gronenborn & Messing, 1978), however, has made it possible to sequence any DNA fragment by the chain termination method, simply by cloning it into the filamentous phage and treating it as single-stranded phage DNA. This is illustrated in Fig. 3.2.

Several filamentous phage vectors have been developed (see Table 3-1; see also Messing, 1979), but the most widely used system is the M13mp (mp2, mp5, mp7, mp8, mp9) series of Messing. This family of phages contains a 780–base-pair piece of *E. coli lac* operon (part of the *i* gene, the *lac* operator and promoter, and the first third of the *z* gene), and the cloning site is within the *z* gene. Using an α complementation system in appropriate host cells and plating in the presence of IPTG inducer and a colored galactose derivative (X-gal) which is split to a blue dye by β-galactosidase, phages with inserted DNA can be distinguished as white plaques from the wild-type phage blue plaques that do not contain an insert (Gronenborn & Messing, 1978).

The other filamentous phage cloning systems (see M13GoRi and entries below in Table 3-1) do not have a method of distinguishing recombinants plaques from nonrecombinants plaques, except by nitrocellulose hybridization blotting methods (Appendix C.1). However, if alkaline phosphatase is used to remove the 3′ phosphate group from the linearized vector (Appendix B.4), T4 DNA ligase cannot circularize the vector and viable molecules can only be formed when a DNA fragment is inserted and donates two such phosphate groups.

The primer used for sequencing inserts in the filamentous phages varies with the system and is generally a restriction enzyme DNA fragment of the vector that primes close to the 5′ end of the insert. In the M13mp series, various primers are available, but the most useful is a 26 b.p. fragment of the phage that primes a few bases away from the *Eco*RI site and which has been cloned into pBR322 plasmid (Anderson et al., 1980) both as an *Eco*RI (PL16) and an *Eco*RI.*Bam*H-I (PL14) fragment and can be obtained in quantity from these sources. Synthetic 17-mer (Duckworth et al., 1981) and other-length primers specific for the M13mp series have been described and are now commercially available.

A MODEL SYSTEM FOR SEQUENCING A 2-kB PIECE OF DNA

Sequencing by the Sanger dideoxy system using the filamentous phage system is random, in that a "shotgun" clone bank of the

Table 3-1. Single-stranded Phage Cloning Vectors

	Reference	Cloning vectors	Method of selection of recombinants
M13mp2	Messing et al. (1977)	EcoR1	
M13mp5	Gronenborn & Messing (1978)	HindIII/EcoR1	
M13mp7	Messing, Crea, & Seebury (1981)	EcoR1/BamH-1/AccI/HindII PstI/SalI	Blue/white plaque assay
M13mp8	Messing & Vieira (1982)	EcoR1/XmaI/BamH-1/HindII/PstI/HindIII	
M13mp9	Messing & Vieira (1982)	HindIII/PstI/HindII/BamH-1/XmaI/EcoR1	
M13mp701	Brownlee (private communication)	EcoR1/BamH-1/PstI/HindII	
M13GoRi	Kaguni & Ray (1979)	EcoR1/XhoI	
M13amp	Ray & Kook (1978)		Ampicillin
fl R199	Boeke, Vovis, & Zinder (1979)	EcoR1	—
fl R299	Boeke (1981)	HindIII	—
M13 His+	Barnes (private communication)		Conversion of cells to his+

DNA is made and the sequences obtained have no relationship to each other until they are analyzed and matched for overlaps in a computer. Various common computer programs are available for this process, which run on microcomputers as well as on mini- and mainframe computers. The original programs were written by R. Staden (1977, 1979, 1980) and, in modified form, can be run on any small laboratory microcomputer (these programs are available from the author). However, several other programs are available (Korn, Queen, & Wegmen, 1977; Godson et al., 1978), and they are relatively simple to write.

Sequencing a 2-kB piece of DNA takes place in several discrete steps. These are:

Step 1. Restriction enzyme digestion and cloning of small pieces of the DNA. The 2-kB piece of DNA is digested with various restriction enzymes to give fragments 300 ± 100 base-pairs long (i.e., by using a restriction enzyme with a four–base-pair recognition site). Sau3A (G↓ATC) is a useful enzyme which gives small DNA fragments that can be cloned in the *BAM*H-I cloning site of M13mp7; but, in general, restriction enzymes that give blunt ends (*Alu*I, AG↓Ct; *Hae*III GG↓CC; *Hind*II, GTR↓YAC; *Rsa*I GT↓AC) are convenient because the fragments can all be cloned into the same vector cloning site (usually the *Hind*II site of M13mp7) by blunt-end ligation (Appendix A.3). In this system it is better to use cleaved vector that has been treated with alkaline phosphatase (Appendix B.4) so that the vector cannot recircularize and all the plaques arise from recombinant phages containing cloned DNA. A useful all-purpose vector is *Hind*III cut M13mp7.

Cloning by blunt-end ligation can be a general procedure. DNA fragments with a 5′ overhang (i.e., *Eco*RI, G↓AATC, *Xho*I, C↓TCGAG, *Bam*H-I G↓GATCC-cleaved DNA) can be filled in with DNA polymerase to give flush ends before cloning (Appendix B.1), and DNA fragments with 3′ overhangs (i.e., *Pst*I, CTGCA↓G, *Hha*I GCG↓C-cleaved DNA) can be trimmed to flush ends with the single-strand specific S1 nuclease (Appendix B.2).

Step 2. Isolation and screening of clones. Filamentous phage plaques containing inserted DNA can be detected either by the blue/white plaque assay (Appendix A.4), blotting using ^{32}P-labeled probes (Appendix C.1), or by accepting all plaques as recombinants when phosphatased vectors are used (Appendix B.4). Single plaques are picked into 1 ml of Luria broth (Appendix D.1) in microtiter plates (1-ml wells) and are grown overnight at 37°C. These can be kept as stocks and used to grow small cultures to prepare single-stranded DNA from the extruded phage (Appendix A.5).

Once a series of single-stranded DNA circles containing fragments of the 2-kB piece of DNA have been obtained, they should be sequenced using a "universal" primer [either the 26-b.p. fragment of the phage vector DNA (Appendix A.9.5) or a synthetic DNA oligonucleotide that primes a few bases away from the 3′ side of the cloning site]. To eliminate duplicate recombinants, it is often useful to grow 24 different recombinants and screen them by sequencing the DNA as single dideoxy A tracks and running them side-by-side on a sequencing gel.

Step. 3 Assembling the final sequence. All the sequences are put into the different data files in the computer, and by sequencing clones from several different restriction enzyme digests of the 2-kB piece of DNA, overlaps and repetitions can be found. The Staden computer programs (Staden, 1977, 1979; Korn, Queen, & Wegmen, 1977) are written for this purpose. In practice, because of the asymptotic nature of accumulation of random sequence data, it is sometimes a problem to fill in the last few gaps in the sequence. In the author's laboratory, it has been found to be better to sequence new DNA clone banks rather than to search exhaustively for missing fragments in a given clone bank. Sequencing the first 12 clones of six different clone banks will usually give sufficient overlapping data to yield a complete and correct sequence of a 2-kB fragment of DNA.

To sequence a 2-kB piece of DNA, it is important to consider carefully the cloning and sequencing strategy as the project evolves and to make use of a series of special diverse strategies in order to complete the sequence in the shortest possible time.

SPECIAL STRATEGIES IN CLONING AND SEQUENCING

If the restriction enzyme cleavage map of a DNA fragment is known before sequencing is attempted, directed cloning can be very powerful in designing the fastest route to obtaining the final complete sequence. Toward the end of a shotgun sequencing procedure, directed cloning, turning cloned fragments around, and *in situ* subcloning can also be very useful in filling in the last gaps in the sequence and confirming doubtful and difficult regions.

(a) Cloning DNA Fragments in a Directed Orientation

Vectors that contain two or more different cloning sites can be used to insert DNA in a known orientation and with little or no background of recircularized vector plaques. M13mp701 is a derivative of M13mp7 in which the duplicate *Eco*RI, *Bam*H-I, *Pst*I cloning sites have been eliminated so that the vector contains only an *Eco*RI, *Bam*H-I, *Pst*I, and *Hind*II cloning site (in that order) in the β-galactoside gene. By cleaving the vector with different combinations of restriction enzymes (i.e., *Eco*RI plus *Hind*II), DNA fragments with different ends (i.e., *Eco*RI plus any blunt end) can be cloned in a known orientation. This is useful for sequencing the 5′ and 3′ fragments of a piece of DNA from the internal restriction site. For instance, if the 2-kB piece of DNA to be sequenced has been excised from the plasmid pBR322 with *Eco*RI, is digested with *Hind*II (or by blunt-end cleavage enzyme), and is cloned into *Eco*RI/*Hind*II-cleaved M13mp701, the two *Eco*RI end fragments will be cloned in such a way that the sequence obtained will be read from the *Hind*II site toward the *Eco*RI site. The sequence reading from the *Eco*RI site toward the *Hind*II site can be obtained by cloning the whole 2-kB *Eco*RI fragment. This strategy is illustrated in Fig. 3-3. Two new M13 vectors, M13mp8 and M13mp9, have now been constructed especially for directed DNA sequencing (Messing & Vieira, 1982), each with the

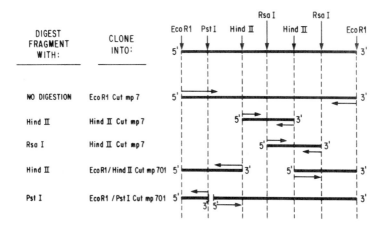

Figure 3-3. Sequencing by selective cloning. By using filamentous phage vectors with several single cloning sites, various regions of the DNA fragment to be sequenced can be selectively cloned and oriented, so that the sequence can be built up selectively rather than at random.

order of cloning sites (*Eco*RI, *Xma*I, *Bam*H-I, *Hind*II, *Pst*I, *Hind*III) in opposite orientations.

(b) Cloning Internal DNA Fragments Only

Clone banks of DNA fragments can be made so that the 5'- and 3'-end fragments are excluded. This is done simply by cleaving the large DNA fragment that has sticky ends (e.g., *Eco*RI) with *Hind*II (or any other blunt-end cleavage enzyme) and ligating the mixture into *Hind*II-cleaved M13mp7. Only the internal fragments with two blunt ends will be cloned. The use of phosphatased vector is recommended in this method. This is also illustrated in Fig. 3-3.

(c) Subcloning in Situ

When a large DNA fragment is cloned into M13mp7, internal fragments and end fragments can be randomly subcloned simply by cleaving the double-stranded recombinant RFI DNA with *Eco*RI (to cut out the insert) and with enzymes that cleave internal to the insert, but not the vector (*Hind*II, *Xho*I, etc.). The double digests are incubated with DNA polymerase to fill in the sticky ends, and the mixture is religated and transfected. The recombinant plaques will contain all the original insert fragments randomly cloned as smaller pieces. All the operations—cleavage, filling in, and ligation—can be carried out in a single reaction mixture. However, care must be taken to use phage DNA that is free of extraneous host cell DNA. This scheme is illustrated in Fig. 3-4.

(d) Turning Cloned Fragments Around

It is sometimes useful to turn a cloned DNA fragment around so that it can be sequenced from the other end. Fragments cloned into M13mp7 can be turned around simply by cleaving the recombinant double-stranded RFI DNA with *Eco*RI (this site flanks all other mp7 cloning sites), religating, and retransfecting the mixture. Vector molecules that have recircularized without an insert will again give blue plaques, and all the white plaques will represent recombinants with the original inserted DNA fragment randomly oriented. The plaques can be screened for orientation of the insert by annealing

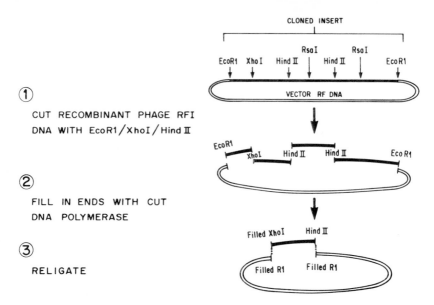

Figure 3-4. Subcloning *in-situ.* When the entire DNA fragment to be sequenced is cloned into a filamentous phage, it is often useful to subclone fragments by digesting the recombinant RFI DNA with various restriction enzymes and religating all in a single series of reactions. In this way, selective and random subcloning can be easily achieved.

heated phage DNA preps and running them on a 0.35% agarose gel (Appendix C.3). Phage DNAs with inserts in the opposite direction will anneal to give partial double-stranded DNA structures which migrate differently from the single-stranded circles.

(e)   Use of Internal Primers

The original methods of dideoxy sequencing did not use this filamentous phage cloning system but used restriction enzyme DNA fragments as primers on a single-stranded DNA template. This is the way that the ϕX174 (Sanger et al., 1978) and G4 genomes were sequenced (Godson et al., 1978). Because of the relatively large size of the DNA restriction enzyme fragments, the 3' extensions were cleaved off the primer with a restriction enzyme before the reaction mixture was put on an acrylamide gel. This allowed the use of strategies to cleave the sequencing extension "forward" to read the sequence further in the 3' direction and "backward" to read the junction properly. It also allows the use of 3' exonuclease-treated primers so that the sequence of the primer restriction enzyme fragment itself could be read to within 15 nucleotides from its 5' end (Appendix B.6).

All this methodology is applicable to large DNA fragments cloned into the filamentous phages; and because clones can be obtained with the large fragment in both orientations, one single internal restriction enzyme DNA fragment can be used to sequence in the 3' and 5' directions from both ends of the fragment, thereby covering a very large sequence. This strategy is illustrated in Fig. 3-5 and described in Appendix B.5.

CHECKING THE SEQUENCE WITH RESTRICTION ENZYMES

Once a nucleotide sequence has been determined, it can be checked for restriction enzyme cleavage sites that are predicted from the sequence, simply by priming and cleaving with restriction enzymes as described above and in Appendix B.5. Cleavage of the chain extension at internal restriction enzyme sites will generate a heavy radioactive band crossing all four dideoxy lanes and migrating at the appropriate size positions in the gel. This strategy is simple and rapid and is often necessary to sort out sequences containing runs

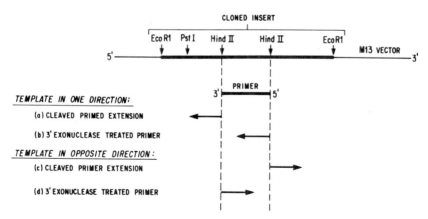

Figure 3-5. Use of internal primers. A single internal primer to a large insert can be effectively used to sequence 250–300 nucleotides on each side of the 3' and 5' ends of the primer and, in addition, by using Exonuclease III digestion of the primer, to sequence 250–300 nucleotides internal from the 3' and 5' ends of the primer. In this way, one primer can generate 500–1,000 bases of sequence. For details see the text.

of G and C residues (i.e., 5′-GGCGCCGG-3′) that are difficult to analyze on the sequence gel data alone.

PROBLEMS OF RANDOM SEQUENCING

It has been found that when DNA is subcloned, the various fragments do not appear in the library at the expected random frequencies. Some fragments appear very often, other fragments appear very rarely. In addition, some DNA fragments do not appear to clone in random orientation and are represented in only one of the two possible orientations, even after repeated subcloning. The basis of this nonrandomness is not known, but it may well be due to the physiology of the phage and the effect of particular inserted nucleotide sequences on phage replication or transcription.

The most effective method of circumventing this nonrandomness is to clone many different restriction enzyme digests of the DNA and to sequence the first dozen or so clones of each digest, all of which are usually unique. In most cases, it is very unrewarding to search exhaustively for a missing member of a clone bank.

An alternative method that appears to generate an unbiased random fragment library is to digest the DNA with DNase I to relatively small fragments and then to separate a size class (300 b.p.) using sucrose density gradients. These fragments are then repaired with DNA polymerase I and cloned into the filamentous phage vector using *Eco*RI linkers (Anderson, 1981). This method, however, requires quite large quantities of starting material (approximately 5 μg) which may not always be available.

However, intelligent use of the accumulated sequence and restriction enzyme mapping data can usually suggest a strategy of crossing missing regions (see Special Strategies, above).

STRATEGIES FOR SEQUENCING LARGE PIECES OF DNA (10–100 kB)

The first large piece of DNA to be sequenced by the dideoxy chain termination procedure was the bacteriophage G4 genome, which is 5,577 nucleotides long (Godson et al., 1978). Since then, several viruses have been sequenced exclusively by the chain termination methods, such as human BK virus (4,396 b.p.; Yang & Wu, 1979) and Cauliflower Mosaic virus (8,031 b.p.; Gardner et al., 1981).

Human mitochondrial DNA (16,000 b.p.; Anderson et al., 1981) has been sequenced in its entirety, and Adenovirus (approximately 35,000 b.p.) and lambda bacteriophage (approximately 48,000 b.p.) are now in the final stages of sequence determination, again using this methodology.

The lessons learned from these large sequencing projects are that restriction enzyme cleavage maps are very important guides to both the accuracy and the strategy of sequencing, that random shotgun sequencing works best on fragments 1,000–2,000 b.p. long, and that continuous computer analysis of the sequence data as it is generated is essential. A single investigator can generate with the chain termination methods up to 1,000 nucleotides of unrelated sequence data a day, but without intelligent selection of sequencing and cloning strategies this data cannot be easily linked up and can be counterproductive.

The most sensible strategy for sequencing a very large piece of DNA, 100–200 kB long, such as T4 bacteriophage or a yeast chromosome, is to digest it to 10–20 kB fragments using partial restriction enzyme digestion and to clone these pieces into lambda vectors. These fragments should then be subcloned as 1–2 kB pieces, using either the filamentous phages or plasmid vectors, and sequenced by the random shotgun procedures described here. In this way, a control is kept on the growth and linkage of the data.

APPENDIXES

Appendix A
Cloning into M13mp7

The protocols described below are written as a series of steps so that they can be followed as a continuous series from cloning the desired DNA to obtaining the final sequence.

A.1. Cleavage of M13mp7 RFI Vector DNA
Ready for Cloning

3 μl   M13mp7 RFI RNA (concentration 1 μg/μl) (D.4)
3 μl   10 × restriction enzyme buffer (10 × Hin buffer [A.7.9] or recommended buffer)
24 μl   H$_2$O
—————
30 μl

Cleave with 3 units of restriction enzyme by incubation at 37°C for 4-6 hr. After digestion, check for completeness of RFI DNA cleavage by running 3 μl of the digest on a 0.35% agarose gel. No RFI orRFII forms of the DNA should be visible. This is important because uncleaved molecules give a background of nonrecombinant plaques in the transfection. After digestion, inactivate the enzyme by heating the mixture at 67°C for 15 min.

This stock solution of 0.1 μg/ml cleaved vector DNA should be kept frozen and is sufficient for approximately 30 ligations.

A.2.   *Cleavage of DNA to Be Cloned*

The DNA fragment to be sequenced is prepared from plasmid or phage DNA by the usual procedures and then cleaved with restriction enzymes into smaller pieces ready for ligation into the linearized vector DNA. The conditions of cleavage are the same as above, but the concentration of the DNA and the number of fragments generated should be approximately known because it is important to maintain a reasonably low ratio of fragment to vector DNA in the ligation step, otherwise multiple fragments are cloned in random order and orientation.

A.3.   *Ligation*

The DNA fragments should be added to give a final ratio of 1:2 fragment DNA molecules per vector DNA molecule.

1 μl	cleaved vector DNA (0.1 μg) (A.1)
1 μl	cleaved fragment DNA (A.2)
2 μl	10 × ligation buffer
2 μl	2 mM ATP (rATP)
14 μl	H_2O

20 μl

add 1 μl T4 DNA ligase

10 × ligation buffer:
200 mM tris-HCl, pH 7.5
100 mM $MgCl_2$
50 mM dithiothreitol

For sticky-end ligation, incubate at 15°C for 2 hr; for blunt-end ligation, incubate overnight at 4°C.

A.4. Transfection

Transfect 1 μl (5 ng) of the ligation mixture from a sticky-end ligation and 3 μl (15 ng) from a blunt-end ligation. Both should give about 500-1,000 recombinant plaques per plate.

Add 1 μl of ligated DNA (A.3) to 0.3 ml of Ca^{2+}-treated cells (see Appendix D.3) in a sterile tube, and stand the tube on ice for 30-40 min. Next, heat-shock the cells by incubation at 45°C for 2 min, and then add 0.3 ml of fresh, uninfected JM103 cells (D.2), 0.05 ml of 2% X-Gal (in DMSO), 0.010 ml IPTG, 3 ml of 0.64% soft agar at 45°C, and overlay immediately on a Luria broth plate (see D.1 for media and chemicals). Incubate the plates overnight at 37°C.

Recombinant phage with inserted DNA will give rise to white plaques; those without inserted DNA, blue plaques. (Caution: this is not the case with HindII-cleaved blunt-end vector because most commercial preparations of HindII enzyme contain a nuclease that digests the ends of the vector and destroys the reading frame and consequently the α complementation reaction.)

Always include transfection control reactions of (a) unligated vector DNA (to measure the background of plaques due to uncleaved vector DNA), (b) ligated vector DNA alone (to measure the efficiency of ligation and the number of white plaques generated due to damaged vector DNA), at the same concentration as the ligated DNA (i.e., 5-15 ng), and (c) no DNA (to measure contamination).

A.5. Growth of Recombinant Phage DNA Ready for Sequencing

White plaques from the transfection plates are picked with sterile toothpicks into 1 ml of Luria broth (D.1) contained in wells of the large 24-well microtiter plates and then incubated overnight at 37°C. After growth, the microtiter plates can be stored at 4°C for months, or frozen for longer storage, without significant loss of phage titer.

To growth phage for DNA sequencing, inoculate 10 ml of Luria broth in a sterile 30-ml screw-cap tube with 0.1 ml of phage suspension from the microtiter well and 0.2 ml of freshly grown JM103 cells (these should always be grown overnight from an inoculum grown on a minimal medium plate [D.1]; after several cycles in unselective media, populations of cells without the F$'$ factor develop, and this results in a reduced filamentous phage yield). Incubate the screw-cap tubes overnight in a roller drum at 37°C.

Next day, pellet the cells in a 15-ml sterile Corex tube by centrifugation at 10,000 rpm for 20 min. Carefully remove 9 ml of supernatant with a 10-ml pipette into a fresh 15-ml Corex tube. Add 1 ml of 25% polyethylene glycol (PEG) and 1 ml of 5 M NaCl, shake to mix, and leave the tube on ice for at least 2 hr (to overnight) to precipitate the filamentous phage (Yamamoto et al., 1970). Recover the PEG precipitate by centrifugation at 10,000 rpm for 15 min, pour off the supernatant, drain and wipe away as much of the residual PEG from the walls of the tube as possible, and resuspend the PEG precipitate in 0.5 ml of water. Transfer this to a 1.5-ml Eppendorf tube for further manipulation. The suspension consists mostly of filamentous phage particles. To extract their single-stranded DNA, add 0.5 ml of redistilled phenol (equilibrated with 0.1 M tris, pH 7.5), shake for 2 min, centrifuge for 10 sec in an Eppendorf microfuge to separate the phases, remove the aqueous phase, and re-extract it with phenol. Remove the residual phenol by extraction with 1-2 volumes of ether, and precipitate the DNA with one-tenth volume 3 M sodium acetate and $2\frac{1}{2}$ volumes of 95% ethanol, at $-20°C$ overnight in a siliconized Eppendorf tube (SS DNA tends to be easily absorbed to plastic surfaces).

Recover the DNA precipitate by centrifugation in an Eppendorf microfuge for 15 min and resuspend the precipitate in 10 μl of water. The concentration and purity of the DNA should be checked by diluting 2.5-250 μl with water (i.e., 1-100 dilution) and measuring the absorption at 260 and 280 mμ. The normal phage DNA yield from 10 ml of cells will give an OD_{260} between 10 and 30. Sequencing requires 1 μl of single-stranded (SS) DNA of $OD_{260} = 15$ (i.e., 0.5 μg/μl).

A.6. Dideoxy Sequencing Procedure

The sequencing reactions are carried out in drawn-out capillaries (melting-point tubes) using graduated capillary pipettes and siliconized $2 \times \frac{1}{2}$ inch glass disposable siliconized tubes (Rochester Scientific) to add, mix, and transfer the microtiter reaction mixture constituents.

Step 1. Annealing the Primer to the Single-stranded DNA Template. The primer normally used to sequence DNA cloned in the filamentous phage M13mp7 series is a short "universal" primer that is complementary to the phage DNA close to the 5' end of the insert (Anderson et al., 1980; Duckworth et al., 1981). A 26-b.p. fragment of M13mp7 cloned into the plasmid pBR322 is a source of such a primer (see A.9.5), but various preparations of synthetic oligonucleotide primers have recently become commercially available. The primer is annealed to the single-stranded DNA in a ratio of 3 to 5 primer molecules per single-stranded DNA template.

The reaction mixture consists of:

1.0 μl	M13mp7 recombinant SS DNA OD$_{260}$ = 15 (i.e. 0.5 μg or an adjusted volume to give this concentration)
1.0 μl	primer DNA (or volume to give SS/primer ratio of 3 to 5)
1.0 μl	10 × Hin buffer (A.7)
7.0 μl	H$_2$O
10.0 μl	

Add each ingredient with a York 5-μl micropipette (graduated in microliters) to a drawn-out capillary. After all the ingredients have been added, mix by blowing the contents of the capillary onto the bottom of a siliconized $2 \times \frac{1}{2}$ inch tube and allow the mixture to return by capillary attraction. Seal the capillary and drop it into boiling water for 3 min to completely denature the DNA. To anneal synthetic single-stranded primer to the template DNA, transfer the boiled capillary to a beaker containing approximately 150 ml of water at 67°C and allow it to cool naturally to room temperature (usually 30 min). To anneal double-stranded DNA primers to the template, incubate the boiled capillary at 67°C for 30 min.

Step 2. Dry 5–10 liter (i.e., 5–10 μCi) of (α-^{32}P)dATP (1.0 mC/ml at a sp. activity of approximately 300 Ci/(m · mi), from Amersham Radiochemicals or New England Nuclear) on the bottom of a siliconized $2 \times \frac{1}{2}$ inch tube. This takes approximately 5 min in a vacuum dessicator using a water pump.

Step 3. While the DNA is annealing and the (α-^{32}P)dATP is drying, prepare the four deoxy/dideoxy reaction mixes. To do this, first make fresh 0.5 mM dATP, dCTP, dGTP, dTTP from the frozen 10 mM stocks (i.e., dilute × 20 in distilled water). Then make the Å, Č, Ğ, and Ť mixes as described below (A.7). Set up four $2\frac{1}{2} \times \frac{1}{2}$ inch siliconized tubes with capillaries in them, in a rack and marked A, C, G, and T. The dideoxy chain extension is done in these reaction capillaries. To each capillary add:

1.0 μl	dideoxy triphosphate (0.25 mM ddATP; 0.33 mM ddCTP; 0.33 mM ddGTP, and 1.0 mM ddTTP)
1.0 μl	appropriate mix (see A.6)

i.e., the reaction mixtures (in μl) so far contain:

A	C	G	T
1.0 ddA	1.0 ddC	1.0 ddG	1.0 ddT
1.0 Å	1.0 Č	1.0 Ǧ	1.0 Ť

Step 4. When the DNA is annealed and the (α-^{32}P)dATP is dry, break the sealed capillary containing the DNA and blow the contents into the siliconized tube containing the dry (α-^{32}P)dATP (it is often better to transfer the DNA solution from the broken capillary to a new capillary before dissolving the (α-^{32}P)dATP: this makes the subsequent transfer easier). Dissolve the (α-^{32}P)dATP in the DNA solution, and check for complete recovery with a hand Geiger counter.

Next, divide the (α-^{32}P)dATP and DNA solution into four 2-μl aliquots by end-to-end transfer from the capillary to four 5-μl graduated York micropipettes (i.e., put 2 μl into each). Transfer 2 μl of the (α-^{32}P)dATP/DNA to each A, C, G and T reaction mixture. The reaction mixtures are now ready for addition of DNA polymerase and chain extension.

Step 5. Chain Extension. In this step diluted DNA polymerase is added to the reaction mixtures and chain extension is allowed to proceed at room temperature for 15 min. The extension is then chased with cold dATP to extend any chains that are prematurely terminated due to the low dATP concentration.

To do this, dilute the DNA polymerase six times with a solution of 0.1 M potassium phosphate pH 8 in 50% glycerol (i.e., add 1 μl of DNA polymerase I Klenow plus 5 μl of dilutent in a capillary, mix, and aliquot the diluted enzyme into four 1-μl portions; Boehringer DNA Pol I Klenow can be diluted up to 20 times and is still active in this reaction). Add 1 μl of the diluted enzyme to each A, C, G, and T reaction mixture and mix thoroughly (by blowing the contents onto the bottom of the siliconized tube and allowing them to return to the capillary, several times, as before). The final reaction mixtures are therefore now as follows:

A	C	G	T
1.0 μl ddA	1.0 μl ddC	1.0 μl ddG	1.0 μl ddT
1.0 μl Å	1.0 μl Č	1.0 μl Ǧ	1.0 μl Ť
2.0 μl DNA/^{32}P	2.0 μl DNA/^{32}P	2.0 μl DNA/^{32}P	2.0 μl DNA/^{32}P
1.0 μl Pol I	1.0 μl Pol I	1.0 μl Pol I	1.0 μl Pol I
5.0 μl	5.0 μl	5.0 μl	5.0 μl

Incubate for 15 min at room temperature, with the reaction mixture still in the capillaries (i.e., not blown out on the bottom of the siliconized tube). After 15 min, chase by adding 1 μl of 0.5 mM dATP and incubate for a further 5–15 min at room temperature.

To make larger chain extensions, incubate the reactions at 37°C in an air oven.

Stop the reaction by adding 10 μl of formamide-dye mix (A.7) to the tube containing the reaction capillary, and blow the contents of the capillary into the formamide; discard the capillary.

Before running the samples on a sequencing gel, they must be denatured. To do this, stand the rack of A, C, G, and T tubes (with the reaction mixtures and formamide-dyes on the bottom) in boiling water for 3 min, and then apply 2 μl of each reaction mixture to the sequencing gel (A.8). This leaves enough sample for several other gel runs. It is usual to make four gel runs per sample to read from 0–300 b.p.

A.7. *Stock Solutions Required for Dideoxy Sequencing*

The following solutions should be made and stored at $-20°$ as basic stock solutions to be used frequently.

1. Deoxytriphosphates.
     10 mM dATP (5.89 mg in 1.0 ml distilled H_2O)
     10 mM dCTP (5.91 mg in 1.0 ml distilled H_2O)
     10 mM dGTP (6.31 mg in 1.0 ml distilled H_2O)
     10 mM TTP (6.24 mg in 1.0 ml distilled H_2O)

     PL Biochemicals dXTP's made up directly from the bottle without further purification are quite satisfactory.

2. Dideoxytriphosphates.
     2.5 mM dideoxy ATP (4 μmol in 1.6 ml distilled H_2O)
     10 mM dideoxy CTP (4 μmol in 0.4 ml distilled H_2O)
     10 mM dideoxy GTP (4 μmol in 0.4 ml distilled H_2O)
     10 mM dideoxy TTP (4 μmol in 0.4 ml distilled H_2O)

     PL Biochemicals supply ddXTP's as 4 μmol per bottle; add 0.4 ml of water (or 1 mM tris-HCl pH 8) and store frozen.

3. DNA Polymerase I dilution buffer. 0.1 M KPO$_4$ pH 8.0 in 50% glycerol.

4. DNA polymerase I Klenow fragment (Boehringer Mannheim) or reverse transcriptase.

5. Formamide sample buffer. Deionize 100 ml of formamide with 5 g Amberlite MB-1 (mixed-bed ion exchange resin) for 30 min and remove resin by filtration. Add 0.3 g xylene cyanol FF, 0.3 g bromophenol blue, and 10 mM Na_2 EDTA pH 7. Store at 4°C.

The following solutions should be made up fresh just prior to sequencing:

6. Deoxytriphosphates (fresh each day).
   0.5 mM dATP in H_2O
   0.5 mM dCTP in H_2O
   0.5 mM dGTP in H_2O
   0.5 mM TTP in H_2O

   i.e., dilute 10 mM stock × 20 in distilled H_2O

7. Mixes (in μl, fresh each day).

	A mix	C mix	G mix	T mix
0.5 mM dCTP	7.5	1	10	10
0.5 mM dGTP	7.5	10	1	10
0.5 mM TTP	7.5	10	10	1
10 × Hin buffer	7.5	7.5	7.5	7.5

8. Dideoxy triphosphates (use until exhausted). The dideoxy triphosphate solutions seem to be more stable than the triphosphate solutions, and it is normal to store the diluted dideoxy solutions frozen and to use them repeatedly until they are exhausted.

   0.25 mM ddATP
   0.33 mM ddCTP
   0.33 mM ddGTP
   1.0 mM ddTTP

   The length of the chain extension, however, depends upon the ratio of dd/d triphosphates and can be controlled by varying the dideoxy concentration and adding the same volume (1 μl) to the reaction mixture.

9. 10 × Hin buffer.
 66 mM tris-HCl pH 7.5
 66 mM MgCl$_2$
 66 mM NaCl
 50 mM dithiothreitol

A.8.   Preparation of Sequencing Gels

The gels used for dideoxy sequencing are usually 8% thin poly-acrylamide gels 0.4 mm × 40 cm × 20 cm, containing 7 M urea and run in 1 × TBE buffer at high voltage (1,200 V and 25–30 mA) so that they run hot (circa 70°C). The slots are 5-mm wide and are separated by 2-mm walls, and only 2–3 μl of sample is applied to each well. These gels are described in Sanger and Coulson (1978).

1. Stock solutions.
  (a) 10 × TBE buffer:
      108 g trizma base
      55 g boric acid (not sodium borate!)
      9.3 g Na$_2$ EDTA

   make up to 1 liter in distilled water, and store at room temperature. The pH should be 8.3.
  (b) 40% acrylamide:
      38 g acrylamide
      2 g *NN*-methylene bis(acrylamide)

   make up to 100 ml in distilled H$_2$O

   Both the acrylamide and bis(acrylamide) should be purified by recrystallization from acetone (dissolve 600 g poly-acrylamide in 1,000 ml acetone at 80°C and crystallize overnight at 4°C, recover the crystals with a Buchner filter funnel, and air dry). The 40% solution should be deionized by adding a few grams of mixed-bed resin (Amberlite MB-1), and stirring gently for 30 min, before removing the resin by filtration. Store the solution at 4°C.
  (c) 1.6% ammonium persulphate (use fresh solutions).
  (d) TEMED (N_1N_1NN-tetramethylethylenediamine).
  (e) Urea (solid), enzyme-grade Schwartz-Mann.

2. Making the Gel.
  (a) The glass plates must be scrupulously clean and grease free. Siliconize one of the plates (the notched one). Use electrical

tape (3MM #56) instead of vaseline to seal the sandwich. Use spacers and combs made from 0.4-mm-thick plastic or teflon sheets.

(b) Make 100 ml of 8% acrylamide (sufficient for 3 gels):

40% acrylamide	20 ml
1.6% ammonium persulphate	3.6 ml
10 × TBE	10 ml
urea	42 g
TEMED	100 μl

Warm the final solution to 25–30°C before pouring. Hold the gel plates at approximately 30°C to pour; when the sandwich is full, insert the comb and clamp it into place with paper clamps. Then lay the gel almost horizontal to set. Polymerization will take 15–20 min. The gel can be used immediately.

3. Running the Gel.   When the walls of the wells are visible, the gel has set. Remove the tape from the bottom of the gel and then carefully pull out the comb and immediately wash the loose acrylamide out of the wells, either with a wash bottle or by quickly attaching the gel to the electrophoresis apparatus, adding the reservoir buffer (1 × TBE), and washing the wells with a Pasteur pipette. Loose acrylamide is a problem only if the gels are used immediately after setting. The gels can be kept overnight and used the next day, as long as they do not dry.

Load the gel with a very finely drawn-out capillary (approximately 3 inch) that has been calibrated to 2–3 μl. The bromophenol blue marker dye migrates in a position that corresponds to DNA chain lengths of 15 nucleotides and the xylene cyanol FF to chain lengths of 80 nucleotides.

4. Autoradiography.   After completion of electrophoresis, remove the gel from the apparatus, and lift the siliconized plate. The thin polyacrylamide sheet will stick to the unsiliconized plate. Drop the gel (still attached to the glass plate) into a bath of 10% acetic acid and fix for 5 min (until the bromophenol blue changes to yellow). Remove the gel from the acetic acid bath, and wash away the excess acetic acid either under a running tap or by dropping the gel and plate into a bath of distilled water. Dry the gel and glass plate with large Kimwipes, wrap it in Handi-wrap, and autoradiograph it at room temperature. An overnight exposure should be sufficient to read the DNA sequence.

Drying the gel (without fixation) using a heated slab gel dryer will give much sharper bands and is preferable to fixation unless too many gels are to be processed.

A.9. Preparation of DNA for Dideoxy Sequencing

Preparation of the DNA is critical for the success of the Sanger enzymatic sequencing methods. The DNA (such as primer fragments) must be free of acrylamide and salts because these can poison the DNA polymerase. The DNA must also be free of RNA fragments (which will give random priming and high backgrounds) and free of contaminating DNA fragments (a single contaminating DNA fragment can give a second sequencing track; random DNA fragments will give spurious bands and high backgrounds). Agarose appears to leach agents that inhibit DNA polymerase and are difficult to remove; the use of agarose gels to separate DNA fragments should therefore be avoided where possible (if agarose is used, however, introduce a hydroxyl apatite column purification step after soaking or electro-eluting the DNA from the gel). Low-melting-point agarose appears to get round these problems.

1. Preparation of DNA Restriction Enzyme Fragments as Primers. Make sure that the DNA is digested with restriction enzyme to completion, because partial digestion products can undetectably contaminate other DNA bands and produce a second sequencing track when the DNA fragment is used for priming. If partial digestion products are visible on an analytical gel, the DNA should be further digested.

2. Extraction of DNA from the Gel. After running the gel, stain it with 0.5 µg/ml of ethidium bromide and then, using UV fluorescence to visualize the DNA, cut out the DNA bands. Lay each gel strip on a sheet of parafilm and chop it into 1-mm cubes with a sharp scalpel or razor blade. (Do not grind the gel or pass it through a syringe needle; these treatments seem to result in polyacrylamide contamination of the final DNA preparation.) DNA fragments of greater than 250 b.p. should be recovered by electroelution.

3. Soaking. Put the chopped gel pieces in a $2 \times \frac{1}{2}$ inch siliconized tube, and add 1.5 ml of 0.3 M sodium acetate, 0.1 M tris-HCl pH 8.5, 0.001 M EDTA, and incubate at $37°$C overnight. Next day, separate the supernatant from the acrylamide by passing it through a glass wool filter. This can be done by punching a hole through the bottom of a small Eppendorf plastic conical centrifuge tube (use a red-hot needle), packing the bottom with glass wool, and standing it in a $2\frac{1}{2} \times \frac{1}{2}$ inch polyallomer SW50 centrifuge tube. Pour the acrylamide and supernatant into the Eppendorf tube and blow the supernatant through the glass wool with a rubber bulb. Measure the volume of the supernatant and add exactly $2\frac{1}{2}$ volumes of 95% ethanol (no salt need be added as it is in there already), and precipitate the DNA overnight at $-20°C$. Next day, recover the DNA by centrifuging at 30,000 rpm for 30 min at $0°C$ in a SW50.1 rotor. Pour off the supernatant, invert the tube to drain, and after 10 min wipe the walls dry of any remaining ethanol. No precipitate should be visible (if there is, it is salt and acrylamide). Dissolve the DNA in 50–100 μl of distilled water and store frozen. The recovery can be checked on an ethidium bromide stained gel. It is convenient to dissolve the fragments in the same number of μl of water as the number of μg originally digested. Assuming 50% recovery, the concentration of fragments can easily be calculated.

4. Electroelution. (This is described by Galibert, Sedat, & Ziff, 1974.) Take a 10-ml Falcon plastic pipette, and cut it off at the 0-ml mark (i.e., 8 inch from the tip). Stand the pipette in a beaker of electroelution buffer (20 mM tris-HCl pH 8.0, 0.002 M EDTA). Soak a small pad of cotton wool in buffer, and lighly pack it into the pipette so that it forms a plug approximately $\frac{3}{4}$ inch from the tip. Make sure that the plug has no air bubbles trapped in it (this is why everything is done wet). Knot a piece of 0.25-inch boiled dialysis tubing and cut it off $2\frac{1}{2}$ inch from the knot. Fill the dialysis tubing with buffer and push it over the outside of the pipette tip, again making sure that no air bubbles are trapped. The pipette is now ready for use. Fill it with buffer and pour in the chopped acrylamide gel pieces. These will sink to the bottom and rest on the cotton-wool plug.

Put the pipette in an old-style disc gel aparatus and electrophorese at 3 mA/tube. DNA fragments 250–700 b.p. will electroelute in 2 hr; fragments up to 3,500 b.p. will electroelute in 4–5 hr. If the DNA has been stained with ethidium bromide, check the gel pieces for fluorescence with a UV lamp after electroelution (there should be none!).

To recover the DNA, remove the dialysis bag from the pipette, pour the contents into a 5-ml polyallomer tube, and proceed as above.

5. *Preparation of the 26-b.p. Universal Primer.* The 25-b.p. fragment of M13 adjacent to the 3′ side of the cloning site has been cloned as a *Bam*H-I *Eco*RI piece in plasmid PL14 and on an *Eco*RI piece in plasmid PL16 (Anderson et al., 1980). To prepare the primer, excise the fragment from 100 μg of the plasmid DNA, separate the fragment on an 8% 2-mm polyacrylamide gel, and recover the DNA by soaking as described above.

Synthetic 15–17 b.p. universal primers for the M13mp7, -mp8, -mp9 series of vectors are now commercially available (Bethesda Research Laboratories; Collaborative Research Labortories; New England Biolabs) and essentially supersede the cloned universal primers.

Appendix B

B.1. Filling in 5′ Sticky Ends Prior to Blunt-end Cloning

When DNA fragments to be cloned have different 5′ overlaps at each end (i.e., *Eco*RI and *Bam*H-I termini) or have 5′ overlapping ends that have no complementary cloning sites in the vector (i.e., *Taq*I, ↓TCGA), it is often convenient to fill in the sticky ends with DNA polymerase to give slush ends and then to clone the fragment into *Hind*II-cleaved M13mp7 vector, by blunt-end ligation. To do this, mix:

1 μl	DNA fragment (0.5 μg)
1 μl	50 μM dATP (i.e., dilute 5 mM dATP stock solution × 100)
1 μl	50 μM dCTP (i.e., dilute 5 mM dCTP stock solution × 100)
1 μl	50 μM dGTP (i.e., dilute 5 mM dGTP stock solution × 100)
1 μl	50 μM dTTP (i.e., dilute 5 mM dTTP stock solution × 100)
1 μl	10 × buffer
4 μl	H_2O
10 μl	

plus 0.5 μl DNA polymerase (Klenow) for 30 min on ice

10 × buffer:
>200 mM tris pH 7.5
>100 mM $MgCl_2$
>10 mM DTT
>10 mM EDTA

These fragments can be ligated directly into *Hind*II-cleaved M13mp7 without further purification.

>1 μl   *Hind*II-cleaved M13mp7 DNA (0.1 μg)
>1 μl   filled-in fragment (0.05 μg)
>2 μl   10 × ligation buffer (A.3)
>2 μl   5 mM ATP
>14 μl H_2O

>――――――
>20 μl

add 1 μl T4 DNA ligase; incubate at room temperature overnight

Transfect 1, 3, and 5 μl as before (A.4).

It is often useful to start the protocol by cleaving first with a restriction enzyme and then filling in and ligating, all in a single series of steps. To do this, digest:

>1 μl   DNA (0.05 μg)
>1 μl   10 × restriction enzyme buffer
>8 μl H_2O

>――――――
>10 μl

add 1 μl restriction enzyme at 37°C for 60 min

Inactivate the enzyme by incubation at 67°C for 15 min and then add 1 μl of a mixture at 50 μM dATP, dCTP, dGTP, and dTTP made up in 10 × polymerase buffer and continue as above.

B.2.   *Removing 3' Overlaps with S1 Nuclease*

3' overlaps cannot be filled in with DNA polymerase and must be removed with a single-stranded specific nuclease, such as S1 nuclease. To do this, incubate:

1 μl   DNA fragment (0.5 μg)
2 μl   10 × buffer
18 μl H$_2$O
_____
21 μl

> add 10 units of S1 nuclease; incubate at room temperature for 30 min

10 × S1 buffer:
> 0.3 M sodium acetate pH 4.5
> 3 M NaCl
> 45 mM ZnCl$_2$

Stop the reaction with 20 μl of 1 M tris pH 9 buffer and phenol extract by adding 40 μl of redistilled phenol (equilibrated with 0.1 M tris pH 8), shaking for 2 min, and then removing the phenol by shaking with 2 ml of ether. Precipitate the DNA by adding one-tenth volume of sodium acetate and 2$\frac{1}{2}$ volumes of 95% ethanol and then standing overnight at $-20°$C. Recover the DNA by centrifuging in an Eppendorf microfuge tube for 15 min, dry the tube, and resuspend the DNA in 10 μl of water. Use 1 μl of DNA solution to clone in the fragments into *Hind*II-cleaved M13mp7 vector, as described above (B.1).

B.3.   Adding Synthetic Linkers

Commercially available synthetic restriction enzyme recognition sites (called synthetic linkers) do not have 5$'$ phosphate groups, and before use they must be phosphorylated. It is normal to phosphorylate linkers, using (γ-^{32}P)-labeled ATP, so that the linkers are labeled and their efficiency of ligation can be checked. It is also useful to have labeled linkers, so that free unattached linkers can be separated after ligation to DNA fragments from attached linkers, using Sephadex columns or polyacrylamide gels.

(a) Phosphorylation of Linkers. Dissolve linkers in water to A$_{260}$ = 10 (i.e., approx. 500 μg/ml). To label the linkers incubate:

10   μl   linkers (i.e., 5.0 μg or 760 pmol)
1.5 μl   0.5 M tris-HCl pH 6.5, 0.1 M MgCl$_2$
0.5 μl   H$_2$O
_____
12.0 μl

> heat to 70°C for 2 min and allow to slow cool to 37°C

Then add

1.5 μl   0.1 M DTT

75   μC ^{32}P (7.5 μl of isotope [New England Nuclear, approx. 3,000 Ci/mmol] dried on bottom of siliconized glass or Eppendorf tube)

1.5 μl   T4 polynucleotide kinase (Boehringer 2,000/units/ 0.44 ml)

_____

15   μl

        incubate at 37°C for 30 min
        heat at 70°C for 2 min
        slow cool to 37°C

Then add

0.5 μl   0.5 M tris-HCl pH 7.5, 0.1 M MgCl$_2$
2   μl   0.1 M DTT
2   μl   5 mM ATP
2   μl   T4 polynucleotide kinase (Boehringer)

_____

21.5 μl

        incubate at 37°C for 30 min

Freeze as a stock of phosphorylated linkers.

(b) Testing the Efficiency of Ligation of Linkers. The phosphorylated linkers should be checked for both efficiency of ligation and efficiency of cleavage with restriction enzyme. To test the linkers incubate:

1   μl   phosphorylated linkers
0.5 μl   0.5 M tris-HCl pH 7.5, 0.1 M MgCl$_2$
0.5 μl   5 mM ATP
0.5 μl   0.1 M DTT
2.5 μl   H$_2$O

_____

5.0 μl

        add 0.5 μl T4 DNA ligase and incubate overnight at room temperature

It has been found necessary to phenol extract the linker before subsequent cleavage in order to kill a residual ligation activity. To do this, add 25 μl H$_2$O and 20 μl of equilibrated phenol (0.1 M tris-HCl pH 7.5), shake for 2 min, and dissolve out the phenol with two 1–2 ml ether washes; then add one-tenth volume of 3 M sodium acetate and precipitate the DNA with $2\frac{1}{2}$ volumes of 95% ethanol overnight at −20°C. Recover the DNA by centrifugation in an

Eppendorf microfuge for 15 min; dry and dissolve the precipitate in 10 μl of restriction enzyme buffer (i.e., *Bam*H-I buffer for *Bam*H-I linkers, etc.). Keep 5 μl for checking ligation on a polyacrylamide gel and digest the remaining 5 μl with the appropriate restriction enzyme. Run both (cut and uncut) samples on an 8% polyacrylamide gel, and stain with 0.5% ethidium bromide to visualize the linker DNA. The ligation ladder should extend to 20–50-mers, and these should all be reduced to 1–2-mers after cleavage.

If this test is positive the linker can be kept frozen and used to exhaustion.

(c) Ligation of Linkers to Blunt-end DNA Fragments. (Either directly from the restriction enzyme cleavaged fragments or after generating flush ends by filling in or S1 nuclease treatment.)

2.5 μl	phosphorylated linkers (i.e., approx. 0.2 μg)
2.0 μl	DNA fragments (approx. 0.5 μg)
1.0 μl	2 mM ATP
4.5 μl	H$_2$O
10 μl	

add 1 μl T4 DNA ligase and incubate overnight at room temperature

(d) Removal of Excess Linkers. After ligation, phenol extract and cleave the excess linkers off the DNA as described above. To remove the excess linker, pass the DNA (dissolved in 25 μl water) over a Sephadex G.100 or a Sepharose CL4 B column made in a 1-ml Falcon plastic pipette using 2 mM tris-HCl pH 7.5, 0.01 mM EDTA as a buffer. The excluded ^{32}P-DNA peak is pooled (usually 4 drops) and lyophilized.

(e) Ligation of Linkered DNA Fragments to the Vector DNA. Redissolve the lyophilized-linkered DNA in 10 μl of 2 × ligation buffer (A.3) and proceed as described in A.3.

10 μl	linkers DNA fragments
1 μl	cleaved vector DNA (0.1 μg)
2 μl	2 mM ATP
7 μl	H$_2$O
20 μl	

add 1 μl T4 DNA ligase and incubate overnight at 37°C

Transfect 1, 3, and 5 μl of ligated DNA as before (A.4).

B.4. Dephosphorylation of Vector DNA Prior to Cloning

It is convenient to dephosphorylate 5–10 μg of *Hind*II-cleaved M13mp7 vector DNA and to resuspend the dephosphorylated DNA at a concentration of 0.1 μg/μl for use as a stock solution ready for cloning (0.1 μg is normally used.)

10 μl	M13mp7 RFI (10 μg)
5 μl	10 × Hin buffer (A.7)
35 μl	H_2O

50 μl

       incubate plus 5 μl *Hind*II (5 units) at 37°C for 2 hr

Check for completeness of cleavage using a 0.35% agarose gel (see A.1).

It is very important to remove all phosphate from the reaction mixture, and the most reliable results are obtained by dialyzing the cleaved DNA. To do this, increase the volume to 200 μl with distilled water and dialyze overnight against 10 mM tris pH 8, 1 mM EDTA. After dialysis, add sufficient 1 M tris pH 7.5 buffer to increase the concentration to 20 mM, and incubate with 1 μl of bacterial alkaline phosphatase (Worthington BAP-F) at 40°C for 60 min. Then phenol-extract three times to remove the phosphatase, wash with ether twice, and precipitate the DNA with salt and ethanol as usual (A.5). Recover the DNA by centrifugation in an Eppendorf microfuge, and dissolve the precipitate in sufficient water to give a concentration of 0.1 μg/ml (i.e., $A_{260} = 2$) ready for cloning.

Before use, check the effectiveness of the dephosphorylation by ligation with and without added restriction enzyme DNA fragments (blunt ended) using the conditions described above (A.4). Five ng of this vector should give about 2 plaques without added DNA fragments and 50–200 plaques with added DNA fragments, all of which contain an insert.

B.5. Use of Internal Primers to Sequence and Cleavage of Chain Extensions from Larger Primers

Restriction enzyme DNA fragments internal to the cloned segment of DNA can be used as primers to sequence from both their 3′ and their 5′ termini (see below). Prepare the DNA fragments as described in A.9.1. Because most of the primer fragments will be larger than 20–30 b.p., the dideoxy-chain extension must be cleaved off so that the sequence can be read from the 3′ end of the primer. The total series of reactions are as follows:

Anneal:

1 μl SS DNA template (approx. 0.5 μg)
1 μl primer fragment (to give a molar ratio of 3 primer:1 template molecule)
1 μl 10 × Hin buffer
2 μl H_2O

_____
10 μl

boil for 3 min in a sealed capillary
anneal by incubation at 67°C for 30 min

Proceed then as described in A.6 (steps 2, 3, 4, and 5); but after the chain extension (15 min at room temperature) and ATP chase (5 min at room temperature), add 0.5 μl of NEB restriction enzyme to each of the dideoxy-chain extension, incubate for 5 min (no longer) at 37°C (put the rack of capillaries and tubes in a 37°C air oven), and then add 10 μl of formamide/dye mix to stop the reaction.

B.6. Exonuclease Treatment of Primers

The sequence of template DNA can be read from the 5′ end of an internal primer, through the sequence of the primer itself, by using Exonuclease III to digest the double-stranded primer DNA fragment (any size) to close to its 5′ termini (Exonuclease III digestion will go past the middle of most double-stranded DNA fragments). The exonuclease digestion is performed prior to annealing to the template DNA. Incubate:

5   μl DNA fragments (approx. 1 pmol; see A.9.3)
1.5 μl 10 × Exonuclease buffer
8.5 μl H_2O

_____
15   μl

add 1 μl Exonuclease III (4 units)

Digest in a sealed drawn-out capillary for 60 min at 37°C and then inactivate the exonuclease by dropping the capillary into boiling water for 3 min.

10 × Exonuclease buffer:
500 mM tris-HCl pH 8
10 mM DTT
10 mM $MgCl_2$

To sequence add:

5 μl	Exonuclease treated primer
1 μl	SS DNA template (0.5 μg)
1 μl	10 × Hin buffer
3 μl	H$_2$O

10 μl

     boil for 3 min
     anneal at 67°C for 30 min

and continue as described in A.3.

Appendix C
Identification of Recombinants

Phage plaques containing recombinant DNA can be identified either by the α-complementation blue/white plaque assay (A.6), by blotting against a ^{32}P-labeled probe, or by using a phosphatased vector that gives only recombinant plaques.

C.1. *Blotting Protocol*

(a) To blot, make a lawn of cells (71–78 or JM 103; see D.2) on a Luria plate by pipetting 1 ml of freshly grown cells onto the plate, swirling until all the surface area is wetted, and then pouring off the excess cells, using a Kimwipe to soak up the dregs. Dry the lawn by leaving the Petri dish open, on a bench.

    (b) When the plate is dry, use a sterile toothpick to spot each of the 1-ml M13 cloned stocks on the lawn in a grid pattern (up to 100 per Petri dish) and incubate the plate at 35°C until the plaques are visible (6 hr to overnight). These grids can be kept for several weeks and still be successfully blotted.

    (c) Lay a circular nitrocellulose filter (Schleicher and Schuell BA85) on the plate (chilled to 4°C), making sure that the contact is complete. Stab a series of holes through the filter and agar with a needle in a pattern that can be used later to orient the filter. Remove the filter (the lawn will not lift if the plates have been properly chilled) and lay it, plaque side uppermost, on a series of 3MM filter papers soaked (no puddles) with the following solutions:

(*a*) 0.5 M NaOH	(7 min, to denature DNA)
(*b*) 1 M tris-HCl pH 7.4	(1 min, to neutralize)

(*c*) 1 M tris-HCl pH 7.4   (1 min, to neutralize)
(*d*) 1.5 M NaCl, 0.05 M tris-HCl pH 7.4   (5 min)

Ordinary household Pyrex glass rectangular baking dishes (8×12 inch) are useful flat surfaces for the solutions. Cut the 3MM paper to fit the bottom of the dish, pour on the solution, and drain. Prepare the complete series of dishes, and transfer the nitrocellulose filters from dish to dish. At each stage, check the pH by touching the nitrocellulose filter with the pH paper, edge on. All solutions should be freshly made up on the day of use. After the final wash, lay the nitrocellulose filter on fresh, dry 3MM paper, plaque side uppermost, and dry to whiteness (approximately 30 min).

(d) When the filter is dry, fill a large Petri dish (20 cm) with ethanol (95%) and drop the filter gently, flatwise, plaques uppermost, onto the surface. Swirl vigorously until the nitrocellulose is completely wet, and then decant off the excess ethanol. Drain the last ethanol by propping the Petri dish on its side. Dry the filter to whiteness; in so doing it should stick to the bottom of the dish (approximately 15 min).

(e) It is normal to take several filters through this procedure together; when they are dry (approximately 30 min), put them between leaves of fresh 3MM paper and stack the papers between covers of corrugated cardboard, taping the sandwich together. Then bake the sandwich in a 67°C oven for 2–24 hr.

(f) When the filters have been baked, wash them in 250–500 ml of prewarmed, prewash solution (see below) and shake them gently at 67°C for 1 hr. Sears-Roebuck plastic sealable "boilable bags" hold these volumes, and up to eight filters can be washed and hybridized together in a single bag. Lay the filled, sealed bag on the bottom of a Pyrex glass baking dish, put the dish in a New Brunswick floor shaker, and shake at 67°C and 30–40 rpm so that the filters gently pass over one another.

(g) To hybridize, transfer the filters to a fresh bag containing 100 ml of prewarmed 6 × SSC-PPFB solution and the ³²P-labeled DNA probe (see below for preparation of a probe). Shake at 30–40 rpm at 67°C as before, but overnight.

(h) When the hybridization is complete, take the filters out of the bag and wash them in 250–500 ml of postwash solution in a fresh bag, shaking at 67°C as before. Repeat once and then air dry the filters on fresh 3MM paper.

(i) Autoradiograph the filter using intensifying screens if necessary.

Solutions for Denaturation. (These should be fresh for each experiment.)

	per 250 ml	
1. 0.5 N NaOH	5 g	
2. 1 M tris-HCl pH 7.4	30.25 g	plus approximately 18 ml conc. HCl (check pH!)
3. 1.5 M NaCl	21.9 g	
0.5 M tris-HCl pH 7.4	15.1 g	plus approximately 9 ml conc. HCl (check pH!)
4. 0.3 M NaCl	4.9 g	

Solutions for Washing and Hybridizing.

per liter

1. 6 × SSC   52.2 g NaCl, 26.5 g Na citrate, $2H_2O$

2. Prewash solution
   (6 × SSC-PPFB solution)

6 × SSC	1 liter
0.02% polyvinyl pyrrollidone (PVP 360, Sigma)	0.2 g
0.02% Ficoll	0.2g
0.02% BSA	0.2 g
0.1% SDS	1.0 g

3. Hybridization solution: same as the prewash solution but with denatured ^{32}P-DNA probe added (see below).

Nick Translation of DNA for ^{32}P-probe.

1 μg	DNA
10 μl	10 × buffer
3 μl	0.1 mM dCTP
3 μl	0.1 mM dGTP
3 μl	0.1 mM dTTP
20–40 μC	(α-^{32}P)dATP (300 Ci/mmol)
84 μl	H_2O
100 μl	

        add 1 μl Boehringer DNA Polymerase I (not Knelow!)
        incubate at room temperature for 60 min

10 × buffer:
    0.5 M tris(HCl) pH 7.8
    0.05 M $MgCl_2$
    0.05 M DDT

After nick translation, pass the 100-μl incubation mixture through a small G.100 Sephadex column (made in a 1-ml Falcon plastic pipette) to remove the unincorporated (α-^{32}P)triphosphates. Use any buffer.

Preparation of the ^{32}P-DNA Probe for Hybridization. E. coli DNA is added to the ^{32}P-DNA probe, and they are denatured in 0.2 N NaOH and treated with diethylpyrocarbonate to inactivate any nucleases present. Add

0.2 ml	^{32}P DNA probe (1–3 × 10^6 cts/min)
0.2 ml	E. coli DNA (1 mg/ml)
0.4 ml	1 N NaOH (fresh)
1.2 ml	H_2O
2.0 ml	

Incubate for 20 min at room temperature, neutralize with HCl, and then treat with 0.1% diethyl pyrocarbonate (i.e., dilute the DEPC 1:10 in 95% ethanol and add 20 μl to the DNA solution) at room temperature for 15 min.

Hybridization. Add the 2 ml of neutralized, denatured ^{32}P-DNA probe to 100 ml of 6 × SSC containing PPFB and 0.1% SDS. This can be used to hybridize up to eight nitrocellulose filters (Petri dish size) in one boilable bag.

Postwash solutions. 3 × SSC (no PPFB/SDS).

C.2. Screening Recombinant Clones for Size at Inserted DNA

The size of the inserted DNA can be assayed directly from the 1-ml stock culture of recombinant plaque, simply by lysing 10–15 μl of the culture and running it on a 0.35% agarose gel. Inserts of 100 b.p. or larger can be detected from the change in mobility of the single-stranded DNA circle. It is useful to remove the host cells from the culture so that the lysate contains only viral DNA. Thus, centrifuge 50 μl of infected cell culture in an Eppendorf microfuge for 2 min.

Lyse the filamentous phage in 15 μl of supernatant by adding 1.5 μl of 10% SDS and 2 μl of 80% glycerol containing 0.3% bromophenol blue dye. Mix and incubate at 67°C for 10 min and electrophorese on a 0.35% agarose gel containing 0.5% ethidium bromide. Use recombinant phage with known-size DNA inserts as size standards.

C.3.   Assay at Orientation of Inserted DNA

The relative orientation of the inserted DNA can be assayed by hybridization and agarose gel analysis.

Mix 10 μl of culture supernatant from two different clones, add 0.2% SDS and dye as described above, and incubate at 67°C for 60 min. Single-stranded DNA circlets containing inserts in the opposite orientation will hybridize to give a partially duplex DNA structure which migrates more slowly on agarose gel electrophoresis. The inserts need not be of equal size, and this method is useful for screening clone banks for region of the genome.

Appendix D

D.1.   Media to Grow Cells and Phage

Recipes for all the media cited in the text are to be found in "Experiments in Molecular Genetics" by Jeffry Miller: Luria broth, p. 433; minimal medial and minimal plates, pp. 431 and 432; X-gal and IPTG, p. 48 or 461; titration of phage, p. 37.

D.2.   Host Cells

The filamentatous phage infect *E. coli* via pili, and therefore the host cells have to be male. Any male *E. coli* strain can be used. Female (F-) cells can be transfected with phage DNA and will generate and release phage into the culture supernatant, but the cells cannot be reinfected. Such a system can be used for biological containment.

For the α-complementation system using M13mp series that contain part of the *lac* operon, cells carry a *lac-pro* deletion in the chromosome, and an F episome (F$'$ *lac* I^q 2 ΔM15 *pro*$^+$) with a proline marker (to maintain it), a deletion of the structural gene for β-galactositase spanning amino acid residues 11–41, and a mutation in the *lac* repressor promoter that causes overproduction of repressor (so that the many copies of the *lac* operator on the M13 phage do not titrate out the repressor). Uninfected cells are therefore *lac⁻*, infected cells *lac⁺*; but infected cells carrying phage with a DNA insert in the β-galactosidase gene are again *lac⁻* (white plaques). The standard *E. coli* host cells for the system are 71–78 {(Δ/*lac,pro*) F' *lac* I^q 2 ΔM15 *pro⁺*} and JM103 {Δ(*lac,pro*), *SupE, StrA,* endA, sbcB15, hsdR4, *thi*, F' *tra* D36 *pro* AB, *lac* I^q 2 M15}. For a full description of the system, refer to the papers by Messing.

D.3. *Preparation of* Ca^{2+}*-treated Cells*

Inoculate 100 ml of Luria broth (D.1) with an overnight culture of JM103 (from a minimal medium plate) and grow at 37°C until the cells are at a density of 5×10^8 cells/ml. Harvest the cells (5,000 rpm in a Sorvall SS34 rotor), resuspend them in 50 ml of sterile 50 mM $CaCl_2$, and leave them on ice for 20 min. Harvest the cells again, and suspend them in 10 ml of 50 mM $CaCl_2$. These cells are now competent to be transfected with DNA.

The efficiency of transfection is approximately 500 plaque/ng of phage RF DNA. This efficiency is strain-dependent and can be increased by different calcium treatment protocols (i.e., using calcium chloride at low pH or by overnight calcium treatment).

D.4. *Preparation of Filamentous Phage RFI DNA*

Inoculate 1 liter of Luria broth with 5 ml of JM103 cells grown overnight to saturation plus 10^{10} p.f.u. of phage particles, and incubate overnight shaking at 37°C. Next day, spin out the cells by centrifuge at 5,000 rpm for 15 min. Keep the supernatant to recover the phage by precipitating with 2% polyethylene glycol and 0.5 M NaCl. Lyse the cells and prepare the RFI DNA by any of the standard plasmid DNA preparation protocols.

Appendix E
Sequencing Double-stranded DNA by the Sanger Methods

The primed synthesis sequencing methods were developed using single-stranded phage DNA as a ready source of single-stranded DNA template, but the methods have been adapted for use directly on double-stranded DNA fragments. Two methods appear to work well.

1. Exonuclease Method. In this method, double-stranded DNA is exhaustively treated with Exonuclease III (3'→5') and the resulting single strands used on a template for restriction enzyme primer DNA fragments. The details of this method are given in Smith (1979).

2. Nick Translation of 5′-terminally Labeled Double-stranded DNA. In this procedure, double-stranded DNA is labeled at its 5′ end with (γ-^{32}P)ATP using T4 polynucleotide kinase and is then randomly nicked with pancreatic DNase to create internal 3′-OH termini. These random termini are then extended with DNA polymerase I and dideoxytriphosphate in the normal extension reaction [but no (α-^{32}P) label]. After denaturation only those chains still attached to the ^{32}P-labeled 5′ terminus will appear on the sequencing gel, and these will give the nucleotide sequence, starting at the 5′ end.

The details of this method are described in Maat and Smith (1978), and a useful variation is given in Sief, Khoury, and Dhar (1980).

ACKNOWLEDGMENTS

The author wishes to acknowledge the time spent in Dr. F. Sanger's laboratory at the Medical Research Council, Cambridge, learning the dideoxy sequencing methodology described here. The other techniques and methods described in this article are a compendium of modifications of standard methods as they were practiced in the author's laboratory until 1979. As most of these methods are well known, reference to the original and subsequent descriptions has been omitted in the interest of clarity and brevity. Thanks are given to Dr. B. L. Smiley and J. Lupsky for critically reading the manuscript and checking the accuracy of the protocols. The work was supported by the NIH grant AI17667A.

REFERENCES

Anderson, S. (1981). DNA sequencing using cloned DNase I-generated fragments. *Nucl. Acids Res.* 9:3016–3027.

Anderson, S., Bankier, A. T., Barrell, B. G., deBruijn, M. H. L., Coulson, A. R., Drouin, J., Eqperon, I. C., Nierlich, D. P., Roe, B. A., Sanger, F., Schreier, P. H., Smith, A. J. H., Staden, R., and Young, I. G. (1981). Sequence and organization of the human mitochondrial genome. *Nature* 290:457–465.

Anderson, S., Gait, M. J., Mayol, L., and Young, J. G. (1980). A short primer for sequencing DNA cloned in the single-stranded phage vector M13mp2. *Nucl. Acids Res.* 8:1731–1743.

Barrell, B. G. (1971). Fractionation and Sequence Analysis of Radioactive Nucleotides. In *Procedures in Nucleic Acid Research*, Vol. 2., eds. G. L. Cantoni and D. R. Davies (New York: Harper & Row), pp. 751–812.

Boeke, J. D. (1981). One and two condon insertion mutants in bacteriophage f1. *Molec. Gen. Genetics* 181:288–291.

Boeke, J. D., Vovis, G. F., and Zinder, N. D. (1979). Insertion mutant of f1 sensitive of EcoR1. *Proc. Natl. Acad. Sci.* 76:2699–2702.

Brownlee, G. G., and Sanger, F. (1969). Chromatography of ^{32}P-labelled oligonucleotides on thin layers of DEAE-cellulose. *Eur. J. Biochem.* 11:395–399.

Duckworth, M. L., Gait, M. J., Goelet, P., Hong, G. F., Singh, M., and Titmas, R. C. (1981). Rapid synthesis of oligodeoxynucleotides VI. Efficient mechanised synthesis of lepta decadeoxyribonucleotides by an improved solid phase phosphotriester route. *Nucl. Acids Res.* 9:1691–1706.

Galibert, F., Sedat, J., and Ziff, E. (1974). Direct determination of DNA nucleotide sequences: Structure of a fragment of bacteriophage ϕX174 DNA. J. Mol. Biol. 87:377–407.

Gardner, R. C., Howarth, A. J., Hahn, P., Brown-Luedi, M., Shepherd, R. J., and Messing, J. (1981). The complete nucleotide sequence of an infectious clone of cauliflower mosaic virus by M13mp7 shotgun sequencing. *Nucl. Acids Res.* 9:2871–2888.

Gingeras, T. R., Milazzo, T. J. P., Sciaky, D., and Roberts, R. J. (1979). Computer programs for the assembly of DNA sequences. *Nucl. Acids Res.* 7:529–545.

Godson, G. N., Barrell, B. G., Staden, R., and Fiddes, J. C. (1978). Nucleotide sequence of bacteriophage G4 DNA. *Nature* 276:236–247.

Gronenborn, B., and Messing, J. (1978). Methylation of single-stranded DNA *in vitro* introduces new restriction endonuclease cleavage sites. *Nature* 272:375–377.

Jeppesen, P. G. N. (1974). A method of separating DNA fragments by electrophoresis in polyacrylamide concentration gradient slab gels. *Anal. Biochem.* 58:195–207.

Kaguni, J., and Ray, D. S. (1979). Cloning of a functional replication origin at phage G4 into the genome at phage M13. *J. Mol. Biol.* 135:863–878.

Korn, L. J., Queen, C. L., and Wegmen, M. N. (1977). Computer analysis of nucleic acid regulatory sequences. *Proc. Natl. Acad. Sci.* 74:4401–4495.

Kornberg, A. (1969). Active center of DNA polymerase. *Science* 163:1410–1418.

Maat, J., and Smith, A. J. H. (1978). A method for sequencing restriction fragments with dideoxynucleoside triphosphates. *Nucl. Acids Res.* 5:4537–4545.

Messing, J., and Vieira, J. (1982). A new pair of M13 vectors for selecting either DNA strand of double-digest restriction fragments. *Gene* 19:269–276.

Messing, J., Crea, R., and Seeburg, P. H. (1981). The system for shotgun DNA sequencing. *Nucl. Acids Res.* 9:309–321.

Messing, J., Gronenborn, B., Mueller-Hill, B., and Hofschneider, P. Y. (1977). Filamentous coliphage M13 as a cloning vehicle: Insertion of a *Hind*II fragment of the *lac* regulatory region in M13 replicative form *in vitro*. *Proc. Natl. Acad. Sci. USA* 74:3642–3646.

Ray, D. S., and Kook, K. (1978). Development of M13 on single stranded cloning vector: Insertion of Tn3 transposon into the genome of M13. In *Single Stranded DNA Phages* (Cold Spring Harbor Laboratory Press), pp. 455–459.

Robertson, H. D., Barrell, B. G., Weith, H. L., and Donelson, J. E. (1973). Isolation and sequence analysis of a ribosome-protected fragment from bacteriophage ϕX174 DNA. *Nature New Biol.* 241:38–40.

Sanger, F., and Coulson, A. R. (1978). The use of this acrylamide gels for DNA sequencing. *FEBS Lett.* 87:107–110.

Sanger, F., and Coulson, A. R. (1975). A rapid method for determining sequences in DNA by primed synthesis with DNA polymerase. *J. Mol. Biol.* 94:441–448.

Sanger, F., and Thompson, E. O. P. (1963). The amino acid sequence in the glycyl chain of insulin. *Biochem. J.* 53:353–374.

Sanger, F., Brownlee, G. G., and Barrell, B. G. (1965). A two-dimensional fractionation procedure for radioactive nucleotides. *J. Mol. Biol.* 13:373–398.

Sanger, F., Nicklen, S., and Coulson, A. R. (1977). DNA sequencing with chain-terminating inhibitors. *Proc. Natl. Acad. Sci. USA* 74:5463-5467.

Sanger, F., Air, G. M., Barrell, B. G., Brown, N. L., Coulson, A. R., Fiddes, J. C., Hutchinson, C. A., III, Slocombe, P. M., and Smith, M. (1977). Nucleotide sequence of bacteriophage ϕX174 DNA. *Nature* (London) 265:687-695.

Sanger, F., Coulson, A. R., Friedmann, T., Air, G. M., Barrell, B. G., Brown, N. L., Fiddes, J. C., Hutchinson, C. A. III, Slocombe, P. M., and Smith, M. (1978). The nucleotide sequence of bacteriophage ϕX174. *J. Mol. Biol.* 125:225-246.

Sanger, F., Donelson, J. E., Coulson, A. R., Kossel, H., and Fischer, D. (1973). Use of DNA polymerase I primed by a synthetic oligonucleotide to determine a nucleotide sequence in phage f1 DNA. *Proc. Natl. Acad. Sci. USA* 70:1209-1213.

Seif, I., Khoury, G., and Dhar, R. (1980). A rapid enzymatic DNA sequencing technique: Determination of sequence alterations in early simian virus 40 temperature sensitive and deletion mutants. *Nucl. Acids Res.* 8:2225-2240.

Smith, A. J. H. The use of exonuclease III for preparing single-stranded DNA for use as a template in the chain terminator sequencing method (1979). *Nucl. Acids Res.* 6:831-848.

Staden, R. (1980). A new computer method for the storage and manipulation of DNA gel reading data. *Nucl. Acid Res.* 8:3673-3694.

Staden, R. (1979). A strategy of DNA sequencing employing computer programs. *Nucl. Acids Res.* 6:2601-2610.

Staden, R. (1977). Sequence data handling by computer. *Nucl. Acids Res.* 4:4037-4051.

Yamamoto, K., Alberts, R. M., Benzinger, R., Lawhorne, L., and Treiber, B. (1970). Rapid bacteriophage sedimentation in the presence of polyethylene glycol with application to large scale virus purification. *Virology* 40:734-744.

Yang, R. C. A., and Wu, R. (1979). BK virus DNA: Complete nucleotide sequence of a human tumor virus. *Science* 206:456-462.

Ziff, E. B., Sedat, J. W., and Galibert, F. (1973). Determination of the nucleotide sequence of a fragment of bacteriophage ϕX174 DNA. *Nature New Biol.* 241:34-37.

4

Nucleotide Sequence of DNA

ALLAN MARSHALL MAXAM

INTRODUCTION

During the early 1970s a number of laboratories began to search in earnest for a direct and effective way to sequence bases in DNA. These pursuits were instigated by developments in both molecular genetics and DNA biochemistry. For some, genetic mapping and physical isolation of the *lac* operator (Sadler & Smith, 1971; Bourgeois & Riggs, 1970; Gilbert, 1972), the λ operators (Maniatis & Ptashne, 1973a; Pirrotta, 1973), and a phage fd promoter (Heyden, Nusslein, & Schaller, 1972) indicated that the primary structure of DNA control elements was within reach. For others, genetic and physical mapping of the proteins and messenger RNAs encoded by the small single-stranded chromosomes of phages fd, f1, and especially φX174 implied that defined genes might be accessible to sequencing.

Throughout this period there also ensued a stream of results obtained with reagents and enzymes that dissect or copy DNA. Pyrimidine tract isostichs were catalogued for the phages φX174, S13, T7, λ, fd, and f1 (Burton & Peterson, 1960; Spencer & Chargaff, 1963; Hall & Sinsheimer, 1963; Cerny, Cerna, & Spencer, 1969; Mushynski & Spencer, 1970a, 1970b; Ling, 1972a, 1972b; Delaney & Spencer, 1973). Experiments introducing ^{32}P-phosphate onto the ends of bacteriophage chromosomes with DNA polymerase or poly-nucleotide kinase and sequencing rather short distances accumulated, resulting in the review of nine in 1974 (Murray & Old, 1974). The

Sanger group reported a sequence of unknown function added to an octanucleotide primer bound somewhere on the phage f1 chromosome (Sanger et al., 1973). Ling (1971) produced an array of discrete phage fd chromosome fragments with the predominantly C-specific T4 Endonuclease IV (Sadowski & Hurwitz, 1969). Danna and Nathans (1971) digested SV-40 DNA with the restriction endonucleases *Hind*II+III and displayed eleven fragments of the chromosome on a gel, followed immediately by Middleton, Edgell, & Hutchison III (1972) with thirteen from ϕX174 with *Hae*III.

With few exceptions, these experiments were not greatly concerned with polymerase, kinase, or restriction enzyme mechanisms, nor very seriously with gene structure (although all are now of prime concern). Most of the experiments just mentioned were, in effect, trying out these enzymes and reagents as tools with which to study the primary structure of DNA. They soon proliferated, and their implication was clear: if existing genetic maps could be phased with the new restriction maps, then restriction enzymes, DNA polymerases, DNA kinase, and certain base-specific reagents might provide access to DNA sequences of known function in chromosomes thousands of base-pairs long.

Not only was this anticipation correct, it was quickly realized for a number of bacterial and phage operons and was recently exceeded when several phage and plasmid chromosomes were sequenced in entirety. This introduction recounts the pursuit and succession of DNA sequencing techniques since 1970, focuses on the three methods that have contributed to the recent flow of DNA primary structural information and, at the end, considers base sequence interpretation.

SEQUENCING BASES IN NUCLEIC ACIDS

The object of DNA or RNA sequencing is to elicit which base— adenine, guanine, cytosine, or thymine/uracil—is attached to each sugar along the backbone. For a unique single strand 100 nucleotides long, this is equivalent to finding out which of 4^{100} (10^{60}) possible configurations of 4 items is at hand. Sequencing techniques introduced over the past ten years have exploited the base specificities of of endonucleases, template strands, and a few chemical reagents to make this a far easier task than it sounds. This section will review some of those techniques in detail. But first, formal solutions to the problem of sequencing bases in nucleic acids will be considered, so that the actual solutions can later be seen in perspective.

Given a unique DNA or RNA, a sequencing method must do two things to solve its covalent primary structure:

1. Distinguish the bases, with agents specific for or properties intrinsic to adenine, guanine, cytosine, and thymine or uracil, and if present, the less common bases.
2. Order nucleotides along the backbone, by specifying an operation that will associate each with the one that precedes it (5'), or the one that follows it (3').

Distinguishing the Bases

Bases can be identified directly or inferentially. The classic approach to RNA sequencing (Sanger, Brownlee, & Barrell, 1965) processes ^{32}P-labeled RNA through one or more rounds of base-specific ribonuclease digestion and oligonucleotide fractionation, followed by alkaline hydrolysis, paper electrophoresis, and autoradiography to reveal the four mononucleotides. It thus identifies bases both directly, by mobility and radioactivity, and by inference, from proximal cleavage by a base-specific ribonuclease. The newer gel sequencing methods distinguish the bases by cleaving strands with agents of known base-specificity (Maxam & Gilbert, 1977), or by terminating synthesis at complementary template positions with single-addition nucleotide precursors (Sanger, Nicklen, & Coulson, 1977). The following is a list of properties that have been used in sequencing techniques to distinguish adenine, guanine, cytosine, and thymine/uracil nucleotides:

Electrophoretic mobility at pH 3.5 (J. D. Smith, 1967)

Displacement from anion exchange layers (Randerath & Randerath, 1967)

Mobility shift of oligonucleotide on loss of terminal nucleotide (variation on above two properties, described in a later section)

Radioactivity, if only one of the four is labeled

Ultraviolet absorption spectra (Beaven, Holiday, & Johnson, 1955)

Recognition by base-specific endonucleases, with proximal strand cleavage (Holley et al., 1965; Sanger, Brownlee, & Barrell, 1965)

Chemical substitution or fission of purine or pyrimidine rings, leading to strand cleavage (reviewed in Kochetkov & Budovskii, 1972, and in D. M. Brown, 1974; Maxam & Gilbert, 1977)

Base-complementarity, by terminating synthesis at specific template positions (Wu & Kaiser, 1968; Englund, 1971, 1972; Sanger & Coulson, 1975; Sanger, Nicklen, & Coulson, 1977)

Ordering Nucleotides

Ordering is the greatest challenge to nucleic acid sequencing methods. Identifying bases merely requires distinguishing four nonidentical heterocyclic rings. But positioning nucleotides means discriminating each of tens to hundreds of identical sugar-phosphate links in a backbone.

Of course, a sequencing method must both identify and order nucleotides to sequence. Thus

$$pNpNpNp \rightarrow pA, pG, pC$$

identifies three bases, and

$$pNpNpN \rightarrow pNp, Np, N$$

certainly reveals a nucleotide order, but neither establishes the base sequence. Only a base analysis *with* an ordering operation will yield the base sequence:

$$pNpNpN \rightarrow pAp, Gp, C = pApGpC$$

Nucleotide order has been established in three fundamentally different ways: by removing nucleotides sequentially from one end, by overlapping oligonucleotide tracts or blocks, or by nesting partial fragments or partial copies. Table 4-1 illustrates these three sequencing strategies with the DNA segments that ideal versions of each would generate.

Stepwise Degradation

Removing nucleotides sequentially from an end is a most intuitive way to sequence the bases in a DNA or RNA strand. Beginning at one end, one removes the terminal nucleotide and identifies its base,

Table 4-1. Unit, Track/Block, and Nested Segment Sequencing Strategies

Sequencing strategy	Pattern of cleavage or synthesis	Base Control	Labeling required	Further analysis required
STEPWISE DEGRADATION	G–A–G–C–T–A–A–C–T	N	Uniform or terminal at steps	Identify nucleosides released at steps
	G–A–G–C–T–A–A–C	+N9 T		
	G–A–G–C–T–A–A	+N8 C		
	G–A–G–C–T–A	+N7 A		
	G–A–G–C–T	+N6 A		
	G–A–G–C	+N5 T		
	G–A–G	+N4 C		
	G–A	+N3 G		
	G	+N2 A		
TRACT OR BLOCK OVERLAP				
(1) Chemical excision tracts	G G–C–T C–T	–A	Uniform or terminal	Sequence and overlap tracts
	A C–T–A–A–C–T	–G		
	G–A–G T–A–A T	–C		
	G–A–G–C A–A–C	–T		
(2) Nuclease incision blocks	G–A G–C–T–A A C–T	+A	Uniform or terminal	Sequence and overlap blocks
	G A–G C–T–A–A–C–T	+G		
	G–A–G–C T–A–A–C T	+C		
	G–A–G–C–T A–A–C–T	+T		

(continued)

Table 4-1. (*Continued*)

Sequencing strategy	Pattern of cleavage or synthesis	Base Control	Labeling required	Further analysis required
NESTED FRAGMENTS OR COPIES				
(1) Chemical excision fragments	G	−A	Terminal for fragments	Phase and separate four sets of fragments or copies by size
	G−A−G−C−T	−A		
	G−A−G−C−T−A	−A		
	G−A	−G	Uniform or terminal for copies	
	G−A−G	−C		
	G−A−G−C−T−A−A	−C		
	G−A−G−C	−T		
	G−A−G−C−T−A−A−C	−T		
(2a) Nuclease incision fragments	G−A	+A	Terminal for fragments	Phase and separate four sets of fragments or copies by size
	G−A−G−C−T−A	+A		
	G−A−G−C−T−A−A	+A		
(2b) Inhibitor terminated copies	G	+G	Uniform or terminal for copies	
	G−A−G	+G		
	G−A−G−C	+C		

	Sequence		
	G–A–G–C–T–A–A–C	+C	Identify terminal nucleotides by mobility shift or end-group analysis
	G–A–G–C–T	+T	
	G–A–G–C–T–A–A–C–T	+T	
(3a) Randomly cleaved fragments	G	+/–N	Terminal for fragments
	G–A	+/–N	
	G–A–G	+/–N	
(3b) Randomly terminated copies	G–A–G–C	+/–N	Uniform or terminal for copies
	G–A–G–C–T	+/–N	
	G–A–G–C–T–A	+/–N	
	G–A–G–C–T–A–A	+/–N	
	G–A–G–C–T–A–A–C	+/–N	
	G–A–G–C–T–A–A–C–T	+/–N	

and then takes what is left of the fragment, removes the second nucleotide, and identifies it, and so on, repeating the cycle as many times as possible. This approach has the advantage that it identifies bases directly and might be automated; its disadvantage is cumulative sequence noise when cleavage is not complete in successive cycles.

Tract or Block Overlap

A second strategy is to cleave the DNA or RNA into sets of discrete tracts or blocks (Table 4-1). Oligonucleotide tracts are what remains of a polynucleotide after exhaustive chemical destruction of one to three of its four component nucleosides ("excision"). Tracts are thus distinguished by their *base composition* (bases not attacked). Oligonucleotide blocks, on the other hand, are pieces of DNA or RNA created by enzymatic strand cleavage proximal to one to three of the four bases ("incision"). Blocks are defined both by their *base composition* (bases not recognized) and their common *terminal nucleotide* (the one recognized at cleavage).

Complete cleavage of a DNA or RNA at every occurrence of a base yields a tract or block set unique to that sequence. Parallel cleavage at another base gives another set. If the original DNA/RNA is small (tens of nucleotides), certain oligonucleotides in one set will overlap uniquely with some in the other. The more restricted the cleavage specificity, the greater the frequency and extent of overlap; single base specificity is ideal. Note also that excision destroys sequence but incision does not, causing blocks to exhibit more overlap in general than tracts (Table 4-1). In either case, strand scission can be partial, followed by isolation of clusters, and then complete, to establish tract or block linkage without reference to a second set.

Oligonucleotides have to be sequenced before they can be overlapped. Stepwise degradation could sequence them, but so can reciprocal block cleavage. Consider a hexanucleotide $N-N-N-N-N-N$ and three endonucleases—one that cleaves after A (Endo A), another after C (Endo C), and a third every nucleotide in turn (Exo N)— and the following two part experiment.

In one part, Endo A cleaves $N-N-N-N-N-N$ to $N-N-N-N$, N. Separation and quantitation reveals $N = 2A$, and in subroutines Endo C splits $N-N-N-N$ into $N-N$, $N-N$, which Exo N in turn breaks down to mononucleotides: $N-N \rightarrow$ A, T and $N-N \rightarrow$ C, G. At this point, working backward through the A/C cleavage specificities in steps will reconstruct the order of these nucleotides:

C, G → G—C   (Endo C)
A, T → T—A   (Endo A)
G—C, T—A → G—C—T—A   (Endo A)
N—N—N—N—N—N = G—C—T—A, A, A

In the other half of the experiment, the two base-specific nucleases are reversed, with first Endo C and then Endo A and Exo *N* breaking the hexanucleotide down in steps. Likewise, systematic reassociation of these C/A products reveals:

N—N—N—N—N—N = A—A—G—C, T, A

Finally, overlapping reciprocal blocks from the two experiments indicates their order and the sequence of *N—N—N—N—N—N*:

Part 1:   A   A   G—C—T—A
Part 2:   A—A—G—C   T   A
Sequence: A—A—G—C—T—A

This nucleic acid sequencing strategy, block-overlap (Table 4-1), is the principle behind the classic RNA sequencing techniques introduced in 1965 (Holley et al., 1965; Sanger, Brownlee, & Barrell, 1965), one of which has been used to sequence RNA copies of DNA (see below).

Nested Fragments or Copies

A third strategy sequences unique DNA by reducing it to a population of nesting segments. Partially cleaving or limited primed copying of a DNA strand will create these segments. Nested DNA segments have a common end and a variable end:

Common ends   *N*
              NN
              NNN
              NNNN   Variable ends
              NNNNN
              NNNNNN

The common end may be one terminus of the original DNA, which was radioactively labeled, or it may be the 5′ end of a primer, or primer extensions, if the primer was clipped off. This shared end relates all the segments. The heterogeneous ends are points at which an agent cleaved the DNA, or at which copying terminated. These variable ends distinguish segments in the nested population.

Nested-segment sequencing methods identify bases by their presence or absence at the variable end. They order those bases by comparing the distance, in nucleotides, from one segment (n) and from the next one ($n + 1$) to a fixed reference point, the common end. This sequencing strategy was proposed for nucleic acids over 15 years ago by Mandeles and Tinoco, Jr. (1963).

When limited internal cleavage generates the nested fragments, the DNA must be labeled at one unique end beforehand. Then, when one double-stranded molecule is cleaved twice—once in one strand and once in the complementary strand—only fragments carrying the original end label will be detected. Pieces of DNA distal to the labeled end and from the other strand are not labeled and therefore do not clutter the nested set. Alternatively, if primed partial copying generates the nested segments, the radioactive label may be introduced either beforehand, on the primer, or during synthesis, in nascent DNA.

Cleaving or copying reactions that are not base-specific will give a *continuous* set of nesting segments. In continuous sets every base in the sequence is represented by a segment that extends from it all the way back to the common end. Neighboring cleavages or terminations give two DNA segments of n and $n + 1$ nucleotides. Polynucleotide pairs of this kind differ only by a small increment of mass and charge, one of the four mononucleotides. Sequencing with continuous sets then consists of assorting all the nested segments by size and identifying the single nucleotide by which each differs from the next smaller one. The procedure for this method is described in a later section on wandering-spot sequencing methods.

Base-specific cleavages or copying terminations, on the other hand, generate *discontinuous* sets of nested segments. In each set, all the variable ends are, or immediately precede, one of the four bases. Four base-specific discontinuous sets are required for sequencing— one each for adenines, guanines, cytosines, and thymines—or some redundant combination thereof. When four populations of discontinuous partial products of this kind are assorted in phase by size, sequencing consists of noting which base-specific agent was responsible for each cleavage or termination in the four-channel succession.

In conclusion, *complete* excision of or incision at one of the four bases gives a set of tracts or blocks. Likewise, *partial* excision or incision will give a unique set of nested segments, as will limited strand copying.

So far, we have considered the above DNA sequencing strategies as though all were always available and all operate with single base specificity. Table 4-1 certainly suggests this, but such was never the case. To see how and in what form these strategies have been realized as techniques, our consideration of DNA sequencing now shifts from a formal to a historical point of view.

PHASE ONE: TRACT AND BLOCK METHODS

The initial phase in the solution to DNA sequencing extended roughly from 1970 to 1973. It began with the first direct determination of an extensive DNA sequence, for the guinea pig α-satellite (Southern, 1970), and ended with the sequences of several promoters and operators. Three techniques emerged in this period which, upon introduction, solved some aspect of DNA sequencing for the first time and could be used on any DNA (as opposed, for instance, to terminal nucleotides). The three techniques, which had both novelty and generality, are pyrimidine tract analysis, copying DNA into RNA, and ribosubstitution. Figure 4-1 summarizes the DNA synthetic and cleavage steps characteristic of each of these techniques.

Realization of all three techniques depended largely on the Sanger-Brownlee-Barrell method for sequencing ^{32}P-labeled RNA (Sanger, Brownlee, & Barrell, 1965). By 1970 this method had been applied to RNA sequencing for five years, demonstrating the utility of block-overlap for nucleic acids. It also led to the perfecting of two-dimensional oligonucleotide fingerprinting systems essential to the three new DNA techniques: electrophoresis at pH 3.5 on cellulose acetate, followed by electrophoresis on DEAE-cellulose paper (Sanger, Brownlee, & Barrell, 1965), or homochromatography on DEAE-cellulose thin layers (Brownlee & Sanger, 1969).

Pyrimidine tracts and ribosubstituion blocks had to be sequenced to be overlapped, but no universal DNA oligonucleotide sequencing technique appeared during this time. Rather, a variety of copying, labeling, and partial digestion tricks, some invented for RNA and carried over to DNA, solved this problem along the way. These included copying with one of the four nucleoside triphosphates ^{32}P-labeled in the α position (Billeter et al., 1969), for transfer of label to 5′ neighbor nucleotides (Josse, Kaiser, & Kornberg, 1961); 5′-phosphorylation of oligonucleotides with polynucleotide kinase and (γ-^{32}P)ATP (Szekely & Sanger, 1969); and partial digestion with spleen and venom exonucleases (reviewed in Barrell, 1971, and in Brownlee, 1972). These will be described in context below.

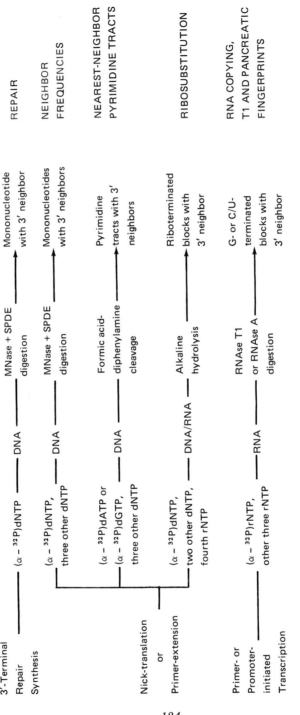

Figure 4-1. Summary of tract and block DNA sequencing techniques. *Abbreviations:* VPDE = venom phosphodiesterase, SPDE = spleen phosphodiesterase, MNase = micrococcal nuclease, d = deoxy, r = ribo, NTP = nucleoside triphosphate.

Pyrimidine Tracts

Pyrimidine tracts are runs of contiguous pyrimidine nucleotides, left after excision of all purine nucleosides from DNA, which can be separated and sequenced. Several crucial depurination experiments preceded the first use of pyrimidine tract analysis to sequence DNA. First, in studying hydrolysis of DNA to bases for compositional analysis, Shapiro and Chargaff found acidic conditions under which purines but not pyrimidines were released from sugars (Shapiro & Chargaff, 1957a, 1957b, 1957c). Next, Burton and Peterson worked up a reaction (66% formic acid, 2% diphenylamine) that completely depurinated and cleaved the DNA backbone, leaving intact clusters of pyrimidine nucleotides with phosphates on both ends (Burton & Petersen, 1957, 1960). Finally, Hall and Sinsheimer applied this reaction to bacteriophage ϕX174 DNA and found a subset of the pyrimidine isopliths obtained from complex calf-thymus DNA, demonstrating genome specificity (Hall & Sinsheimer, 1963).

The stage was thus set for a pyrimidine tract DNA sequencing experiment. This took place in 1970 when Southern (1970) separated the strands of guinea pig α-satellite DNA, subjected each to the Burton-Peterson reaction (above), and resolved the pyrimidine oligo-nucleotide products with a version of Sanger two-dimensional electrophoresis. Fingerprints from each strand exhibited one strong spot embedded in a simple background, T—T from the heavy strand and C—C—C—T from the light strand, with one- or two-base changes from C—C—C—T accounting for most spots in its background. Juxta-position of the two tracts on opposite strands then suggested that guinea pig α-satellite DNA is predominantly a simple, repeating hexanucleotide:

$$\begin{matrix} \text{—C—C—C—T—} & & \text{—C—C—C—T—A—A—} \\ \text{—T—T—} & \rightarrow & \text{—G—G—G—A—T—T—} \end{matrix}$$

This pyrimidine tract sequencing experiment is notable for having produced the first unique DNA fingerprint, and for having first determined the core sequence of a specialized DNA.

Two years later, Ling published a high-resolution pyrimidine "fingerprint" of a phage chromosome (Ling, 1972a). He depurinated and cleaved DNA from phage fd with the Burton-Peterson reaction, electrophoresed the pyrimidine tracts on cellulose acetate at pH 3.5, transferred them to a thin layer of DEAE-cellulose, and then sub-jected them to displacement by oligonucleotides of equal charge ascending in chromatographic fronts. This two-dimensional fractionation system, electrophoresis → homochromatography (Brownlee & Sanger, 1969), resolved all but the isomeric fd tracts. Ling sequencd many of the larger tracts, including the longest, 20 consecutive pyrimidine nucleotides, with a new partial exonuclease digestion/ mobility-shift technique (described below in the section on wandering-spot sequencing methods).

Pyrimidine tract analysis was then applied to protein binding sites. That same year, Heyden, Nusslein, and Schaller (1972) produced pyrimidine tracts from a 40–base-pair fragment of fd DNA protected against DNAse I digestion by RNA polymerase *in vitro*, and the same group later worked out an fd promoter sequence with only pyrimidine tract data (Schaller, Gray, & Hermann, 1975). In 1973, Maniatis and Ptashne found nested pyrimidine tract sets in λ operator fragments of increasing size, protected against DNAse I by repressor of increasing concentration, and on this account proposed multiple operators in phage λ (Maniatis & Ptashne, 1973b).

Copying DNA into RNA

In its time, transcribing DNA into an RNA copy for sequencing purposes amounted to a retreat from the effort to sequence DNA directly, back to the certainty of Sanger pancreatic and T1 ribo-nuclease fingerprints (Sanger, Brownlee, & Barrell, 1965). However, in converting DNA into a replica nucleic acid which could be sequenced by a sure method, RNA copying succeeded, which is the prime requirement of any DNA sequencing method. For unambiguous sequencing, an RNA copy of DNA must be a uniform transcript of one strand. Hence, use of this DNA sequencing technique has usually required priming and/or trimming to limit the copy to the sequence of interest.

One strategy first trims the DNA and then copies it into RNA. Several kinds of defined DNA fragments have been copied *in vitro* with RNA polymerase and sequenced: the template for a unique 6S transcript of λ bacteriophage DNA (Lebowitz, Weissman, & Radding, 1971); a 29-base strand encoding part of $tRNA^{tyr}$ produced by organic synthesis (Terao, Dahlberg, & Khorana, 1972); 27–base-pair fragments protected against DNAse I digestion by *lac* repressor (Gilbert & Maxam, 1973); 40-45 base-pair DNA protected against DNAse by RNA polymerase (Walz & Pirrotta, 1975; Sugimoto et al., 1975); 48- and 200-base strands released from φX174 DNA by T4 endonuclease IV (Blackburn, 1975, 1976); and restriction fragments of various lengths (Dhar, Zain, et al., 1974; Fiers et al., 1974). In most of these cases, the RNA was synthesized off both complementary strands (when both were present), initiated and terminated at multiple but nonrandom points, included some end-to-end transcripts originating at the last template pyrimidine, and initiated repeatedly, giving net synthesis. Oligonucleotide primers plus noninitiating nucleoside triphosphate concentrations (Downey & So, 1970) have restricted initiation to one or a few points on small templates, and transcripts of one strand have been resolved from those of the other by gel electrophoresis and by hybridization to a viral plus strand (Gilbert & Maxam, 1973; Pirrotta, 1975; Walz & Pirrotta, 1975; Sugimoto et al., 1975). Some have found complementary sequences in tandem at the 3' ends of transcripts, where only the first half was known to be present in the DNA template; this suggests that transcribing RNA polymerases can double back or jump from one template to another (Terao, Dahlberg, & Khorana, 1972; Volckaert et al., 1977).

A second strategy copies the DNA into RNA first and then trims the RNA. There are three steps: First, messenger RNA is transcribed *in vitro* off one strand of a chromosomal DNA (usually phage), having the target DNA sequence embedded in it somewhere downstream from the promoter. Second, this copy RNA is hybridized to a second DNA, which lacks regions that flanked the target in the original template, and digested with ribonuclease. Finally, the target-protected RNA segment is separated from the hybrid and is fingerprinted.

Pribnow used copy DNA/trim RNA to sequence physically defined DNA segments, the second and third phage T7 early promoters (Pribnow, 1975a, 1975b). In the first experiment (Pribnow, 1975a) he initiated messenger RNA *in vitro* at the first two promoters (A1 and A2) and elongated through the third (A3), and prepared third-promoter DNA, protected against DNAse I digestion by RNA polymerase. Hybridizing the RNA and DNA protected an RNA copy of one strand of the A3 promoter against ribonuclease, which he then isolated and sequenced.

Barnes et al. (1975) used copy DNA/trim RNA to isolate a genetically defined DNA target, the *lac* control region. They prepared RNA copies of each strand separately, initiated at promoters in two transducing phage carrying opposite downstream orientations of the *lac* operon, and *lac* phage DNAs finely deleted for on *i* or the other *z* genes flanking the promoter-operator region. Sequential rounds of hybridization and ribonuclease digestion to the two phage DNAs cut each RNA down to a copy of one strand of the control region. Each strand was sequenced (Dickson et al., 1975) by the traditional method (Sanger, Brownlee, & Barrell, 1965).

RNA copying in both trimming modes amassed extensive and significant DNA sequence during its time, including that of the *lac* operator (Gilbert & Maxam, 1973) and promoter (Dickson et al., 1975), part of the left operator region in phage λ (Pirrotta, 1975), and the promoters in the phages T7 (Pribnow, 1975a, 1975b), fd (Sugimoto et al., 1975), and λ (Walz & Pirrotta, 1975) and SV40 DNA (Dhar et al., 1974b; Zain et al., 1974). During and after this period, the Fiers and Weissman groups used the same strategy to sequence portions of SV40, protecting randomly or uniquely initiated copy RNA with physically defined restriction fragments (Dhar, Zain et al., 1974; Fiers et al., 1974).

Ribosubstitution

The ribosubstitution DNA sequencing technique grew out of the observation by Berg, Fancher, and Chamberlin (1963) that replacing Mg^{2+} with Mn^{2+} and supplying three deoxy- and one ribonucleoside triphosphates would induce DNA polymerase I to incorporate ribonucleotides into DNA. Ribosubstitution is employed in DNA sequencing as follows. A DNA repair reaction is set up, containing DNA polymerase I and a single-stranded template, to which is hybridized a complementary endogenous or exogenous primer. To this reaction is added manganese chloride, one ribonucleoside triphosphate, and the other three deoxynucleoside triphosphates. DNA polymerase extends the primer, inserting ribo-C opposite template guanines, and deoxy-A, deoxy-G, and deoxy-T nucleotides at all other positions. This gives primer extensions fully substituted with ribo-C, which ribonuclease or alkali will break at every —CpN— linkage into C-terminal blocks.

The ribonucleoside triphosphate can be ATP, GTP, or CTP, and any one of the triphosphates (ribo or deoxyribo) may be α-labeled with ^{32}P, for transfer to nearest neighbors after incorporation and cleavage. After ribonuclease digestion or alkaline hydrolysis and fingerprinting, the block oligonucleotides are eluted and sequenced.

Salser et al. (1972) tested ribosubstitution repair reactions on sheared M13 DNA, concluded from fingerprint analysis that they copied DNA faithfully, and so used them on a rat satellite DNA (Fry et al., 1973). Some have questioned the fidelity of DNA polymerase I copying with Mn^{2+} when rGTP is substituted for dGTP

(Van de Sande, Loewen, & Khorana, 1972), but all G-terminal ribosubstitution blocks from the left λ operator region (Maniatis et al., 1974) were later confirmed by direct DNA sequencing. Sanger et al. (1973) sequenced 50 nucleotides in phage f1 by hydrolyzing a nested series of ribosubstituted primer extensions and sequencing the oligonucleotides (described in the next section). Much later, Barnes (1978) made limited ribosubstitution the basis of a nested-segment DNA sequencing technique.

PHASE TWO: WANDERING-SPOT METHODS

Despite successes, oligonucleotide block and tract methods were considered temporary solutions to DNA sequencing. First, without a pure site-specific DNA binding protein, sets of mapped deletions, or an appropriate oligonucleotide primer there was no obvious route from chromosomal DNA to an appropriate small copy RNA. Second, pyrimidine tracts—the only chemically homogenous DNA tracts then available—contain only two of the four bases and exhibit unambiguous overlaps only with very small or highly repetitious DNAs. Third, although straightforward, sequencing and overlapping ribonuclease A and T1 oligonucleotides of copy RNA was slow and tedious. Fourth, restriction enzymes of increasing variety were cleaving phage and plasmid DNAs into distinct segments which existing methods were not fully exploiting. What was needed was a way to sequence DNA restriction fragments directly and continuously. A quite different sequencing method emerged in 1973 which began to meet this need.

The new, direct DNA sequencing technique became known as two-dimensional mobility-shift analysis, or the wandering spot. It can be traced back to the cytosine- and thymine-distinguishing electrophoresis → homochromatography mobility shifts exploited by Ling to sequence the long pyrimidine tracts of phage fd (Ling, 1972a), and farther back to sequencing oligonucleotides by partial exonuclease digestion and DEAE-cellulose chromatography (Razzell & Khorana, 1961; Holley, Madison, & Zamir, 1964); the latter gives a one-dimensional array of nested segments, each (n) differing from the next smaller one ($n-1$) by a single terminal nucleotide. When similar partial exonuclease products are electrophoresed on DEAE-cellulose paper, the interval between two successive segments (of chain length n and $n-1$) was often found to be characteristic of that nucleotide and was expressed as an "*M* value" (Sanger, Brownlee, & Barrell, 1965; Brownlee & Sanger, 1967):

$$M = \frac{\text{Mobility}_{n-1}}{\text{Mobility}_n}$$

M values for nested oligonucleotides electrophoresed on DEAE-cellulose at pH 3.5 follow the order $G \geqslant A > U > C$ (Brownlee, 1972) and thus indicate whether two successive fragments differ by a purine, uracil, or cytosine nucleotide.

The wandering-spot method sequences into a DNA strand from one end. That end can be 5' or 3', but it must be uniform (all the molecules must end with the same nucleotide in the sequence), such as the fixed termini of tracts, blocks, primers, or restriction fragments. If the DNA is double-stranded it must be labeled in or on only one strand.

The wandering-spot technique works as follows: First, either venom ($3' \rightarrow 5'$) or spleen ($5' \rightarrow 3'$) exonuclease partially digests the DNA strand to within 1–20 nucleotides of the uniform end. The variable ends in this nested set thus correspond to points at which exonucleases ceased removing nucleotides from the distal end. This population is electrophoresed at pH 3.5, transferred to a thin layer of DEAE-cellulose, homochromatographed (Brownlee & Sanger, 1969), and autoradiographed.

The autoradiogram reveals a winding trail of spots with mobilities characteristic of the sequence. Each spot in the trail is longer than the next higher spot by one nucleotide. Which nucleotide that is—dTMP, dGMP, dAMP, or dCMP—determines where the second spot lies with respect to the first.

Thymine, guanine, adenine, and cytosine nucleotides have negative charges at pH 3.5 which decrease from about -1 to -0.1 in that order (J. D. Smith, 1967). Their individual charge-to-mass ratios are, however, more significant for electrophoretic mobility. At pH 3.5 dTMP and dGMP both have greater charge:mass ratios than that of an oligonucleotide of random base composition; hence loss of either decreases mobility. dAMP and dCMP, on the other hand, have below-average charge:mass ratios, and loss of either of these nucleotides increases mobility. The second dimension distinguishes purines from pyrimidines: loss of G or A increases homochromatographic mobility about twice as much as loss of C or T, possibly due to stronger interaction between purines and cellulose.

The composite two-dimensional change in mobility on loss of one nucleotide is called a "shift." Roughly, a short 10-o'clock shift indicates loss of C; a long 11-o'clock shift, loss of A; a long 1-o'clock shift, loss of G; and a short 2-o'clock shift, loss of T:

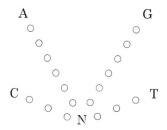

The succession of such shifts indicates the sequence of the DNA. Five accounts of this technique appeared almost simultaneously in 1973.

Ziff, Sedat, and Galibert (1973) provided details of partial venom and spleen exonuclease sequencing of uniformly labeled T4 Endonuclease IV (Sadowski & Hurwitz, 1969) fragments, derived the sequence of a 48-nucleotide fragment from ϕX174, and later fully developed the mobility shift method (Galibert, Sedat, & Ziff, 1974).

Sanger et al. (1973) reported a second-order version of the same technique which mapped, prepared, and then sequenced oligo-nucleotides contained in nested primer extensions. First, the primer was extended on its template for some tens of nucleotides with ^{32}P-labeling and ribosubstitution at cytidines. Second, the limited primer extensions from this reaction were resolved into a long, wandering trail on two-dimensional electrophoresis-homochromato-graphy, and the longer extensions were extracted. Third, each of these was cleaved into C-terminal blocks with ribonuclease and electrophoresed in one dimension on DEAE paper beside the others. Extensions of increasing length from the synthetic reaction released extra oligonucleotide blocks at the cleavage step, which were resolved by electrophoresis, and thus ordered with respect to the primer end. Fourth and finally, all the oligonucleotides were sequenced and strung together according to the block map into a 50-base sequence adjacent to the site of primer hybridization.

Murray and Murray (1973) and Weigel et al. (1973) also des-cribed a two-dimensional mobility-shift sequencing technique with two very important differences. One was that they end-labeled DNA with polynucleotide kinase and $(\gamma\text{-}^{32}P)$ATP at 5′ ends (Richardson, 1965) or with DNA polymerase and $(\alpha\text{-}^{32}P)$dNTP at 3′ ends. The other difference was that they partially digested with an *endo*nuclease, DNAse I, rather than an *exo*nuclease. This approach gave sequences for the ends of phages λ, 421, and 21.

Wu, Tu, and Padmanabhan (1973) also reported wandering-spot sequencing of ^{32}P-end-labeled DNA, a 12-base oligodeoxynucleotide produced by organic synthesis. This group employed, in addition to polynucleotide kinase for 5′-labeling, terminal deoxynucleotidyl transferase and $(\alpha\text{-}^{32}P)$rNTP for 3′-labeling, and the venom and spleen exonucleases for generating nested fragments.

The experimental tactic these five reports demonstrated in 1973 was to map and sequence DNA continuously from one end. The wandering spot produced the first complete sequence of a λ operator the following year, by limited primer extension with ribosubstitution and partial exonuclease/mobility-shift analysis of the component oligonucleotides (Maniatis et al., 1974). The Wu group then improved the range and definition of the technique (Jay et al., 1974) and contributed a quantitative analysis of the mobility-shift phenomenon (Bambara, Jay, & Wu, 1974; Tu et al., 1976).

PHASE THREE: BASE–SPECIFIC
NESTED–SEGMENT METHODS

The first long-range DNA sequencing technique appeared in 1975 under the title "A Rapid Method for Determining Sequences in DNA by Primed Synthesis with DNA Polymerase," otherwise known as the plus-minus method (Sanger & Coulson, 1975). The second, in 1977, was "A New Method for Sequencing DNA," since known as the chemical method (Maxam & Gilbert, 1977). The third technique came out later in 1977 with the title "DNA Sequencing with Chain-Terminating Inhibitors" (Sanger, Nicklen, & Coulson, 1977).

The first two of these techniques arose independently, but as we shall see, both had origins in the mobility-shift techniques of the previous section. The ribosubstitution version of the wandering spot (Sanger et al., 1973) ordered blocks by limiting primer extension, anticipating the plus-minus primed synthesis method which does the same. The endonuclease version of the wandering spot (Murray &

Murray, 1973; Weigel et al., 1973), on the other hand, labeled an extant DNA end with ^{32}P and then reduced the DNA to nested, end-labeled fragments by partial internal cleavage, thus anticipating the chemical method. What the new methods did that was different was to replace single-channel, random nested fragments with four-channel, base-specific nested segments—the plus-minus partially terminating and the chemical partially cleaving at each (or at most two) of the four bases—and to extend the sequencing range by replacing DEAE-homochromatography with gel electrophoresis.

In the discussion below it will also be apparent that the chemical sequencing method descended from pyrimidine and purine tract techniques (excision tracts) and from extensive knowledge of alkylation, depurination, and hydrazinolysis of DNA. Likewise, the plus-minus and chain-termination methods have roots in DNA repair synthesis, which had been used to sequence the cohesive ends of bacteriophages λ (Wu & Kaiser, 1968; Wu & Taylor, 1971) and T7 (Englund, 1971, 1972), and to supplement chemical with enzymatic DNA synthesis (Kleppe et al., 1971). That dideoxynucleoside triphosphates terminate growing DNA chains was also known at the time (Atkinson et al., 1969).

The Chemical Method

A detailed determination of the nucleotide sequence of a DNA molecule is beyond our present means, nor is it likely to occur in the near future. . . . Even the smallest functional DNA varieties seen, those occuring in certain small phages, must contain something like 5000 nucleotides in a row. We may, therefore, leave the task of reading the complete nucleotide sequence of a DNA to the 21st Century which will, however, have other worries.

As I have pointed out before . . . no chemical solution of the sequence problem is likely to occur in the near future.

Erwin Chargaff, *What Really is DNA?*
(Chargaff, 1968)

Ten years later, a chemical DNA sequencing technique derived complete base sequences for the bacterial plasmid pBR322 (Sutcliffe, 1978), the bacteriophage fd (Beck et al., 1978), and the animal virus SV-40 (Fiers et al., 1978; Reddy et al., 1978), all about 5,000 nucleotides in size. There is incidental irony in the fact that the hydrazinolysis of pyrimidines, a reaction elucidated by this master of nucleic acid chemistry and his colleagues (see below), provided half the base cleavage specificity needed to sequence these three chromosomes.

How does the chemical technique sequence DNA? It can be used on any discrete DNA fragment, either single-stranded or double-stranded, which is labeled on one end with ^{32}P-phosphate and has the desired sequence situated within several hundred nucleotides of the end. *In vitro* restriction enzyme cutting and ^{32}P-end-labeling routines will rapidly reduce DNA purified from cells to fragments of this form. Two chemical steps are then used to map each of the bases along one of these fragments. The first step is a reaction that modifies one (or at most two) of the four bases at a random position in each molecule. Thus in 200 bases proximal to the labeled end of each DNA molecule, only 1 base out of 50 others of its kind is attacked. However, all 50 ultimately react in the molecule population. The second step consists of complete reactions that break the DNA wherever a base reacted appropriately in the first step.

Four two-step, base-specific cleavage procedures are used to sequence the four bases in an end-labeled DNA. One reaction with dimethylsulfate and piperidine cleaves the DNA at guanines (G); another with acid and then piperidine cleaves at both guanines and adenines (G + A); a third reaction with hydrazine and then piperidine breaks the DNA at cytosines and thymines (C + T); and a fourth reaction with hydrazine in the presence of an inorganic salt followed by piperidine breaks at cytosines (C). Phased electrophoresis through a polyacrylamide gel assorts partial cleavage products from the four reactions by size, and autoradiography registers those that are end-labeled.

The result is four adjoining ladders of bands on a sheet of X-ray film. All these bands are images of DNA fragments which have have a common and fixed ^{32}P-labeled end, created at the outset by a restriction enzyme, and a unique and variable broken end, created by one of the four base-specific chemical reactions. Each band identifies and locates a base in the DNA strand: the lane(s) it appears in reveals whether the base was G, G/A, C/T, or C; how far it moved down the gel measures, in nucleotides, how close the base was to the labeled end. Examples of these sequencing band patterns can be found in the sections below.

What were the origins of the chemical DNA sequencing method? It evolved from an experiment conducted by Walter Gilbert early in 1975 to find out whether *lac* repressor occupies the major, the minor, or both grooves of the *lac* operator. That experiment evolved in turn from those of Mirzabekov and Kolchinsky (1974) and Mirzabekov and Melnikova (1974) probing for histones and several small ligands in the large and small grooves of DNA, and from an endonuclease version of the wandering-spot sequencing technique for ^{32}P-end-labeled DNA.

It was known at the time that the 7-nitrogen of guanine and the 3-nitrogen of adenine are targets for methylation by dimethylsulfate in native DNA (Lawley & Brookes, 1963), and that the guanine N7 in the major groove reacts about seven times faster than the adenine N3 in the minor groove (Lawley & Brookes, 1963; Lawley & Thatcher, 1970). It was further known that 7-methyl guanine and 3-methyl adenine have unstable glycosidic bonds and can be released from DNA at neutral pH (Zoltewicz et al., 1970), and that alkali will break this depurinated DNA (Tamm et al., 1953; Pochon & Michelson, 1967). Thus methylation, release of methylated purines, and alkali cleavage were established DNA reactions.

Mirzabekov and Kolchinsky probed the large and small DNA grooves for histones by methylating calf-thymus DNA and calf-thymus chromatin with [3H]-dimethylsulfate in parallel reactions. After purifying and acid-hydrolyzing the methylated DNAs, they separated and quantitated 7-methyl guanine, 3-methyl adenine, and unreacted bases. The analysis revealed that guanine 7-nitrogens in the major groove are 10% less reactive in chromatin as in naked DNA, while adenine 3-nitrogens in the minor grooves are equally reactive. Further experiments designed to characterize the distribution of purines methylated in chromatin removed 7-methyl guanine and 3-methyl adenine by heating at pH 7, cleaved the DNA at the empty sugars at pH 12, and sized the products by centrifuging through alkaline sucrose gradients (Mirzabekov & Kolchinsky, 1974). In effect, these experiments cleaved DNA at specific bases by methylation, depurination, and phosphate elimination.

Finally, single-end-labeled restriction fragments (Maniatis, Jeffrey, & Kleid 1975) were routinely available. I excised a 95 base-pair *Alu*I (Roberts et al., 1976) fragment from the *lac* UV5 promoter-operator region (Gilbert et al., 1975). After 5′-end-labeling with polynucleotide kinase and (γ-^{32}P)ATP (Richardson, 1965), this piece was split asymmetrically with *Hap*II (Sugimato et al., 1975), and 40- and 55-base-pair end-pieces were isolated on polyacrylamide gels. I used a combined exonuclease/endonuclease wandering-spot method (Galibert, Sedat, & Ziff, 1974) to sequence these and other ^{32}P-end-labeled restriction fragments during this time. As explained above, the wandering spot is a nested-segment DNA sequencing technique.

Against this background, Gilbert performed the following experiment to screen the two operator grooves for occupation by the *lac* repressor. He began with a restriction fragment having the *lac* operator embedded in it and ^{32}P-phosphate on one end, the 55-base-pair segment mentioned above. This operator DNA was divided into two portions, and repressor was added to one. He then reacted both portions with dimethylsulfate such that about one purine in each labeled molecule became methylated, heated the DNA at neutral pH to release 7-methyl guanines and 3-methyl adenines, cleaved strands with alkali at the empty sugars, electrophoresed the products on a polyacrylamide gel which resolves oligonucleotides (Maniatis, Jeffrey, & Van de Sande, 1975), and developed an autoradiogram.

On the autoradiogram were two ladders of bands of two intensities, dark ones from strand cleavage at sugars occupied by fast-methylating 7-methyl guanines, and light ones from slower-reacting 3-methyl adenines. This succession of dark and light bands was readily correlated with the succession of guanines and adenines in the labeled operator strand. Moreover, when the minus- and plus-repressor ladders were compared, it was apparent that the repressor protected some but not all operator guanine N7's and adenine N3's against methylation, 4/20 on the top strand and 3/15 on the bottom strand. This suggested specific interaction of these seven operator purines with *lac* repressor amino acids (Gilbert, Maxam, & Mirzabekov, 1976). Repetition of this probing experiment indicated that it would reproducibly cleave DNA at guanines and adenines.

It should be noted in passing that Sverdlov et al. (1973) had proposed limited base-specific chemical cleavage for sequencing end-labeled DNA in 1973. As a demonstration, this group 5'-phosphorylated the synthetic tetranucleotide TpGpTpG with polynucleotide kinase and (γ-^{32}P)ATP, partially depurinated and cleaved it with the Burton reaction (Burton & Petersen, 1957, 1960), chromatographed the products on DEAE-cellulose, and identified peaks for the two nested fragments pTp and pTpGpTp.

Another restriction fragment, from the β-galactosidase gene, contained a second *lac* repressor binding site (Gilbert et al., 1975). Its sequence was not known, but purine methylation protection seemed to me an appropriate way to locate the binding site within it. This fragment was (5'-^{32}P)-labeled, split, methylated in the presence and absence of repressor, cleaved, electrophoresed, and autoradiographed, as before. This time the repressor suppressed methylation of three guanines. On the complementary strand, it protected a single guanine in the same vicinity. Overall, the repressor blocked dimethylsulfate alkylation of only four purines, all within 20–30 base-pairs of one end of the fragment (Gilbert, Majors, & Maxam, 1976; A. M. Maxam & W. Gilbert, unpublished). This defined the *lac* pseudo-operator by *lac* repressor contacts with its purines, and centered it about 25 base-pairs from a restriction enzyme cleavage site which had been localized to several hundred base pairs inside the *lac z* gene (Gilbert et al., 1975).

This mapping of the interaction of a protein with DNA of unknown sequence had special significance. As revealing as the N7 and N3 protections were, their immediate effect was to stimulate sequencing of the pseudo-operator. We believed that aligning the two operator structures would reveal an homologous sequence essential for repressor recognition. We also wanted to tie the secondary operator to the *lac* control region, by aligning their sequences with the β-galactosidase amino acid sequence being worked out by Fowler and Zabin (1977).

Clearly, the methylation/depurination/cleavage patterns provided purine sequence information. The one dark-G/light-A ladder was nevertheless not a DNA sequencing method. (Strictly speaking, locating guanines and adenines on both complementary strands would give a sequence, and this was attempted on the pseudo-operator fragment; the problem with this approach is that it requires another not-too-distant restriction enzyme cleavage site and two preparations of end-labeled DNA.) It appeared to me that what was required for sequencing, as opposed to probing, was a reagent that would do for pyrimidines what dimethylsulfate did for purines. Then, if cleavage products from both were electrophoresed in phase, C and T bands would mesh with the G and A bands. This added the concept of a simultaneous second (and later, third and fourth) analytic channel to distinguish the DNA bases.

Hydrazinolysis was an obvious first choice for breaking DNA at cytosines and thymines. Hydrazine was found to split the uracil ring over fifty years ago (Levene & Bass, 1926), the cytosine ring twenty years ago (Baron & Brown, 1955), and DNA pyrimidines a few years later, giving strand cleavage (Takemura, 1958, 1959; Habermann, 1961, 1962, 1963a, 1963b). Then, throughout the 1960s, others used hydrazinolysis followed by various strand-cleaving reactions in attempts to elicit chemically defined purine tracts from DNA which, by base complementarity, had to match the corresponding pyrimidine tracts (Chargaff et al., 1963; Sedat & Sinsheimer, 1964; Cape & Spencer, 1968; Vanyushin & Bur'yanov, 1969; Turler, Buchowicz, & Chargaff, 1969).

My first attempts at sequencing both purines and pyrimidines in DNA by partial chemical cleavage were structured as follows. I treated a portion of end-labeled DNA with dimethylsulfate, heated at neutral pH, and alkali, as in the repressor protection experiment, and reacted other portions with hydrazine. Hydrazinolysis conditions in these first experiments were much like those Turler and Chargaff had suggested for preparation of apyrimidinic acid (Turler & Chargaff, 1969). The reaction was run at lower temperature for partial rather than complete destruction of pyrimidine rings. After removing excess hydrazine, I cleaved the DNA with alkali, and electrophoresed the products on a gel beside those from the methylation/depurination/ cleavage reaction above.

Patterns on the first autoradiograms were promising in that lanes from some hydrazine reactions exhibited zones of ^{32}P-labeled DNA. These zones interdigitated with bands in the G/A lane, as would be expected of partial C/T products. Methylation/depurination/alkali-elimination cleaved strongly at G and weakly at A, while hydrazinolysis/transamination/piperidine-elimination was breaking the DNA equally at C and T. This at least determined the nucleotide spacing between purines.

It was then necessary to find other conditions or reactions that would distinguish C from T and, with greater reliability, A from G. Many of those who first studied the hydrazine-pyrimidine reaction observed that uracil reacts much faster than cytosine, which in turn reacts slightly faster than thymine (Baron & Brown, 1955; Habermann, 1961; Temperli et al., 1964; Budovskii, Haines, & Kochetkov, 1964; Lingens & Schneider-Bernlohr, 1966; Ellery & Symons, 1966; Hayes & Hayes-Baron, 1967). However, the difference in rate is at best only several-fold and in these preliminary experiments was neither great, uniform, nor reproducible enough to discriminate reliably between C and T.

Then what might be called a lucky accident revealed a way to suppress the reaction of hydrazine with thymine. Hoping to isolate the cause of the C/T irreproducibility just mentioned, I wondered if it derived from variable amounts of salt carried over to the hydrazinolysis reaction in the ^{32}P-labeled DNA. Hence I introduced salt in one reaction, containing 50% hydrazine in water and a few micrograms of DNA (less than a micromole of nucleotides), and after cleaving and electrophoresing observed the near absence of T-bands on the autoradiogram. Sodium chloride greatly diminished the rate of reaction of hydrazine with thymines in DNA, but not with cytosines. Other inorganic salts, including the chloride of the divalent magnesium cation, had the same qualitative effect.

Although hydrazine with and without salts was able to distinguish cytosines from thymines, hydrazine-alkali cleavage products moved in electrophoresis as blurred rather than sharp bands. This suggested that they were heterogeneous in charge or mass, or both. I soon realized that this heterogeneity in individual partial cleavage products was recapitulating hydrazinolysis aberrations observed by others for years.

First, hydrazine reactions were said to yield "apyrimidinic acid," but when the presumably empty portion of the sugar-phosphate backbone was hydrolyzed, it failed to release the expected amounts of deoxyribose and inorganic phosphate (Habermann, 1961; Chargaff et al., 1963; Temperli et al., 1964; Cashmore & Peterson, 1968). This implied that not all the pyrimidine nucleotides or interpyrimidine nucleoside phosphates cleaved off the purine clusters; or, if they did, some of the nucleoside phosphate was in a form that did not break down to deoxyribose and orthophosphate.

Treating hydrazinolysis products with benzaldehyde before cleaving with alkali (Baron & Brown, 1955) was known to boost the yield of deoxyribose and orthophosphate and also bring the yield of purine and pyrimidine clusters from calf-thymus DNA into closer accord (Chargaff et al., 1963). This suggested that the "apyrimidinic acid" contained sugar hydrazones incompatible with strand cleavage by alkali-catalyzed β-elimination. On these grounds, I introduced a benzaldehyde reaction, said to be effective in regenerating sugar aldehydes from hydrazones (Temperli et al., 1964), between the limited hydrazine reaction and alkali cleavage. However, this failed to unify partial pyrimidine cleavage products, as judged by band sharpness on autoradiograms.

Second, ^{14}C-labeled pyrimidine ring fragments had been detected on sugars in the "apyrimidinic acid" isolated after mild hydrazine reactions (Cashmore & Petersen, 1968). I had been using diluted hydrazine solutions (50% in water) at low temperature, so these base fragments were likely present at some level. If they were interfering with β-elimination of phosphates, that would account for heterogeneity in the products. Holding hydrazine-treated DNA at neutral or acidic pH before alkali cleavage (Cashmore & Peterson, 1968) failed to release these fragments, if they were indeed present.

It was known that primary amines would also cleave DNA at sugars without bases, probably as the Schiff base formed with the sugar aldehyde (Livingston, 1964; Coombs & Livingston, 1969). Proceeding from this observation, Vanyushin and Bur'yanov obtained "apyrimidinic acid" from calf-thymus DNA with hydrazine and treated it with aniline hydrochloride at pH 4.5 (Vanyushin & Bur'yanov, 1969). A 4-hr reaction at $37°C$ gave the appropriate yields of interpyrimidine orthophosphate and of purine tracts (as required by complementarity with the pyrimidine tracts).

I then treated hydrazinolysis products of the end-labeled restriction fragment with aniline hydrochloride and salts of other aromatic primary amines (Turchinskii et al., 1970). In most cases, fragments cleaved at pyrimidines now electrophoresed as diffuse bands rather than broad zones. However, each band was a doublet, retarded approximately 1 and 2 nucleotide intervals behind pyrimidine gaps between purine fragments. Reaction with benzaldehyde after aniline gave single, 1-nucleotide-retarded bands. If the methylation/depurination/alkali-cleavage partial products were of the form

$$^{32}\text{P-phosphate-[nucleotide]}_n\text{-phosphate}$$

then hydrazinolysis/amine-displacement/amine-cleavage products would most likely be

$$^{32}\text{P-phosphate-[nucleotide]}_n\text{-phosphate-}\frac{\text{ribose}}{\text{derivative}}\text{-aniline}$$

and

$$^{32}\text{P-phosphate-[nucleotide]}_n\text{-phosphate-}\frac{\text{ribose}}{\text{derivative}}$$

with the latter predominating when a benzaldehyde reaction is included after aniline cleavage.

Apparently, aniline could displace the pyrimidine glycosidic nitrogen (after hydrazinolysis), and also hydrazine itself, from the deoxyribose $C1'$, and then catalyze elimination of the $3'$- but not the $5'$-phosphate. This interpretation was consistent with the detection of ^{14}C-aniline on fractionated purine tracts (Vanyushin & Bur'yanov, 1969), the slower but tighter electrophoretic mobility of the doublets, and the observation that benzaldehyde converted these to retarded singlets.

At this point in our account of the chemical DNA sequencing method it is useful to review what the hydrazine experiments were trying to achieve. Doing so at the time led directly to a final solution to the problem of cleaving DNA at pyrimidines.

Breaking DNA at each of the four bases in separate reactions was certainly the overall objective. Also, modifying bases, removing the modified bases from their sugars, and cleaving the DNA backbone at those sugars was the productive route. Moreover, it was probable that this reaction sequence could be initiated by disrupting electron resonance in these unsaturated heterocyclic bases in at least two ways: ring substitution and ring fission.

The reactions cleaving DNA at purines were conducted in distinct steps, the first of which was adding a substituent to the adenine or guanine ring. Either protonation (acid depurination) or alkylation (dimethylsulfate methylation) of ring nitrogens would weaken the purine glycosidic bond in DNA. Ionization of that bond (Shapiro & Danzig, 1972) then released the purine base, leaving the sugar intact as a link in the backbone, but with its 1′ carbon in the free aldehyde form. Finally, sequential alkali-catalyzed β-elimination (Linstead, Owen, & Webb, 1953) of the 3′-phosphate and the 5′-phosphate (Brown & Todd, 1955) excised the sugar and broke the strand.

Nearly all other chemical reactions known to cleave DNA (reviewed by D. M. Brown, 1974, and by Kochetkov & Budovskii, 1972) involve ring fission. The aggregate result of investigations into the reaction of hydrazine with DNA pyrimidines (Kochetkov & Budovskii, 1972) suggest that it proceeds as follows. In mild reactions, a hydrazine molecule attacks cytosine or thymine at C6 and then at C4, where it splits the pyrimidine ring. This cyclizes half the base (C4—C5—C6) into a new five-membered ring containing the hydrazine nitrogens, leaving the other half (N1—C2—N3) as a ureido group still attached, at N1, to both the new ring and the sugar:

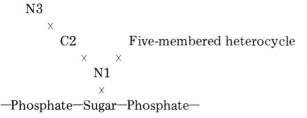

An intermediate in the reaction of this type has been found (Lingens, 1964), and ^{14}C-urea is released slowly at neutral pH from the products of mild hydrazine reactions (Cashmore & Petersen, 1968). At higher temperatures (60°C) anhydrous hydrazine reacts with the above product further, releasing urea and methyl-pyrazolone from thymine, or urea and amino-pyrazole plus ureido-pyrazole from cytosine, leaving the sugar as the hydrazone (Temperli et al., 1964; Cashmore & Peterson, 1968, 1978).

The products of limited hydrazine reactions used in DNA sequencing probably include all the above, leaving the glycosidic carbon attached to the glycosidic nitrogen as a secondary amine, as a tertiary amine, and as the hydrazone. As mentioned, when I reacted end-labeled DNA, first with aqueous hydrazine at room temperature, then with benzaldehyde (to liberate sugar aldehydes from hydrazones), and finally with alkali or aniline (for β-elimination), it did not break cleanly at pyrimidines. Apparently, alkali fails to eliminate either phosphate from some of the hydrazinolysis products, while the amine eliminates only the 3′-phosphate.

I then asked what sort of reactions would be likely to carry all these products of hydrazine-pyrimidine reactions through to discrete DNA cleavages at thymines and cytosines. First, the ureido group, the five-membered heterocycle, or both would have to be displaced from the glycosidic carbon (C1′). As noted, aniline would do this, as would to some extent any primary or secondary amine capable of forming a Schiff base with the sugar aldehyde. Second, if hydrazine is present on the sugar as the hydrazone, it too would have to be removed. Since purine tracts obtained with hydrazine and aniline were indistinguishable from those obtained with hydrazine, benzaldehyde, and aniline (Vanyushin & Bur'yanov, 1969), it appeared that amines might displace hydrazine from the hydrazone. Third, the 3′- and then the 5′-phosphate would have to be eliminated from the sugar to cleave the backbone in the manner that methylation/depurination/alkali-cleavage does. Here alkali would not suffice because alkali-catalyzed β-elimination apparently requires a sugar carbonyl at C1′; nor would aniline, since in a Schiff base with C1′ it fosters elimination of only the 3′-phosphate.

Although successive reactions of DNA with hydrazine, aniline, benzaldehyde, and alkali were likely to satisfy all three of these requirements, it appeared that a secondary amine as the free base might also do so but with fewer steps. I wondered whether the cyclic secondary amine piperidine might displace ring-opened forms of the glycosidic nitrogen, by transamination, and form a Schiff base (enamine) with the sugar, the piperidone. A piperidone (Capon, 1965) has a covalently fixed positive charge on quarternary nitrogen which, in the presence of excess piperidine as the free base, should catalyze efficient β-elimination of both phosphates from the sugar. Whatever the mechanisms, piperidine worked promptly. When I reacted one of the end-labeled *lac* operon restriction fragments with aqueous hydrazine at room temperature and then with aqueous piperidine at 90°C and electrophoresed, bands in the pyrimidine lane meshed with those in the purine lane. These bands were sharp and located only at positions not occupied by purine bands.

As mentioned earlier, limited methylation/depurination/alkali-cleavage provides an average 7-fold greater intensity of G-bands over A-bands on sequencing autoradiograms. Unfortunately, this difference is not uniform. It was therefore necessary to develop a partial cleavage reaction that would distinguish A from G independently of the original G > A reaction. Margison and O'Connor (1973) observed the release of 3-methyl adenines but not 7-methyl guanines from methylated DNA at low pH. Even at neutral pH, 3MeA is released 4- to 6-fold faster than 7MeG (Lawley & Brookes, 1963; Kriek & Emmelot, 1964). Hence, the first reaction I employed increased the relative intensity of A-bands by partial release of methylated purines at pH 2, followed by the usual alkali-cleavage. The second employed normal alkali at 90°C to open adenine rings about 10-fold faster than cytosine rings (Jones, Mian, & Walker, 1966), followed by displacement and strand cleavage with piperidine (A > C).

The five reactions for cleaving DNA at specific bases (G > A, C + T, C, A > G, A > C), plus one more (G) described below, were published in 1977 (Maxam & Gilbert, 1977). Although useful during 1976-78, some of these reactions had inherent problems. The main problem was that the A > G and A > C reactions were inadequate for long-distance sequencing. Blurriness of bands from both increased with chain length, suggesting cumulative heterogeneity in charge/mass due to side reactions.

The final solution distinguishing adenines from guanines substituted all the early purine-specific cleavage reactions with one breaking only at G and another breaking equally at G and A. G-cleavage begins with the usual limited methylation of DNA with dimethylsulfate, but then incorporates ring-opening of 7-methyl guanine (Haines, Reese, & Todd, 1962; Lawley & Brookes, 1963) on the DNA, followed by transamination and phosphate-elimination, all with piperidine as the free base. The G + A cleavage is a conventional but limited acid depurination (Tamm, Hodes, & Chargaff, 1952) followed by phosphate-elimination with piperidine. The four partial cleavage reactions now used for DNA sequencing have base specificities G/G + A/C + T/C. How we conduct these reactions is described in detail in Maxam and Gilbert (1980). They are superior to the original A > G/G > A/C + T/C set in several ways. First, all four employ piperidine for backbone scission, which is removed by evaporation to leave the cleaved DNA salt-free, so that it may be dissolved in a few microliters of formamide for loading on thin (0.3 mm) sequencing gels (Sanger & Coulson, 1978). (Originally, chemically cleaved DNA was of necessity loaded in tens of microliters of 0.1 N sodium hydroxide.) Second, the new reactions minimize side-reaction damage and hence charge/mass heterogeneity in the DNA, to the extent that end-labeled DNA can now be sequenced out to several hundreds of nucleotides with the chemical method.

Third, the G/G + A/C + T/C base cleavage specificities make reading sequence from autoradiograms easier by being 1-2-2-1 symmetric, and by providing a central axis (between G + A and C + T) on one or the other side of which will be a band for every base in the sequence.

COMPARISON AND ASSESSMENT OF THE CHEMICAL AND ENZYMATIC METHODS OF DNA SEQUENCING

It is always interesting to compare two good methods for achieving the same experimental result. At issue in comparing the chemical and chain-termination DNA sequencing techniques are the answers to five questions:

1. Does the method require a special DNA configuration and if so how can I convert my DNA to that form?
2. How much DNA does the method require for sequencing?
3. How much labor do I have to invest, overall, to derive the sequence?
4. How accurate will the sequence I get with this method be?
5. Will the method identify naturally occurring DNA ends and modifications?

Let us begin by comparing sequencing modes, or how the two methods deal with different DNA configurations.

Double-stranded DNA

In the majority of cases the DNA one wants to sequence will be double-stranded. The chemical method will sequence double-stranded DNA if it is ^{32}P-labeled on only one end. There are two ways to make such DNA. One, cut the DNA with a restriction enzyme, ^{32}P-label all 5' or 3' ends on all the fragments, digest the mixture with a second restriction enzyme, and electrophorese beside a small portion of the DNA that was not cut with the second enzyme. Any fragment in the preparative but not the marker lane is singly end-labeled and ready for sequencing. Two, cut the DNA with a restriction enzyme, label all the 5' or 3' ends as before, denature the DNA, and electrophorese on a strand-separation gel beside a small portion of the labeled fragment mixture which was not denatured. Any fragment in the preparative but not the marker lane is potentially an end-labeled single strand and is ready for sequencing.

Double-stranded DNA has been a problem for primer-extension sequencing methods, although solutions are now appearing. Indeed, because a decision on whether or not to use the dideoxy method usually rests on finding a way to do this, it has received considerable attention. At present, it seems that double-stranded DNA can be converted to the primer-template configuration required for chain-termination sequencing by one of three maneuvers: (1) use an end-label/nick-translation version of the chain-termination technique on double-stranded restriction fragments; (2) chew 5' or 3' ends back extensively with an exonuclease, to expose single-stranded template DNA; or, (3) clone the DNA into a single-stranded phage.

An end-label/nick-translation version of the chain-termination method was published by Maat and Smith (1978). The usual round of $(5'-^{32}P)$-labeling, secondary restriction enzyme cleavage, and gel electrophoresis provides singly end-labeled restriction fragments. One of these fragments is mixed with the Klenow DNA polymerase, three deoxynucleoside triphosphates, one dideoxytriphosphate, and a small, carefully titrated amount of DNAse I. Synthesis begins at a random nick, put in one strand by the DNAse, and proceeds just a few nucleotides to a template slot for the dideoxynucleotide, which always incorporates and terminates repair. The result is a standard chain-termination sequencing pattern, with the exception of considerable variation in band intensity, which Maat and Smith attribute to DNAse I sequence preferences. Restoring the missing fourth deoxynucleoside triphosphate to some level might even-out these patterns.

In general, end-label/nick-translation versions of the plus-minus and chain-termination methods have not yet been widely used. This aversion may be due to difficulty in controlling the pattern and extent of DNAse nicking. More likely, these variants are used infrequently because they of necessity relinquish one of the most appealing features of their parent methods: continuous ^{32}P-labeling during primer extension.

Another double-stranded DNA conversion for dideoxy sequencing cuts back both 3' or both 5' ends with an exonuclease. This exposes single-stranded DNA for use as template (Smith, 1979). A Cold Spring Harbor group has persisted with this approach in sequencing the adenovirus-2 chromosome (R. J. Roberts, personal communication; Weissman, 1979). They chew back the 3' ends of large adenovirus restriction fragments with *E. coli* Exonuclease III, uncovering much of each strand, to which they hybridize a smaller restriction fragment from the same region. This restriction fragment becomes a primer in an individual chain-termination sequencing run, followed by others in subsequent runs.

One can imagine several pitfalls. The 3′ ends left on both strands after exonuclease digestion are primers in their own right for extension with labeling and chain termination. If the exonuclease terminated randomly over a long distance, there will be no problem; but if termination is nonrandom, interference may build up in the sequence band pattern. Also, with chromosomes as large as that of adenovirus, this strategy relies on the ability of exonucleases to traverse long distances and gel electrophoresis to resolve a multitude of restriction fragments. Apparently, these hypothetical problems did not materialize or were overcome.

A third tactic for sequencing double-stranded DNA with the chain-termination technique segregates the two strands biologically. The essence of this strategy is to subclone restriction fragments into the double-stranded replicative form of a single-stranded DNA phage, and then to grow and extract viral minus strand and sequence on its single-stranded insert. Gronenborn and Messing (1978) have constructed a derivative of phage M13 which they suggest as suitable for this purpose. This phage, M13mp2, has in it a 789-base fragment of the *lac* operon which encodes the last 31 amino acids of the *lac* repressor, includes the *lac* promoter and operator, and encodes the first 146 amino acids of β-galactosidase (Maxam, Chapman, et al., unpublished). A single *Eco*RI cloning site has been created in the *lac* region in this phage at codons 5–6 of β-galactosidase (Gronenborn & Messing, 1978). An up-to-date account of the status of enzymatic sequencing methods is provided in Chapter 3.

One idea is to isolate the insert from an existing double-stranded plasmid or phage clone, cut it into average 256–base-pair segments with a tetranucleotide-recognizing restriction enzyme, and shotgun subclone all of these into an M13 phage such as the one just mentioned. A part of the strategy is to also prepare and stockpile a cloning-site-proximal M13 restriction fragment for use as a universal primer. One would then sequence by extending the M13 primer with chain terminators into each subcloned region—in one direction on one strand in one clone and, for confirmation, in the other direction on the other strand in an opposite-orientation sibling clone. That M13 does not lyse but exudes from infected cells with a high titer, 10^{12} ml^{-1} (Marvin & Horn, 1969), makes this source of single-stranded DNA for chain-termination sequencing attractive. Centrifuging cells out of a 1-ml M13-producing culture will leave 10^{12} phage particles in the supernatant, from which about 1 pmol of template DNA can be extracted, enough for ten sequencing experiments (Schreier & Cortese, 1979).

The shortcomings of this strategy are that it requires subcloning and that internal regions in longer subcloned sugments will not be accessible with the external M13 primer.

Single-stranded DNA

The chemical method will sequence single-stranded DNA directly once it has been labeled on a 5' end with polynucleotide kinase and $(\gamma\text{-}^{32}\text{P})\text{ATP}$, or at the 3' end with terminal transferase and $(\alpha\text{-}^{32}\text{P})\text{ATP}$. Chain termination is a copy technique and hence does not sequence single-stranded DNA directly. Both techniques will sequence primed copies of single-stranded DNA. With chain termination, copying and sequencing are one and the same process, and primed synthesis is its original and most natural mode. When it is to be sequenced by the chemical method, the 5' end of the primer must be labeled before or after primer extension on the template. Extensive copying, of course, produces extensive double-stranded DNA, which can be treated as such (see above).

Amount and Specific Activity of DNA

The point that the chemical method has most in its favor is its ability to sequence DNA directly at any singular end that can be labeled with ^{32}P. This end can be created *in vitro* by restriction enzyme cleavage or *in vivo* by a DNA replication step. However, to its disadvantage, the chemical method can tolerate only a few ^{32}P-phosphates at this end. The strong point of the chain-termination method is that it continuously labels the DNA which it is sequencing by copying. It therefore requires considerably less DNA for sequencing than the chemical method.

Consider one DNA segment, 200 nucleotides long, in a nested chemical or chain-termination sequencing array. In a chemical cleavage set, that fragment has one ^{32}P-phosphate at one end, transferred from γ-labeled ATP, at a specific activity of about 3,000 Ci/mmol. In a chain-termination set, the same fragment, when separated from its primer, can have up to 199 ^{32}P-phosphates at 300–3,000 Ci/mmol. Taking the lower precursor specific activity (300) as economically feasible, the 200-nucleotide primer-extension would have a net specific activity of 60,000 Ci/mmol. Thus for sequence autoradio-grams of equal intensity and exposure time, the chemical method needs 2 pmol of DNA for every 0.1 pmol required by the chain-termination method.

Sequencing Labor

Each method as presently constituted has steps that require con-siderably more labor than others. These are not necessarily intrinsic to the methods and are therefore not irreplaceable. In the chemical method, ethanol precipitations after base-modification reactions involve repeated manipulations, and lyophilizations after strand scission reactions consume considerable time. (I have a number of ideas on how to eliminate some of these steps and intend to pursue them.) This labor is balanced, in the chain-termination method, by the work and time required to generate two kinds of DNA, the primer and a single-stranded template.

Sequence Accuracy

By what criteria are we to judge the accuracy of a DNA sequence? The single, most important criterion is independent sequence con-firmation. The same sequence derived in the same way on another occasion is not an independent sequence. A base sequence obtained from the corresponding region in the complementary strand is an independent sequence, as are messenger RNA and protein sequences, except where RNA splicing (reviewed by Crick, 1979, and by Darnell, 1978), TGA coding for tryptophan (reviewed by Hall, 1979), or genetic code ambiguity is involved. Sutcliffe noted that about half of his pBR322 sequence matched the available DNA and protein sequence of others and, more importantly, that 75% of the chromo-some had been sequenced on both complementary strands. With an error rate of zero in the confirmed regions, he considered those totally correct and, by induction, the other 25% probably correct. Although the whole sequence was not independently confirmed, the three-fourths that was demonstrated that it is practical for one person working alone to confirm all of a 5,000-base sequence.

Aside from the general need for independent confirmation of DNA sequences, there are specific areas where it is needed most. One is where nested segments become long enough to fold into stable hairpins, which move faster and thus pile up at one position in the gel. This problem, called "compression" or "collapse," and its remedies (running sequencing gels at higher temperature, or better, sequencing the same region on the complementary strand), have been exhibited and discussed (Brown & Smith, 1977; Sutcliffe, 1978; Barrell, 1978).

Other regions most in need of confirmation are the extremities of a sequence obtained from one sequencing run. Bands may be missed at one end because very short fragments ran off the end of the gel, and at the other end as well, where very long fragments are barely resolved. This is true for both sequencing methods, chemical and chain termination. All extremities should be confirmed with sequences that cross them. Overlapping sequences can be derived from other nearby restriction enzyme sites, and these also serve to link blocks of sequence.

Biologically Significant DNA Ends and Modifications

The chemical method will sequence unique RNA/DNA junctures. Short RNA primers are commonly synthesized and then extended as DNA at origins of replication, and replication intermediates containing both RNA and DNA have been isolated from a variety of cells. The chemical method will work on the RNA/DNA transition point after hydrolysis of the RNA and ^{32}P-labeling of the exposed DNA $5'$ end. In this way Tomizawa, Ohmori, and Bird (1977) sequenced the Colicin E1 plasmid replication origin, Gillum and Clayton (1978) sequenced the $5'$ end of the nascent strand in a mouse mitochondrial D-loop, and Haseltine, Maxam, and Gilbert (1977) sequenced the tRNA^{trp}-primed reverse transcript of Rous sarcoma virus.

The chemical method will sequence unique nicks and gaps in double-stranded DNA. The ϕX174 gene A protein cuts one strand of its own gene to create an origin for positive strand synthesis (reviewed by Sims, Koths, & Dressler, 1979). Purified A-protein will put this nick in ϕX174 RF DNA *in vitro*, the DNA ends it creates have been sequenced (Langeveld et al., 1978), matched to the complete ϕX174 sequence (Sanger et al., 1977). This located the positive strand origin precisely on the chromosome. Another example is the plasmid ColE1 relaxation gap, which was defined by Bastia (1978) by labeling and sequencing the flanking DNA ends, as well as the whole region on the complementary strand.

The chemical method will also detect one of the two common modified bases in DNA. 5-Methyl cytosine can be tentatively assigned to a position in a DNA sequence by its failure to react with hydrazine under conditions otherwise useful for detecting cytosine, and confirmed by finding a guanine at the corresponding position in the complementary strand (Ohmori, Tomizawa, & Maxam, 1978).

REFERENCES

Atkinson, M. R., Deutscher, M. P., Kornberg, A., Russell, A. F., and Moffatt, J. G. (1969). Enzymatic synthesis of deoxyribonucleic acid. XXXIV. Termination of chain growth by a $2',3'$-dideoxyribonucleotide. *Biochem.* 8:4897–4904.

Bambara, R., Jay, E., and Wu, R. (1974). DNA sequence analysis: A formula to predict electrophoretic mobilities of oligonucleotides on cellulose acetate. *Nucl. Acids Res.* 1:1503–1520.

Barnes, W. M. (1978). DNA sequencing by partial ribosubstitution. J. Mol. Biol. 119:83–99.

Barnes, W. M., Reznikoff, W. S., Blattner, F. R., Dickson, R. C., and Abelson, J. (1975). The isolation of RNA homologous to the genetic control elements of the *lac* operon. *J. Biol. Chem.* 250:8184–8192.

Baron, F., and Brown, D. M. (1955). Nucleotides. Part XXXIII. The structure of cytidylic acids a and b. *J. Chem. Soc.* 1955:2855–2860.

Barrell, B. G. (1978). Sequence analysis of bacteriophage *phi*X174 DNA. In *Biochemistry of Nucleic Acids*, ed. B. F. C. Clark (Baltimore University Park Press), pp. 125–179.

Barrell, B. G. (1971). Fractionation and sequence analysis of radioactive nucleotides. In *Procedures in Nucleic Acid Research*, ed. G. L. Cantoni and D. R. Davies (New York: Harper & Row), pp. 751–779.

Bastia, D. (1978). Determination of restriction sites and the nucleotide sequence surrounding the relaxation site of ColE1. *J. Mol. Biol.* 124:601–639.

Beaven, G. H., Holiday, E. R., and Johnson, E. A. (1955). Optical properties of nucleic acids and their components. In *The Nucleic Acids*, Vol. I, ed. E. Chargaff and J. N. Davidson (New York: Academic Press), pp. 493–553.

Beck, E., Sommer, R., Auerswald, E. A., Kurz, Ch., Zink, B., Osterburg, G., Schaller, H., Sugimoto, K., Sugisaki, H., Okamoto, T., and Takanami, M. (1978). Nucleotide sequence of bacteriophage fd DNA. *Nucl. Acids Res.* 5:4495–4503.

Billeter, M. A., Dahlberg, J. E., Goodman, H. M., Hindley, J., and Weissman, C. (1969). Sequence of the first 175 nucleotides from the 5′ terminus of Q*beta* RNA synthesized *in vitro*. *Nature* 224:1083–1086.

Blackburn, E. H. (1976). Transcription and sequence analysis of a fragment of bacteriophage *phi*X174 DNA. *J. Mol. Biol.* 107:417–431.

Blackburn, E. H. (1975). Transcription by *Escherichia coli* RNA polymerase of a single-stranded fragment of bacteriophage *phi*X174 DNA 48 residues in length. *J. Mol. Biol.* 93:367–374.

Berg, P., Fancher, H., and Chamberlin, M. (1963). The synthesis of mixed polynucleotides containing ribo- and deoxyribonucleotides by purified preparations of DNA polymerase from *Escherichia coli*. In *Informational Molecules*, eds. H. J. Vogel, V. Bryson, and J. O. Lampen (New York: Academic Press), pp. 467–483.

Bourgeois, S., and Riggs, A. D. (1970). The *lac* repressor-operator interaction: Assay and purification of operator DNA. *Biochem. Biophys. Res. Commun.* 38:348–354.

Brown, D. M. (1974). Chemical reactions of polynucleotides and nucleic acids. In *Basic Principles of Nucleic Acid Chemistry*, ed. P. O. P. T'so (New York/London: Academic Press), pp. 1–90.

Brown, D. M., and Todd, A. R. (1955). Evidence on the nature of the chemical bonds in nucleic acids. In *The Nucleic Acids*, Vol. I, eds. E. Chargaff and J. N. Davidson (New York: Academic Press), pp. 409–445.

Brown, N. L., and Smith, M. (1977). The sequence of a region of bacteriophage *phi*X174 DNA coding for parts of genes *A* and *B J. Mol. Biol.* 116:1–30.

Brownlee, G. G. (1972). Determination of sequences in RNA. In *Laboratory Techniques in Biochemistry and Molecular Biology*, Vol. 3, Part I, eds. T. S. Work and E. Work (Amsterdam/NewYork: North-Holland/Elsevier).

Brownlee, G. G., and Sanger, F. (1969). Chromatography of [32] P-labeled oligonucleotides on thin layers of DEAE-cellulose. *Eur. J. Biochem.* 11:395–399.

Brownlee, G. G., and Sanger, F. (1967). Nucleotide sequences from the low molecular weight ribosomal RNA of *Escherichia coli. J. Mol. Biol.* 23:337–353.

Budovskii, E. I., Haines, J. A., and Kochetkov, N. F. (1964). Hydrazinolysis of pyrimidine nucleosides and DNA. *Dokl. Akad. Nauk. USSR* 158:874–876.

Burton, K., and Petersen, G. B. (1960). The frequencies of certain sequences of nucleotides in deoxyribonucleic acid. *Biochem. J.* 75:17–27.

Burton, K., and Petersen, G. B. (1957). The quantitive distribution of pyrimidine nucleotides in calf thymus deoxyribonucleic acid. *Biochim. Biophys. Acta* 26:667–668.

Cape, R. E., and Spencer, J. H. (1968). Nucleotide clusters in deoxyribonucleic acids. I. Isolation of purine oligonucleotides. *Can. J. Biochem.* 46:1063–1073.

Capon, B. (1965). Mechanism in carbohydrate chemistry. *Acc. Chem. Res.*

Cashmore, A. R., and Petersen, G. B. (1978). The degradation of DNA by hydrazine: Identification of 3-ureidopyrazole as a product of the hydrazinolysis of deoxycytidylic acid residues. Nucl. Acids Res. 5:1485–2491.

Cashmore, A. R., and Petersen, G. B. (1968). The degradation of DNA by hydrazine: A critical study of the reaction for the quantitative determination of purine nucleotide sequences. *Biochim. Biophys. Acta* 174:591–603.

Cerny, R., Cerna, E., and Spencer, J. H. (1969). Nucleotide clusters in deoxyribonucleic acids. IV. Pyrimidine oligonucleotides of bacteriophage S13 *suN15* DNA and replicative form DNA. *J. Mol. Biol.* 46:145–156.

Chargaff, E. (1968). What really is DNA? Remarks on the changing aspects of a scientific concept. In *Progress in Nucleic Acid Research and Molecular Biology*, eds. J. N. Davidson and W. E. Cohn (New York/London: Academic Press), pp. 297–333.

Chargaff, E., Rust, P. Temperli, A., Morisawa, S., and Danon, A. (1963). Investigation of the purine sequences in deoxyribonucleic acids. *Biochim. Biophys. Acta* 76:149–151.

Chow, L. T., Gelinas, R. E., Broker, T. R., and Roberts, R. J. (1977). An amazing sequence arrangement at the 5' ends of adenovirus 2 messenger RNAs. *Cell* 12:1–8.

Coombs, M. M., and Livingston, D. C. (1969). Reaction of apurinic acid with aldehyde reagents. *Biochim. Biophys. Acta* 174:161–173.

Crick, F. (1979). Split genes and RNA splicing. *Science* 204:264–271.

Danna, K., and Nathans, D. (1971). Specific cleavage of simian virus 40 DNA by restriction endonuclease of *Haemophilus influenzae*. *Proc. Natl. Acad. Sci. USA* 68:2913–2917.

Darnell, Jr., J. E. (1978). Implications of RNA–RNA splicing in evolution of eukaryotic cells. *Science* 202:1257–1260.

Delaney, A. D., and Spencer, J. H. (1973). Sequence determination of oligonucleotides from S13 *suN15* DNA. *Fed. Proc.* 32:2536 (Abstract).

Dhar, R., Weissman, S. M., Zain, B. S., Pan, J., and Lewis, A. M. jun. (1974). The nucleotide sequence preceding an RNA polymerase initiation site on SV40 DNA. Part 2. The sequence of the early strand transcript. *Nucl. Acids Res.* 1:595–613.

Dhar, R., Zain, B. S., Weissman, S. M., Pan, J. and Subramanian, K. N. (1974). Nucleotide sequences of RNA transcribed in infected cells and by *Escherichia coli* RNA polymerase from a segment of simian virus 40 DNA. *Proc. Natl. Acad. Sci. USA* 71:371–375.

Dickson, R. C., Abelson, J., Barnes, W. J. M., and Reznikoff W. S. (1975). Genetic regulation: The *lac* control region. *Science* 187:27–35.

Downey, K. M., and So, A. G. (1970). Studies on the kinetics of ribonucleic acid chain initiation and elongation. *Biochem.* 9:2520–2525.

Ellery, B. W., and Symons, R. H. (1966). Loss of adenine during the hydrazine degradation of deoxyribonucleic acid. *Nature* 210:1159–1160.

Elson, E., and Jovin, T. M. (1969). Fractionation of oligodeoxynucleotides by polyacrylamide gel electrophoresis. *Anal. Biochem.* 27:193–204.

Englund, P. T. (1972). The 3β-terminal nucleotide sequences of T7 DNA. *J. Mol. Biol.* 66:209–224.

Englund, P. T. (1971). Analysis of nucleotide sequences of 3' termini of duplex deoxyribonucleic acid with the use of the T4 deoxyribonucleic acid polymerase. *J. Biol. Chem.* 246:3269–3276.

Federoff, N. V., and Brown, D. D. (1978). The nucleotide sequence of oocyte 5S DNA in *Xenopus laevis*. I. The AT-rich spacer. *Cell* 13:701–716.

Fiers, W., Contreras, R., Haegeman, G., Rogiers, R., Van De Voorde, A., Van Heuverswyn, H., Van Herreweghe, J., Volckaert, G., and Ysebaert, M.

(1978). Complete nucleotide sequence of SV40 DNA. *Nature* 273:113–120.

Fowler, A. V., and Zabin, I. (1977). The amino acid sequence of β-galactosidase of *Escherichia coli. Proc. Natl. Acad. Sci. USA* 74:1507–1510.

Fry, K., Poon, R. Whitcombe, P. Idriss, J., Salser, W., Mazrimas, J., and Hatch, F. (1973). Nucleotide sequence of HS-*beta* satellite DNA from kangaroo rat *Diplodomys ordii. Proc. Natl. Acad. Sci. USA* 70:2642–2646.

Galibert, F., Sedat, J., and Ziff, E. (1974). Direct determination of DNA nucleotide sequences: Structure of a fragment of bacteriophage *phi*X174 DNA. *J. Mol. Biol.* 87:377–407.

Ganoza, M. C., Fraser, A. R., and Neilson, T. (1978). Nucleotides contiguous to AUG affect translation initiation. *Biochem.* 17:2769–2775.

Gilbert, W. (1972). The *lac* repressor and the *lac* operator. In *Polymerization in Biological Systems* (Amsterdam/New York: North-Holland/Elsevier), pp. 245–259.

Gilbert, W., and Maxam, A. (1973). The nucleotide sequence of the *lac* operator. *Proc. Natl. Acad. Sci. USA* 70:3581–3584.

Gilbert, W., Majors, J., and Maxam, A. (1976). How proteins recognize DNA sequences. In *Organization and Expression of Chromosomes*, eds. V. G. Alfrey, E. K. F. Bautz, B. J. McCarthy, R. T. Schimke, and A. Tissieres (Berlin: Dahlem Konferenzen), pp. 167–178.

Gilbert, W., Maxam, A., and Mirzabekov, A. (1976). Contacts between the *lac* repressor and DNA revealed by methylation. In *Control of Ribosome Synthesis*, eds. N. O. Kjeldgaard and O. Maaloe (Copenhagen: Munksgaard), pp. 139–148.

Gilbert, W., Gralla, J., Majors, J., and Maxam, A. (1975). Lactose operator sequences and the action of *lac* repressor. In *Protein-Ligand Interactions, eds. H. Sund and G. Blauer (Berlin: Walter de Gruyter), pp. 193–210.*

Gillum, A. M., and Clayton, D. A. (1978). Displacement-loop replication initiation sequence in animal mitochondrial DNA exists as a family of discrete lengths. *Proc. Natl. Acad. Sci. USA* 75:677–681.

Gordon, M. P., Weliky, V. S., and Brown, G. B. (1957). A study of the action of acid and alkali on certain purines and purine nucleosides. *J. Amer. Chem. Soc.* 79:3245–3255.

Gronenborn, B., and Messing, J. M. (1978). Methylation of single-stranded DNA *in vitro* introduces new restriction endonuclease cleavage sites. *Nature* 272:375–377.

Habermann, V. (1963a). Studies on deoxyribonucleic acids. II. Degradation of deoxyribonucleic acids to purine nucleotide sequences. *Coll. Czech. Chem. Commun.* 28:510–517.

Habermann, V. (1963b). Participation of purine and pyrimidine nucleotide sequences in the molecule of calf thymus deoxyribonucleic acid. *Nature* 200: 782–783.

Habermann, V. (1962). The degradation of apyrimidinic deoxyribonucleic acid in alkali: A method for the isolation of purine nucleotide sequences from deoxyribonucleic acid. *Biochim. Biophys. Acta* 55:999–1001.

Habermann, V. (1961). Studies on deoxyribonucleic acids. I. The preparation of apyrimidinic deoxyribonucleic acids. *Coll. Czech. Chem. Commun.* 26: 3147–3156.

Habermann, V., Habermannova, S., and Cerhova, M. (1963). The distribution into pyrimidine and purine nucleotide clusters in the polynucleotide chain from *Escherichia coli* C. *Biochim. Biophys. Acta* 76:310–311.

Haines, J. A., Reese, C. B., and Todd, A. R. (1962). The methylation of guanosine and related compounds with diazomethane. *J. Chem. Soc.* 1962:5281–5288.

Hall, B. D. (1979). Mitochondria spring surprises. *Nature* 282:129–130.

Hall, J. B., and Sinsheimer, R. L. (1963). The structure of the DNA of bacteriophage *phi*X174. IV. Pyrimidine sequences. *J. Mol. Biol.* 6:115–127.

Haseltine, W. A., Maxam, A. M., and Gilbert, W. (1977). Rous sarcoma virus genome is terminally redundant: The 5′ sequence. *Proc. Natl. Acad. Sci. USA* 74:989–993.

Hayes, D. H., and Hayes-Baron, F. (1967). Hydrazinolysis of some purines and pyrimidines and their related nucleosides and nucleotides. *J. Chem. Soc.* 1967:1528–1533.

Heyden, B., Nusslein, C., and Schaller, H. (1972). Single RNA polymerase binding site isolated. *Nature New Biol.* 240:9–12.

Holley, R. W., Apgar, J., Everett, G. A., Madison, J. T., Marquisee, M., Merrill, S. H., Penswick, J. R., and Zamir, A. (1965). Structure of a ribonucleic acid. *Science* 147:1462–1464.

Holley, R. W., Madison, J. T., and Zamir, A. (1964). A new method for sequence determination of large oligonucleotides. *Biochem. Biophys. Res. Comm.* 17:389–394.

Jay, E., Bambara, R., Padmanabhan, R., and Wu, R. (1974). DNA sequence analysis: A general, simple, and rapid method for sequencing large oligodeoxyribonucleotide fragments by mapping. *Nucl. Acids Res.* 1:331–354.

Jones, A. S., Mian, A. M., and Walker, J. T. (1966). The action of alkali on some purines and their derivatives. *J. Chem. Soc.* 1966C:692–695.

Josse, J., Kaiser, A. D., and Kornberg, A. (1961). Enzymatic synthesis of deoxyribonucleic acid. VII. Frequency of nearest neighbor base sequences in deoxyribonucleic acid. *J. Biol. Chem.* 236:864–875.

Klenow, H., and Henningsen, I. (1970). Selective elimination of the exonuclease activity of the deoxyribonucleic acid polymerase from *Escherichia coli* B by limited proteolysis. *Proc. Natl. Acad. Sci. USA* 65:168–175.

Kleppe, K., Ohtsuka, E., Kleppe, R., Moulineaux, I., and Khorana, H. G. (1971). Studies on polynucleotides XCVI. Repair replication of short synthetic RNAs as catalyzed by DNA polymerases. *J. Mol. Biol.* 56:341–361.

Klessig, D. F. (1977). Two adenovirus mRNAs have a common 5′ terminal leader sequence encoded at least 10 kb upstream from their main coding regions. *Cell* 12:9–21.

Kochetkov, N. K., and Budovskii, E. E. (1972). *Organic Chemistry of Nucleic Acids*, Part B. (London/New York: Plenum).

Krayev, A. S. (1981). The use of diethylpyrocarbonate for sequencing adenines and guanines in DNA. *FEBS Lett.* 130:19–22.

Kriek, E., and Emmelot, P. (1964). Methylation of deoxyribonucleic acid by diazomethane. *Biochim. Biophys. Acta* 91:59–66.

Langeveld, S. A., Van Mansfeld, A. D. M., Baas, P. D., Jansz, H. S., Van Arkel, G. A., and Weisbeck, P. J. (1978). Nucleotide sequence of the origin of replication in bacteriophage *phi*X174 DNA. *Nature* 271:417–420.

Lawley, P. D., and Brookes, P. (1963). Further studies on the alkylation of nucleic acids and their constituent nucleotides. *Biochem. J.* 89:127–138.

Lawley, P. D., and Thatcher, C. J. (1970). Methylation of deoxyribonucleic acid in cultured mammalian cells by N-methyl-N′-nitro-N-nitrosoguanidine. *Biochem. J.* 116:693–707.

Lebowitz, P., Weissman, S. M., and Radding, C. M. (1971). Nucleotide sequence of a ribonucleic acid transcribed *in vitro* from λphage deoxyribonucleic acid. *J. Biol. Chem.* 246:5120–5139.

Levene, P. A., and Bass, L. W. (1926). The action of hydrazine hydrate on uridine. *J. Biol. Chem.* 71:167–172.

Ling, V. (1972a). Fractionation and sequences of the large pyrimidine oligonucleotides from bacteriophage fd DNA. *J. Mol. Biol.* 64:87–102.

Ling, V. (1972b). Pyrimidine sequences from the DNA of bacteriophages fd, f1, and *phi*X174. *Proc. Natl. Acad. Sci. USA* 69:742–746.

Ling, V. (1971). Partial digestion of ^{32}P-fd DNA with T4 endonuclease IV. *FEBS Lett.* 19:50–54.

Lingens, F., and Schneider-Bernlohr, H. (1966). Uber die wirkung von hydrazin und methylsubstituierten hydrazinen auf nucleoside, nucleotide und ribonucleinsaure. *Biochim. Biophys. Acta* 123:611–613.

Linstead, R. P., Owen, L. N., and Webb, R. F. (1953). Elimination reactions of esters. Part I. The formation of *alpha,beta*-unsaturated acids from *beta*-acyloxy compounds. *J. Chem. Soc.* 1953:1211–1218.

Livingston, D. C. (1964). Degradation of apurinic acid by condensation with aldehyde reagents. *Biochim. Biophys. Acta* 87:538–540.

Maat, J., and Smith, A. J. H. (1978). A method for sequencing restriction fragments with dideoxynucleoside triphosphates. *Nucl. Acids Res.* 5:4537–4545.

Mandeles, S., and Tinoco, Jr., I. (1963). A general method for determination of nucleotide sequences in nucleic acids. *Biopolymers* 1:183–190.

Maniatis, T., and Ptashne, M. (1973a). Structure of the *lambda* operators. *Nature* 246:133–136.

Maniatis, T., and Ptashne, M. (1973b). Multiple repressor binding at the operators in bacteriophage *lambda*. *Proc. Natl. Acad. Sci. USA* 70:1531–1535.

Maniatis, T., Jeffrey, A., and Kleid, J. (1975). Nucleotide sequence of the right-hand operator of phage *lambda*. *Proc. Natl. Acad. Sci. USA* 72:1184–1188.

Maniatis, T., Jeffrey, A., and Van De Sande, H. (1975). Chain length determination of small double- and single-stranded DNA molecules by polyacrylamide gel electrophoresis. *Biochem.* 14:3787–3794.

Maniatis, T., Ptashne, M., Barrell, B. G., and Donelson, J. (1974). Sequence of a repressor-binding site in the DNA of bacteriophage *lambda*. *Nature* 250: 394-397.

Margison, G. P., and O'Connor, P. J. (1973). Biological implications of the instability of the *N*-glycosidic bond of 3-methyldeoxyadenosine in DNA. *Biochim. Biophys. Acta* 331:349-356.

Marvin, D. A., and Hohn, B. (1969). Filamentous bacterial viruses. *Bacteriol. Rev.* 33:172-209.

Maxam, A. M., and Gilbert, W. (1980). Sequencing end-labeled DNA with base-specific chemical cleavages. In *Methods in Enzymology*, No. 65, eds. L. Grossman and K. Moldave (New York: Academic Press), pp. 499-560.

Maxam, A. M., and Gilbert, W. (1977). A new method for sequencing DNA. *Proc. Natl. Acad. Sci. USA* 74:560-564.

Maxam, A. M., Tizard, R., Skryabin, K. G., and Gilbert, W. (1977). Promoter region for yeast 5S ribosomal RNA. *Nature* 267:643-645.

Middleton, J. H., Edgell, M. H., and Hutchison, III, C. A. (1972). Specific fragments of *phi*X174 deoxyribonucleic acid produced by a restriction enzyme from *Haemophilus aegyptius*, endonuclease Z. *J. Virol* 10:42-50.

Miller, J. R., Cartwright, E. M., Brownlee, G. G. Federoff, N. V., and Brown D. D. (1978). The nucleotide sequence of oocyte 5S DNA in *Xenopus laevis*. II. The GC-rich region. *Cell* *13*:717-725.

Mirzabekov, A. D., and Kolchinsky, A. M. (1974). Localisation of some molecules within the grooves of DNA by modification of their complexes with dimethyl sulphate. *Molec. Biol. Rep.* 1:385-390.

Mirzabekov, A. D., and Melnikova, A. F. (1974). Localisation of chromatin proteins within DNA grooves by methylation of chromatin with dimethyl sulphate. *Molec. Biol. Rep.* 1:379-384.

Murray, K., and N. E. Murray (1973). Terminal nucleotide sequences of DNA from temperate coliphages. *Nature New Biol.* 243:134-139.

Murray, K., and Old, R. W. (1974). The primary structure of DNA. In *Progress in Nucleic Acid Research and Molecular Biology*, Vol. 14, ed, W. E. Cohn, (New York/London: Academic Press), pp. 117-185.

Mushynski, W. E., and Spencer, J. H. (1970a). Nucleotide clusters in deoxyribonucleic acids. V. The pyrimidine oligonucleotides of strands *r* and *l* of bacteriophage T7 DNA. *J. Mol. Biol.* 52:91-106.

Mushynski, W. E., and Spencer, J. H. (1970b). Nucleotide clusters in deoxyribonucleic acids. VI. The pyrimidine oligonucleotides of strands *r* and *l* of bacteriophage *lambda* DNA. *J. Mol. Biol.* 52:107-120.

Ohmori, H., Tomizawa, J., and Maxam, A. M. (1978). Detection of 5-methylcytosine in DNA sequences. *Nucl. Acids Res.* 5:1479-1485.

Peacock, A. C., and Dingman, C. W. (1967). Resolution of multiple ribonucleic acid species by polyacrylamide gel electrophoresis. *Biochem* 6:1818-1827.

Pirrotta, V. (1975). Sequence of the O_R operator of phage *lambda*. *Nature* 254:114-117.

Pirrotta, V. (1973). Isolation of the operators of phage *lambda*. *Nature New Biol.* 244:13-16.

Pochon, F., and Michelson, A. M. (1967). Polynucleotides. IX. Methylation of nucleic acids, homopolynucleotides and complexes. *Biochim. Biophys. Acta* 149:99-106.

Pribnow, D. (1975a). Nucleotide sequence of an RNA polymerase binding site at an early T7 promoter. *Proc. Natl. Acad. Sci. USA* 72:784-788.

Pribnow, D. (1975b). Bacteriophage T7 early promoters: Nucleotide sequences of two RNA polymerase binding sites. J. Mol. Biol. 99:419-443.

Randerath, K., and Randerath, E. (1967). Thin-layer separation methods for nucleic acid derivatives. In *Methods in Enzymology*, Vol. XIIA, eds. L. Grossman and K. Moldave (New York/London: Academic Press), pp. 323-347.

Razzell, W. E., and Khorana, H. G. (1961). Studies on polynucleotides, X. Enzymic degradation. Some properties and mode of action of spleen phosphodiesterase. *J. Biol. Chem.* 236:1144-1149.

Reddy, V. B., Thimmappaya, B., Dhar, R. Subramanian, K. N., Zain, B. S., Pan, J., Ghosh, P. K., Celma, M. L. and Weissman, S. M. (1978). The genome of simian virus 40. *Science* 200:494-502.

Richardson, C. C. (1965). Phosphorylation of nucleic acid by an enzyme from T4 bacteriophage-infected *Escherichia coli. Proc. Natl. Acad. Sci. USA* 54:158-165.

Roberts, R. J., Myers, P. A., Morrison, A., and Murray, K. (1976). A specific endonuclease from *Arthrobacter luteus. J. Mol. Biol.* 102:157–165.

Sadler, J. R., and Smith, T. F. (1971). Mapping of the lactose operator. *J. Mol. Biol.* 62:139–169.

Sadowski, P. D., and Hurwitz, J. (1969). Enzymatic breakage of DNA. II. Purification of properties of endonuclease IV and T4 phage-infected *Escherichia coli, J. Biol. Chem.* 244:6192–6198.

Saf, A., Coulson, A. R., Friedmann, T., Air, G. M., Barrell, B. G., Brown, N. L., Fiddes, J. C., Hutchinson, III, C. A., Slocombe, P. M., and Smith, M. (1978). The nucleotide sequence of bacteriophage *phi*X174. *J. Mol. Biol.* 125:225–246.

Salser, W., Fry, K., Brunk, C., and Poon, R. (1972). Nucleotide sequencing of DNA: Preliminary characterization of the products of specific cleavages at guanine, cytosine, or adenine residues. *Proc. Natl. Acad. Sci. USA* 69: 238–242.

Sanger, F., and Coulson, A. R. (1978). The use of thin acrylamide gels for DNA sequencing. *FEBS Lett.* 87:107–110.

Sanger, F., and Coulson, A. R. (1975). A rapid method for determining sequences in DNA by primed synthesis with DNA polymerase. *J. Mol. Biol.* 94:441–448.

Sanger, F., Brownlee, G. G., and Barrell, B. G. (1965). A two-dimensional fractionation procedure for radioactive nucleotides. *J. Mol. Biol.* 13:373–398.

Sanger, F., Nicklen, S., and Coulson, A. R. (1977). DNA sequencing with chain-terminating inhibitors. *Proc. Natl. Acad. Sci. USA* 74:5463–5467.

Sanger, F., Air, G. M., Barrell, B. G., Brown, N. L., Coulson, A. R., Fiddes, J. C., Hutchinson, III, C. A., Slocombe, P. M., and Smith, M. (1977). Nucleotide sequence of bacteriophage *phi*X174 DNA. *Nature* 265:687–695.

Sanger, F., Donelson, J. E., Coulson, A. R., Kossel, H., and Fischer, D. (1973), Use of DNA polymerase I primed by a synthetic oligonucleotide to determine a nucleotide sequence in phage f1 DNA. *Proc. Natl. Acad. Sci. USA* 70: 1209–1213.

Schaller, H., Gray, C., and Hermann, K. (1975). Nucleotide sequence of an RNA polymerase binding site from the DNA of bacteriophage fd. *Proc. Natl. Acad. Sci. USA* 72:737–741.

Schreier, P. H., and Cortese, R. (1979). A fast and simple method for sequencing DNA cloned in the single-stranded bacteriophage M13. *J. Mol. Biol.* 129: 169–172.

Sedat, J., and Sinsheimer, R. L. (1964). Structure of the DNA of bacteriophage *phi*X174. V. Purine sequences. *J. Mol. Biol.* 9:489–497.

Shapiro, H. S., and Chargaff, E. (1957a). Characterization of nucleotide arrangement in deoxyribonucleic acids through stepwise degradation. *Biochim. Biophys. Acta* 23:451–452.

Shapiro, H. S., and Chargaff, E. (1957b). Studies on the nucleotide arrangement in deoxyribonucleic acids. I. The relationship between the production of pyrimidine nucleoside $3',5'$-diphosphates and specific features of nucleotide sequence. *Biochim. Biophys. Acta* 26:596–608.

Shapiro, H. S., and Chargaff, E. (1957c). Studies on the nucleotide arrangement in deoxyribonucleic acids. II. Differential analysis of pyrimidine nucleotide distribution as a method of characterization. *Biochim. Biophys. Acta* 26:608–623.

Shapiro, R., and Danzig, M. (1972). Acid hydrolysis of deoxycytidine and deoxyuridine derivatives. The general mechanism of deoxyribonucleoside hydrolysis *Biochem.* 11:23–29.

Sims, J., Koths, K. E., and Dressler, D. H. (1978). Single-stranded phage replication: Positive- and negative-strand DNA synthesis. *Cold Spring Harbor Symp. Quant. Biol.* 43:349–365.

Smith, A. J. H. (1979). The use of exonuclease III for preparing single stranded DNA for use as a template in the chain terminator sequencing method. *Nucl. Acids Res.* 6:831–848.

Smith, J. D. (1967). Paper electrophoresis of nucleic acid components. In *Methods in Enzymology*, Vol. XII, Part A, eds. L. Grossman and K. Moldare (New York/London: Academic Press), pp. 350–361.

Southern, E. M. (1970). Base-sequence and evolution of guinea-pig *alpha*-satelite DNA. *Nature* 227:794–798.

Spencer, J. H., and Chargaff, E. (1963). Studies on the nucleotide arrangement in deoxyribonucleic acids. VI. Pyrimidine nucleotide clusters: Frequency and distribution in several species of the AT-type. *Biochim. Biophys. Acta* 68:18–27.

Sugimoto, K., Okamoto, T., Sugisaki, H., and Takanami, M. (1975). The nucleotide sequence of an RNA polymerase binding site on bacteriophage fd DNA. *Nature* 253:410–414.

Sutcliffe, J. G. (1978). Complete nucleotide sequence of the *Escherichia coli* plasmid pBR322. *Cold Spring Harbor Symp. Quant. Biol.* 43:77-90.

Sverdolv, E. D., Monastyrskaya, G. S., Chestukhin, A. V., and Budovskii, E. I. (1973). The primary structure of oligonucleotides. Partial apurinization as a method to determine the positions of purine and pyrimidine residues. *FEBS Lett.* 33:15-17.

Szekely, M., and Sanger, F. (1969). Use of polynucleotide kinase in fingerprinting non-radioactive nucleic acids. *J. Mol. Biol.* 43:607-617.

Takemura, S. (1959). Hydrazinolysis of nucleic acids. I. The formation of deoxy-riboapyrimidinic acid from herring sperm deoxyribonucleic acid. *Bull. Chem. Soc. Japan* 32:920-926.

Takemura, S. (1958). Hydrazinolysis of herring sperm deoxyribonucleic acid. *Biochim. Biophys. Acta* 29:447-448.

Tamm, C., Hodes, M. E., and Chargaff, E. (1952). The formation of apurinic acid from the desoxyribonucleic acid of calf thymus. *J. Biol. Chem.* 195: 49-63.

Tamm, C., Shapiro, H. S., Lipshitz, R., and Chargaff, E. (1953). Distribution density of nucleotides within a desoxyribonucleic acid chain. *J. Biol. Chem.* 203:673-688.

Temperli, A., Turler, H., Rust, P., Danon, A., and Chargaff, E. (1964). Studies on the nucleotide arrangement in deoxyribonucleic acids. IX. Selective degradation of pyrimidine deoxyribonucleotides. *Biochim. Biophys. Acta* 91:462-476.

Terao, T., Dahlberg, J. E., and Khorana, H. G. (1972). Studies on polynucleo-tides. CXX. On the transcription of a synthetic 29-unit-long deoxyribo-polynucleotide. *J. Biol. Chem.* 247:6157-6166.

Tomizawa, J., Ohmori, H., and Bird, R. E. (1977). Origin of replication of colicin E1 plasmid DNA. *Proc. Natl. Acad. Sci. USA* 74:1865-1869.

Tomlinson, R. V., and Tener, G. M. (1963). The effect of urea, formamide, and glycols on the secondary binding forces in the ion-exchange chroma-tography of polynucleotides on DEAE-cellulose. *Biochem.* 2:697-702.

Tu, C. D., Jay, E. Bahl, C. P., and Wu, R. (1976). A reliable mapping method for sequence determination of oligodeoxyribonucleotides by mobility shift analysis. *Anal. Biochem.* 74:73-93.

Turchinski, M. F., Gus'Kova, L. I., Khazai, I., Budovskii, E. I., and Kochetkov,

N. K. (1970). Chemical method of specific degradation of ribonucleic acids with selectively removed bases. 3. Cleavage of phosphodiester bonds in ribose-2- and 3-phosphates catalyzed by amines. *Molec. Biol.* (English translation of *Molekulyarnaya Biologiya)* 4:343-348.

Turler, H., and Chargaff, E. (1969). Studies on the nucleotide arrangement in deoxyribonucleic acids. XII. Apyrimidinic acid from calf thymus deoxyribonucleic acid: Preparation and properties. Biochim. Biophys. Acta 195: 446-455.

Turler, H., Buchowicz, J., and Chargaff, E. (1969). Studies on the nucleotide arrangement in deoxyribonucleic acids. XIII. Frequency and composition of purine isostichs in calf thymus deoxyribonucleic acid. *Biochim. Biophys. Acta* 195:446-455.

Van De Sande, J. H., Loewen, P. C., and Khorana, H. G. (1972). Studies on polynucleotides. CXVIII. A further study of ribonucleotide incorporation into deoxyribonucleic acid chains by deoxyribonucleic acid polymerase I of *Escherichia coli. J. Biol. Chem.* 247:6140-6148.

Vanyushin, B. F., and Bur'yanov, Y. I. (1969). Production and alkaline hydrolysis of apyrimidinic acid. *Biochem.* (English translation of *Biokhimiya)* 34:546-555.

Walz, A., and Pirrotta, V. (1975). Sequence of the P_R promoter of phage lambda. Nature 254: 118-121.

Weigel, P. H., Englund, P. T., Murray, K., and Old, R. W. (1973). The 3'-terminal sequences of bacteriophage *lambda* DNA. *Proc. Natl. Acad. Sci. USA* 70: 1151-1155.

Weissman, S. M. (1979). Curent approaches to DNA sequencing. *Anal. Biochem.* 98:243-253, 1979.

Wu, R., and Kaiser, A. D. (1968). Structure and base sequence in the cohesive ends of bacteriophage λDNA. *J. Mol. Biol.* 35:523-537.

Wu, R., and Taylor, E. (1971). Nucleotide sequence analysis of DNA. II. Complete nucleotide sequence of the cohesive ends of bacteriophage λDNA. *J. Mol. Biol.* 57:491-511.

Wu, R., Tu, C. D., and Padmanabhan, R. (1973). Nucleotide sequence analysis of DNA. XII. The chemical synthesis and sequence analysis of a dodecadeoxynucleotide which binds to the endolysin gene of bacteriophage *lambda. Biochem. Biophys. Res. Comm.* 55:1092-1099.

Zain, B. S., Weissman, S. M., Dhar, R., and Pan, J. (1974). The nucleotide sequence preceding an RNA polymerase initiation site on SV40 DNA. Part 1. The sequence of the late strand transcript. *Nucl. Acids Res.* 1: 577–594

Ziff, E., Sedat, J., and Galibert, F. (1973). Determination of the nucleotide sequence of a fragment of bacteriophage *phi*X174 DNA. *Nature New Biol.* 241:34–37.

Zoltewicz, J. A., Clark, D. F., Sharpless, T. W., and Grahe, G. (1970). Kinetics and mechanism of the acid-catalyzed hydrolysis of some purine nucleosides. *J. Amer. Chem. Soc.* 92:1741–1750.

ADDENDUM BY SHERMAN M. WEISSMAN, M.D.

Subsequent to the development of the new and powerful methods of gel sequencing, the factors limiting the speed of data generation became the resolution obtained on sequencing gels, identification of suitably spaced sites on a DNA molecule from which to initiate sequencing reactions, and the handling of the raw data obtained from the sequence films. The third topic is not dealt with in the present volume, but many workers have evolved sequence-handling data systems and an excellent initial source for this material is provided by the series of articles published in *Nucleic Acids Research* of January, 1982.

Useful methods have been suggested for both the enzymatic sequencing and the DNA sequencing methods that rapidly provide convenient sites for sequence initiation along an inserted DNA molecule. These methods depend essentially on random creation of double-stranded breaks in the DNA molecules by use of nonspecific deoxyribonuclease in the presence of manganese chloride. The constructions are arranged so that the individual molecules can be cloned and sequencing initiated near the site of the original double-stranded break. The initial DNA is cloned either in single-stranded phage (Hong, 1982; Poncz et al., 1982) for enzymatic sequencing or in plasmids such as pBR322 (Frischauf, Garoff, & Lehrach, 1980) for DNA sequencing. The clone is treated with DNAse and manganese under prescribed conditions, so as to obtain principally uncut and singly cut molecules. The ends generated with DNAse are then converted to blunt ends by treatment with DNA polymerase and/or S1 nuclease.

In the proposal to use double-stranded plasmid clones for chemical sequencing, a representative scheme was described for sequencing a DNA molecule cloned at the *Pst* site pBR322 (inactivating the ampicillin gene). To do this, an RI linker was inserted at the site of random DNAse cleavage in the original cloned DNA. The DNA molecules were then redigested with *Eco*RI, separating the nonessential fragment of plasmid between the *Pst* and RI sites and also a portion of the insert closest to the plasmid RI sites (see Fig. 4-A-1). The cleaved molecules were fractionated on gels, and molecules of various-sized lengths

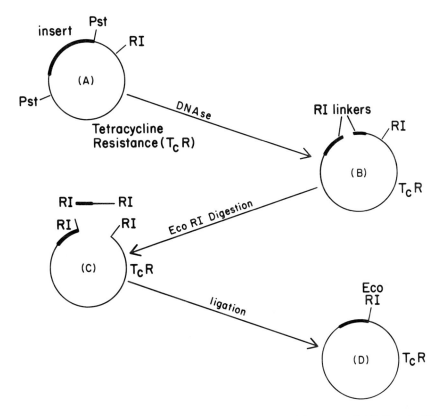

Figure 4-A-1. Method for systematic sequencing of inserts at *Pst* I site of pBR or equivalent vector by the chemical sequencing method. Light segments of circle indicate plasmid; dark segment indicates insert to be sequenced. Pst indicates *Pst* I sites; RI, the *Eco*RI site. *A*: Initial plasmid with approximate localization of tetracycline resistance region TcR after treatment with manganese and DNAse I produces molecules that have been linearized by random double-stranded cuts. *B*: Such molecules are illustrated in which the cut has occurred within the insert RI linkers and ligated with *Hind* insert as indicated; after *Eco*RI digestion one excises the small fragment between RI linkers and regenerates RI ends containing the bulk of plasmid, including tetracycline resistance. *C*: Religation generates a plasmid in which a unique RI site is located adjacent to the initial DNAse I cleavage site within the insert and provides a suitable starting point for sequencing.

corresponding to the large *Eco*RI to *Pst* fragment of the vector, plus varying amounts of insert, were excised from the gel and ligated to form circles. These molecules of course will also contain other molecules in which the random DNAse nick and the RI linker has been inserted elsewhere in the plasmid. However, most of these latter molecules will lack the tetracycline resistance gene and so would not score as colonies on plating on tetracycline plates. The gel fractionation step permits the selection of a sampling of molecules with RI linkers inserted at different positions throughout the insert (Fig. 4-A-1). The circularized DNA fragments are then cloned into bacteria. Sequencing can thus be accomplished simply by cutting DNA from each new clone at the single *Eco*RI site, labeling at the RI site, and then recutting with those enzymes (*Cla*I or *Hind*III) that have unique cleavage sites close to the *Eco*RI site of the plasmid. Thus, one will obtain a very short labeled fragment from plasmid and a long fragment from the genomic insert. In principle, a mixture could be sequenced directly without gel purification, although the first nucleotides would be ambiguous. Alternatively, the large labeled fragment could be purified by a short gel run. The general procedure is of course applicable with other combinations of inserts and selectable markers.

The method described for enzymatic sequencing is similar in principle, except that the initial DNAse cleavage is done on a cloned DNA fragment in an appropriate M13 vector designed for sequencing from a synthetic oligonucleotide primer. The clone is designed in such a way that there is a site complementary to the synthetic primer located adjacent to one side of the insertion site for the cloned DNA in the M13 phage (Fig. 4-A-2). Between the primer binding site and the site of insertion of the cloned DNA is a unique restriction endonuclease cleavage site (*A*). After DNAse cleavage and "polishing" of the cleavage site, one cuts at the unique restriction site (*A*) between the primer binding site and the insert and then fuses the DNAse cleavage site in the insert to one end of the restriction site (*A*). The origin of replication of the phage lies close to the site of the inserted DNA on the side away from the primer binding site so that initial DNA cleavages within the vector will generally remove the origin of insert and produce molecules that will not generate phage plaques. DNAs prepared from miniscale lysates of individual clones are themselves sized. The DNA from clones of various sizes can be prepared and sequenced from the unique primer site to read across the random DNAse cleavage site into different segments of the cloned fragment. This procedure avoids the major problem in shotgun M13 sequencing of the increasing redundancy of the work as one approaches completion.

In addition to the excellent reactions described by Maxam and Gilbert, Krayev (1981) has described the use of diethylpyrocarborate to obtain selective cleavage of DNA at purines (reaction at pH 3) or G C (reaction at pH 8).

REFERENCES

Frischauf, A. M., Garoff, H., Lehrach, K. (1980). A subcloning strategy for DNA sequence analysis. *Nucl. Acids Res.* 8:5541–5549.

Hong, G. F. (1982). A systematic DNA sequencing strategy. *J. Mol. Biol.* 158: 539-49.

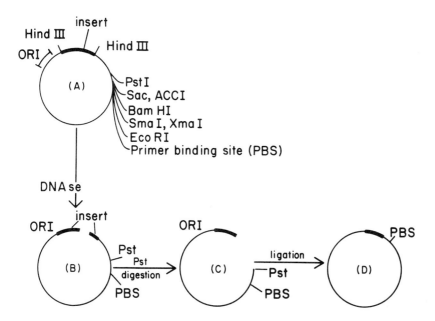

Figure 4-A-2. Method for systematic generation of clones for sequencing by the enzymatic primer extension method. The method is illustrated for the plasmid insert between *Hind*III sites and M13. ORI indicates the region of origin of replication of M13. The light arcs of the circle indicate the phage DNA sequences, and the dark line indicates the insert. *Pst*I, *Sac*, ACCI, *Bam*HI, *Sma*I, *Xma*I, and *Eco*RI indicate adjacent cleavage sites for each of those restriction enzymes. The primer binding site (PBS) is the site at which the oligonucleotide primer can be annealed in order to initiate the sequencing reaction. After random DNAse I cleavage and S1 treatment, one obtains a variety of linearized molecules after cleavage with *Pst*I removes the S1 segment containing a portion of the insert and of phage DNA at the *Pst* site. If the random cutting had occurred counterclockwise to the ORI region, two nonviable fragments of phage would have been generated, one containing an incomplete set of phage genes, the other lacking an origin. After cutting with *Pst* and S1 nuclease, the phage can be religated to generate a new vector in which a primer binding site is located close to the initial random DNAse cleavage site within the insert.

Krayev, A. S. (1981). The use of diethylpyrocarbonate for sequencing adenines and guanines in DNA. *FEBBS Lett.* 130:19–22.

Poncz, M., Solowiejczyk, D., Ballantine, M., Scwartz, E., and Surrey, S. (1982). "Nonrandom" DNA sequence analysis in bacteriophage M13 by the dideoxy chain termination method. *Proc. Natl. Acad. Sci. USA* 79:4298–4302.

5

Selected Postlabeling Procedures for Base Composition and Sequence Analysis of Nucleic Acids

ERIKA RANDERATH and KURT RANDERATH

I. RADIOACTIVE DERIVATIVE (POSTLABELING) METHODS FOR STRUCTURAL ANALYSIS OF NUCLEIC ACIDS

Structural analysis is not only a prerequisite for our understanding of the functions of nucleic acids in normal cells and tissues, but may also provide insights into the etiology of diseases as well as into the mechanism of action of certain drugs. For example, it may reveal carcinogen- or drug-induced alterations of nucleic acids and help to characterize structurally altered nucleic acids in cancer cells.

For the structural analysis of nucleic acids in the mammalian organisms, highly sensitive base composition and sequence analysis methods are required due to the limited availability of nucleic acids from mammalian sources. To achieve such high sensitivity, radioactive derivative (postlabeling) methods have been systematically developed during the past 13 years. These methods combine the advantages of using nonradioactive nucleic acids and isotopic labeling by introducing radioactivity into the nucleic acids after their isolation. They have now largely displaced the older, much less sensitive sequencing methods based on spectrophotometry, as well as *in vivo* labeling methods, and thus have made systematic structural studies on mammalian nucleic acids possible. Postlabeling methods for nucleic acids can be grouped as follows:

A. Methods for base analysis of nucleic acids
  (i) ^{3}H and ^{125}I derivative methods for base analysis of RNA based on the enzymatic digestion to nucleosides and chemical ^{3}H- and ^{125}I-labeling.
  (ii) ^{32}P derivative method for base analysis of RNA entailing the enzymatic digestion of the nucleic acid to ribonucleotides and their enzymatic ^{32}P-labeling.
  (iii) ^{32}P derivative method for base analysis of DNA based on the same principles as (ii).
B. Methods for sequence analysis of nucleic acids
  (i) ^{3}H derivative method for RNA based on mapping of oligonucleotide digests and direct determination of each position in individual oligonucleotides.
  (ii)   Combined ^{3}H/^{32}P derivative method for RNA entailing the same principles as (i).
  (iii) ^{32}P derivative method for RNA based on fingerprinting combined with mobility-shift analysis of individual oligonucleotides.
  (iv) ^{32}P gel readout method for RNA entailing partial enzymatic digestions of end-labeled RNA.
  (v) ^{32}P gel readout method for RNA based on partial chemical degradations of end-labeled RNA.
  (vi) ^{32}P thin-layer readout method entailing single-hit chemical cleavages of RNA.
  (vii) Special methods for large RNAs.
  (viii) Enzymatic ^{32}P-sequencing methods for DNA based on primed synthesis.
  (ix)   Chemical ^{32}P-sequencing method for DNA entailing partial chemical degradation of DNA.

Radioactive derivative methods originated in early experiments by Khorana (1959), Ralph, Young, and Khorana (1963), and RajBhandary, Young, and Khorana (1964), in which radioactive label was incorporated into end groups of nucleic acids by chemical means. For the past 14 years, our laboratory has been involved in the development and application of postlabeling methods for base composition and sequence analysis of nucleic acids. Thus, ^{3}H derivative methods for base composition analysis (Randerath, K., & Randerath, 1968, 1971, 1973; Randerath, K., Gupta, & Randerath, 1980; Randerath, E., Yu, & Randerath, 1972) and sequence analysis of RNA (Randerath, K., 1973; Randerath, K., & Randerath, 1973; Randerath, K., et al., 1974a, 1974b; Randerath, K., Gupta, & Randerath, 1980; Sivarajan et al., 1974; Gupta, Randerath, E., and Randerath, 1976a;

Gupta & Randerath, 1977a, 1977b; Randerath, E., et al., 1979) were developed in our laboratory. Such methods were used to compare tRNA from normal and neoplastic mammalian tissues (Randerath, K., 1971; Randerath, K., Mackinnon, & Randerath, 1971; Randerath, K., & Randerath, 1973; Randerath, E., et al., 1974; Chia et al., 1976) and to investigate effects of anticancer drugs on tRNA (Lu et al., 1976a, 1976b; Lu, Tseng, & Randerath, 1979; Tseng, Medina, & Randerath, 1978; Randerath, K., 1979) and tRNA modifying enzymes (Tseng, Medina, & Randerath, 1978; Lu, Tseng, & Randerath, 1979; Randerath, K., 1979; Lu & Randerath, 1979). Our laboratory first sequenced a nonradioactive nucleic acid, tRNA$_{UAG}^{Leu}$ from yeast, by postlabeling methods (Randerath, K., et al., 1975).

The ^{3}H-derivative method for base analysis of RNA (for references, see above) is both a qualitative and a quantitative method. It requires as little as 0.5 μg of a low molecular weight RNA, such as tRNA, and enables the identification and quantitation of most base-modified constituents known to occur in tRNA. The newly developed ^{125}I derivative method for base analysis of RNA (Randerath, K., 1981), which is considerably more sensitive than the ^{3}H method, has so far only been used for the determination of the four major nucleosides. The ^{32}P derivative method for base analysis of RNA (Silberklang et al., 1977; Silberklang, Gillum, & RajBhandary, 1979; Nishimura, 1972) based on the (γ-^{32}P)ATP/T$_4$ polynucleotide kinase reaction (Székely & Sanger, 1969; Székely, 1971; Richardson, 1971; Simsek et al., 1973), which is more sensitive than the ^{3}H-derivative method, has not been standardized for quantitative analysis of modified nucleosides. Conditions for quantitative determination of the four major ribonucleosides by ^{32}P-labeling have been described recently (Davies et al., 1979). For this purpose, as little as 0.01 μg of RNA is required. The ^{32}P method is particularly useful for the detection of certain modified nucleosides, such as ms^{2}i^{6}A, which cannot be analyzed by the ^{3}H derivative method. The ^{3}H and ^{125}I derivative methods for base analysis are described in this chapter, and the ^{32}P derivative method is treated in Chapter 6.

An enzymatic ^{32}P derivative method for base analysis of mutagen/carcinogen-modified DNA has recently been developed in our laboratory (Reddy, Gupta, & Randerath, 1981; Randerath, K., Reddy, & Gupta, 1981). A detailed account of this method, which is much more sensitive and versatile than conventional *in vivo* labeling methods, will be given in this chapter.

The first postlabeling methods for sequence analysis of RNA developed by us (B.i and B.ii above) (Randerath, K., 1973; Randerath,

K., & Randerath, 1973; Randerath, K., Randerath, et al., 1974; Randerath, K., et al., 1975; Randerath, K., 1979; Randerath, Gupta, & Randerath, 1980; Sivarajan et al., 1974; Gupta, Randerath, & Randerath, 1976a, 1976b; Gupta & Randerath, 1977a, 1977b) and others (B.iii) (Simsek et al., 1973; Gillum et al., 1975; Silberklang, Gillum, & RajBhandary, 1979) entailed conventional fingerprinting of complete and partial RNase T_1 and A digests of the RNA. While less sensitive than the ^{32}P thin-layer readout method (B.vi) developed later (Gupta & Randerath, 1979), they are much more sensitive than conventional spectrophotometric methods, requiring 10-120 μg of a tRNA for complete sequence analysis, whereas conventional methods require mg amounts of the tRNA. Some limiting factors of the earlier postlabeling methods are detection of unlabeled fragments on fingerprints, the relatively low specific activity of ^3H, or the requirement for additional methods for modified constituents that cannot be identified by mobility-shift analysis. An advantage of the ^3H or combined $^3H/^{32}P$ derivative methods is the direct identification of each nucleotide, including modified ones, in the polynucleotide chain.

The ^{32}P derivative methods (B.iv to B.vi) developed later are more rapid than the earlier methods since they do not involve time-consuming fingerprinting techniques and yield a direct readout of the RNA sequence. The two gel readout methods (B.iv and B.v) are based, similarly to the DNA sequencing method first described by Maxam and Gilbert (1977) (B.ix), on the analysis of nested sets of partial fragments, which are obtained either by base-specific enzymatic cleavages (Gupta & Randerath, 1977a, 1977b; Donis-Keller, Maxam, & Gilbert, 1977; Simoncsits et al., 1977; Lockard et al., 1978; Randerath, E., et al., 1979; Silberklang, Gillum, & RajBhandary, 1979; RajBhandary, 1980; Randerath, K., Gupta, & Randerath, 1980) or base-specific chemical (Peattie, 1979) degradations of 5'- or 3'-end-labeled RNA. They are suitable for RNAs of chain lengths of up to about 150 nucleotides and require only a few μg of RNA. They do not afford the identification of modified constituents and therefore require complementary methods for this purpose. Complete RNase digests of such RNAs may be prepared, and the resulting oligonucleotides may be analyzed by the combined $^3H/^{32}P$ derivative method (B.ii) for the positions of the modified constituents. Overlaps for establishing the complete sequence will then be provided by gel readout sequencing. However, the method of choice for sequence analysis of RNAs containing modified constituents is the thin-layer readout method (B.vi) (Gupta & Randerath, 1979; Randerath, K., Gupta & Randerath, 1980), which is more straightforward, rapid, and sensitive than the other methods. It is based on the direct ^{32}P analysis of the 5' termini of fragments generated by single-hit chemical cleavages of RNA (Stanley & Vassilenko, 1978) and requires as little as 0.2–0.5 µg of a tRNA for complete sequence analysis, providing an unambiguous identification and location in the chain of most modified nucleotides known to occur in tRNA. An additional advantage of this method is that it clearly distinguishes between C and U residues, which is not the case for the enzymatic gel readout method (B.iv).

As to RNA sequencing, we will mainly concentrate on the description of the thin-layer readout method (B.vi) since it proved most successful in our hands. We will also include a description of the enzymatic ^{32}P gel readout method. The ^{3}H and the combined ^{3}H/^{32}P-sequencing methods will only be outlined briefly, as they are more time-consuming and the former method is also considerably less sensitive than any of the methods developed more recently. Regarding the ^{32}P derivative method for sequence analysis of RNA based on fingerprinting and mobility shift analysis (B.iii), the reader is referred to a comprehensive article by Silberklang, Gillum, and RajBhandary (1979). Methods for the sequence analysis of large RNAs will only be summarized briefly in this chapter. The chemical ^{32}P gel readout method (B.v) as well as enzymatic (reviewed by Godson, 1980) and chemical (Maxam & Gilbert, 1977)[32]P-sequencing methods are the subjects of Chapters 7, 3, and 4, respectively, and therefore will not be covered here.

II. FLUOROGRAPHY AND SCREEN-INTENSIFIED AUTORADIOGRAPHY FOR ENHANCED DETECTION OF RADIOISOTOPES (^{3}H, ^{32}P, ^{125}I) ON CHROMATOGRAMS OR GELS

A. General Aspects

The development of fluorographic and screen-intensified autoradiographic procedures for the enhanced detection of radioisotopes, such as ^{3}H, ^{32}P, and ^{125}I, have greatly contributed to the high sensitivity of postlabeling methods for structural analysis of RNA and DNA. In particular, the large enhancement of ^{3}H detection has made this isotope suitable for analytical studies on nucleic acids.

While high-energy β-particles or γ-ray-emitting isotopes, such as ^{32}P and ^{125}I, respectively, can be readily detected by autoradiography, the low-energy β-particle emitter ^{3}H requires extremely long film exposure times. However, the speed of ^{3}H for thin-layer chromatograms can be enhanced greatly, i.e., about 200-fold (Randerath, K., 1970), by fluorographic methods (Wilson, 1958; Randerath, K., 1969, 1970), as the incorporation of a scintillator, such as 2,5-diphenyloxazole (PPO), into the chromatogram or gel converts the energy of the ^{3}H particles into light and thereby produces an image in the photographic emulsion. The principles underlying the fluorographic process and the factors that influence this process, such as amount of scintillator incorporated, temperature of film exposure, and choice of film, were first investigated systematically by K. Randerath (1969, 1970) for ^{3}H detection on thin-layer chromatograms. Similar studies were subsequently also conducted for ^{3}H detection on polyacrylamide gels (Bonner & Laskey, 1974; Laskey & Mills, 1975; Laskey, 1980). While ^{3}H is thus best detected by fluorography, the autoradiographic detection of ^{32}P and ^{125}I can be enhanced considerably by the use of an X-ray intensifying screen (Villareal & Berg, 1977; Benton & Davis, 1977; Laskey & Mills, 1977; Swanstrom & Shank, 1978; Laskey, 1980). The effect of the intensifying screen is similar to that of the scintillator in fluorography in that the majority of events recorded on the film are long-wavelength UV photons resulting from fluorescence of the screen, which in turn is caused by the radiation of the isotope. Factors that influence screen-enhanced autoradiography, such as type of screen and film, as well as temperature of exposure, have been investigated systematically (Laskey & Mills, 1977; Swanstrom & Shank, 1978; Laskey, 1980).

We will now describe procedures for the enhanced detection of ^{3}H, ^{32}P, and [125]I. For background information on the procedures, the reader is referred to the literature quoted above.

B. Materials

X-Omat AR X-ray film, X-Omatic cassettes, GBX developer, and GBX fixer from Eastman-Kodak Co.

2,5-Diphenyloxazole (PPO) and Omnifluor from New England Nuclear Corp. (Boston, Massachusetts).

Cronex Lighting Plus Intensifying Screens from DuPont (Wilmington, Delaware).

[14]C-ink from Schwarz/Mann (Orangeburg, New York). ^{99}Tc-ink [prepared by diluting (^{99}Tc)NH$_4$TcO$_4$ (New England Nuclear) with regular Indian ink].

#SE 540 Compact Slab Gel Dryer (for gels up to 18 × 34.5 cm) or #SE 1140 Large Slab Gel Dryer (for gels up to 35 × 44 cm) from Hoefer Scientific Instruments (San Francisco, California). Similar equipment may also be obtained from Bio-Rad Laboratories (Richmond, California).

C. Fluorography for the Detection of ^{3}H on Thin-layer Chromatograms

[3]H-Labeled compounds, such as nucleoside trialcohols (Section III) and oligonucleotide 3'-dialcohols (Sections VI and VII), are detected on thin-layer chromatograms by fluorography (Randerath, K., 1969, 1970). Optimal results are obtained if the appropriate amount of scintillator (see below) and a blue-sensitive screen-type X-ray film, such as Kodak X-Omat AR X-ray film, are used and the film exposure is carried out at −70 to −80°C. This film is about as sensitive as Kodak RB-54 Royal Blue Medical X-ray film which was used previously (Randerath, K., 1970) but is no longer available commercially.

Procedure. As rapidly as possible, pour a 7% (w/v) solution of PPO in diethyl ether over the chromatogram containing the ^{3}H-labeled compounds. Distribute scintillator evenly by tilting; complete this operation in 1–3 sec, depending on the size of the layer. Bring chromatogram into a vertical position, allow excess solution to drain off, and agitate until the ether has evaporated. The volume of PPO solution used should be 35–40 μl/cm^2, i.e., 14–16 ml for a 20 × 20 cm layer. It is important that this treatment is carried out so that no scintillator crystals build up on the surface of the thin layer, otherwise the light emission process will be disturbed, resulting in decreased sensitivity.

Apply ^{14}C-ink to a few points on the layer for later alignment of chromatogram and film, if counting of the compounds on the chromatogram is intended.

Conduct further operations in a darkroom under proper lighting conditions. Place the layer in contact with the film and keep the assembly in a cassette at −70 to −80°C (deep-freeze of Dry Ice). Alternatively, if cassettes are not available, keep thin-layer sheet and film between two glass plates held together with adhesive tape and wrap assembly in sufficient aluminum foil to keep it dark. If several films are to be exposed simultaneously for the same period of time, assemble the whole set interleaved with cardboard or aluminum sheets to prevent penetration of light. Expose films for the time period required to visualize the labeled compounds. Under the conditions specified, the fluorographic method allows one to visualize 2–3 nCi of $^3H/cm^2$ in 24 hr (about 1 nCi/day for an average thin-layer spot).

Develop films automatically or, for optimal results, manually. For manual development, use Kodak GBX developer and Kodak GBX fixer (5–10 min each), then rinse films with water and allow them to dry.

If the ^3H-labeled compounds are to be counted, mark compound areas by superimposing film and chromatogram and perforating the film around spots with a pin. A darkened area on the film usually does not completely coincide with the area actually occupied by the radioactive compound on the layer. Spots at the sites of major or minor components may, respectively, be overexposed or underexposed. Thus, in the latter case, cut a somewhat larger area from the chromatogram than indicated by the darkening of the film. For semiquantitative estimation of radioactivity, place cutouts in vials, add 2 ml of scintillation fluid, and count in a liquid scintillation counter.

The scintillation fluid consists of 0.3% Omnifluor (w/v) and 25% Triton X-100 (v/v) in xylene.

For quantitative determinations, elute cutouts, layer side down, in beakers or vials with 1–2 ml of eluent, the eluent depending on the type of compounds to be eluted and the thin-layer material. For example, extract ^3H-labeled nucleoside trialcohols from cellulose thin layers with 2 N ammonia at room temperature for at least 1–2 hr with agitation. Use 0.5 M LiCl solution for silica gel layers. Transfer 500-μl aliquots to vials, mix with 10 ml of scintillation fluid, and count in a liquid scintillation counter.

D.   Detection of ^3H in Polyacrylamide, Mixed Polyacrylamide/Agarose, or Agarose Gels by Fluorography

^3H-Labeled nucleic acids in polyacrylamide, polyacrylamide/agarose, or agarose gels may also be detected by fluorography (Bonner & Laskey, 1974; Laskey & Mills, 1975; Laskey, 1980).

Procedure.   Polyacrylamide gels are treated as follows. Immediately after electrophoresis or after staining, soak the gel in about 20 times its volume of dimethylsulfoxide (DMSO) for 30 min (or longer for gels greater than 3 mm thick) followed by a second 30-min immersion in fresh DMSO. It is essential to remove all water from the gel this way or PPO will not enter the gel. Retain the separate tanks of used DMSO for reuse in the same sequence. Immerse the gel in 4 volumes of a 22% (w/v) solution of PPO in DMSO for 3 hr. Longer times may result in diffusion of bands.

Caution: Avoid contact of DMSO solution with skin. Do not reuse residual PPO solution. Recover PPO from residual solution by precipitation with 3 volumes of 10% ethanol.

Soak the impregnated gel in excess water for 1 hr (or longer for gels greater than 3 mm thick) to remove DMSO. Dry the gel under vacuum for 1 hr. Use a commercially available slab gel drier for this purpose.

Treat plain agarose gels differently from the polyacrylamide gels, as they dissolve in DMSO. Before impregnation with PPO, equilibrate the agarose gel with ethanol by soaking it in 20 volumes of absolute ethanol for at least 30 min, followed by a second (and if necessary third) 30-min immersion in fresh ethanol to remove water. Soak gel in a 3% (w/v) solution of PPO in ethanol for at least 3 hr. Finally, immerse gel in water for up to 1 hr to remove ethanol, and subsequently dry it as above. The gel can be gently heated during drying. However, if it is heated too much before most of the water has been removed, the agarose will melt.

For mixed polyacrylamide/agarose gels, determine whether DMSO or ethanol should be used as the solvent by soaking a small piece of gel in each solvent; only if it dissolves in DMSO should ethanol be used. For gels not soluble in DMSO and containing less than 5% acrylamide, use a 10% (w/v) solution of PPO in DMSO.

Mark dried gel with ^{99}Tc-ink. Place X-Omat AR X-ray film in contact with the dried gel and expose at $-70°C$ using a cassette or glass plates and aluminum foil. To enhance the sensitivity of ^3H detection further, sensitize the film by preexposure to a light flash prior to the actual exposure (Laskey & Mills, 1975; Laskey, 1980). To do this, expose the film immediately before use to a single instantaneous flash of light from an electronic photographic flash unit or stroboscope. Adjust the intensity of the flash so as to increase the absorbance of the film to 0.15 (A_{540}) above the absorbance of the unexposed film. Note that it is essential that the duration of the flash is short, i.e., in the order of 1 msec. For more technical details, consult the above references.

Using "untreated" RP Royal X-Omat film, which appears similar in performance to X-Omat AR film, Bonner and Laskey (1974) could detect 3,000 disintegrations of ^3H/min in a gel band (1 × 0.1 cm) at $-70°C$ in 24 hr. With preexposed film, the sensitivity was found to be 10 times increased over this value (Laskey & Mills, 1975).

E. Detection of ^{32}P and ^{125}I on Chromatograms or Gels by Screen-intensified Autoradiography

Autoradiography of ^{32}P- or ^{125}I-labeled compounds on chromatograms or gels is enhanced by the use of calcium tungstate (Cronex Lighting Plus) intensifying screens (Laskey & Mills, 1977; Swanstrom & Shank, 1978; Laskey, 1980). Optimal results are obtained by using a blue-sensitive screen-type medical X-ray film, such as Kodak X-Omat AR X-ray film, and exposing the film at −70 to −80°C. The sensitivity of screen-enhanced detection is at least 5 times greater at −70°C than at 23°C.

Procedure. Mark chromatogram or dried gel with ^{99}Tc-ink. Place film (preferably X-Omat AR) between the sample (chromatogram or dried gel) and the screen, and expose the assembly in a cassette or between glass plates wrapped in aluminum foil at −70°C.

Cover gels containing water with a thin plastic foil, e.g., Saran wrap, before film exposure. Mark small pieces of tape attached to the Saran wrap with ^{99}Tc-ink. As nondenaturing gels will crack when they are frozen, autoradiograph above 0°C. Gels containing urea, on the other hand, can be exposed to temperatures below 0°C.

Again, to increase the sensitivity of detection further, the film may be preexposed to a brief light flash (see above).

Comments. Using Kodak RP Royal X-Omat film, Swanstrom and Shank (1978) found that a calcium tungstate intensifying screen at −70°C produces an 8- to 10-fold enhancement in the detection of ^{32}P and a 30- to 40-fold enhancement in the detection of ^{125}I. Thus, 10-20 dpm of ^{32}P and 50-100 dpm of ^{125}I in an area of 10 mm^2 are detectable in a 20-hr exposure.

III. ^{3}H DERIVATIVE METHOD FOR BASE ANALYSIS OF RNA

A. General Aspects

The ^{3}H derivative method for base analysis of RNA (Randerath, K., & Randerath, 1968, 1971, 1973; Randerath, E., Yu, & Randerath,

1972; Randerath, K., Gupta, & Randerath, 1980) makes possible the quantitative determination of the four major ribonucleosides and of the majority of base-modified nucleosides known to occur in tRNA (see also Section D below). It has also proved useful for the characterization of unknown modified constituents, such as a hyper-modified uridine in mammalian mitochondrial tRNAs (Randerath, E., Agrawal, & Randerath, 1981b, and unpublished experiments). Purified RNAs, as well as unfractionated tRNA and oligonucleotides extracted from thin-layer chromatograms or gels, can be analyzed by the ^{3}H derivative method. A minimum of 0.5 μg of a purified tRNA or 2–6 pmol of an oligonucleotide is required for this purpose. One minor component in about 1,000 nucleotides can be detected by this method in 24 hr if a total of 2–5 μCi of ^{3}H-labeled digest is chromatographed.

B.   Outline of Method

The ^{3}H derivative method for base analysis of RNA entails the enzymatic degradation of the RNA to nucleosides. The RNA digest is then treated with periodate, followed by tritiated borohydride so as to convert the nucleosides, via dialdehydes, to ^{3}H-labeled nucleoside trialcohols:

$$\tag{1}$$

Nucleoside (R = H)	Nucleoside dialdehyde (R = H)	Nucleoside trialcohol (R = H)

The mixture of the nucleoside trialcohols is separated by two-dimensional thin-layer chromatography on cellulose. Labeled compounds on chromatograms are detected by low-temperature solid scintillation fluorography (Randerath, K., 1969, 1970; see Section II above), eluted, and counted. Since the count rate of each nucleoside trialcohol has been shown to be directly proportional to the original concentration of the parent nucleoside (Randerath, K., & Randerath, 1969, 1971; Randerath, E., Yu, & Randerath, 1972), the base composition can be calculated from the nucleoside trialcohol count rates.

C. Materials

N,N-Bis(2-hydroxyethyl)glycine (bicine) from Calbiochem-Behring Corp. (San Diego, California). Aqueous solutions can be stored up to 4 weeks at 4°C. Do not freeze and thaw repeatedly (Randerath, K., Randerath, et al., 1974).

RNase A (code R4875) and RNase T_2 (code R3751) from Sigma Chemical Co. (Saint Louis, Missouri), snake venom phosphodiesterase (code VPH) and *E. coli* alkaline phosphatase (electrophoretically purified, code BAPF) from Worthington Biochemical Corp. (Freehold, New Jersey). Alkaline phosphatase is dialyzed against distilled water (Randerath, K., Randerath, et al., 1974) and then kept at −18°C.

Potassium borohydride pellets from Alfa Inorganics (Beverly, Massachusetts).

[3]H-Labeled potassium borohydride (code TRK.293; specific activity >3 Ci/mmol) and sodium borohydride (code TRK.45; specific activity > 20 Ci/mmol) from Amersham Corp. (Arlington Heights, Illinois).

Cellulose sheets (20 x 20 cm) from E. Merck (EM Laboratories, Inc., Elmsford, New York; #5502).

Chromatographic markers: Prepare solutions of labeled (0.05 mM, about 1 Ci/mmol) and unlabeled (2 mM) nucleoside trialcohols by periodate oxidation of nucleosides, followed by borohydride reduction, as described in K. Randerath and Randerath (1971, 1973) and E. Randerath, Yu, & Randerath (1972).

D. Procedure

The standard procedure, as well as two scaled-down versions of the standard procedure, will be described. While the standard procedure and scaled-down version 1 have been applied in our laboratory mainly to the analysis of unfractionated tRNA and of various tRNA species, respectively, scaled-down version 2 has been utilized for the analysis of oligonucleotide 3′-dialcohols, as obtained by the combined ^{3}H/^{32}P-sequencing method (see Section VII).

(i) Standard Procedure

Enzymatic Digestion of RNA. Prepare enzyme/buffer mixture from 39 μl of water, 0.5 μl of 1 M MgCl$_2$, 2.5 μl of 0.6 M bicine-Na, pH 8.0, and 8 μl of enzyme solution A.

> Enzyme solution A (stored at $-18°$C) contains 1.25 μg of RNase A, 1.25 μg of snake venom phosphodiesterase, and 1 μg of alkaline phosphatase per μl.

Add enzyme/buffer mixture (1 μl per μg of RNA) to the dried RNA residue. Use a minimum of 15 μg of RNA (based on 1 A$_{260}$ unit = 40 μg of RNA). Incubate the mixture at 37°C for 6 hr. Process the digest immediately, or keep it frozen at $-80°$C. Avoid repeated freezing and thawing of digest (Randerath, E., et al., 1974). Reincubate frozen digest for 5-10 min at 37°C and vortex briefly before further processing.

^{3}*H-Labeling of Enzymatic Digest.* Add 40 μl of water and 10 μl of freshly prepared 9 mM NaIO$_4$ to 15 μl of the above digest. Incubate the reaction mixture at 23°C for 2 hr in the dark, and then cool briefly on ice. Add 1 μl of 1 M potassium phosphate, pH 6.8, vortex briefly, then immediately add 5 μl of borohydride solution A. It is important to keep the pH of the reaction mixture between 7 and 8 during the reduction.

> Prepare borohydride solution A (2-6 Ci/mmol) as follows: Dissolve (^{3}H)KBH$_4$ in freshly dissolved 0.1 N KOH (CO$_2$-free) at a concentration of 0.1 M. If desired, adjust specific activity by addition of unlabeled KBH$_4$. Store solution in portions of 20-50 μl at -70 to $-80°$C. To minimize losses of reducing capacity, especially of high specific-activity preparations, lyophilize the portions of the solution and store the dried residues at -70 to $-80°$C. Reconstitute solution immediately before use by adding the original volume of 0.01 N KOH. Perform all operations involving (^{3}H)KBH$_4$ under a well-ventilated hood.

Incubate the reaction mixture at 23°C for 2 hr in the dark. Destroy excess borohydride by the addition of 100 μl of 1 N acetic acid. Keep the solution in the open tube for 5-10 min at room temperature, and then evaporate it in a stream of air. Take the dry residue up in 50 μl of 0.1 N formic acid. Note that the specific activity of the nucleoside trialcohols is half that of the (^{3}H)KBH$_4$ used. Store the final solution at $-18°$C for a maximum of 8 weeks; decomposition by self-radiolysis occurs on more prolonged storage. Decomposition of some radioactive background material takes place during the first few hours after preparation of the labeled trialcohols; thus, better results are obtained if the labeled digest is chromatographed after this time period only.

Chromatography of Labeled Digest.   Apply labeled digest corresponding to 2-5 μCi to a cellulose thin-layer sheet at 2.5 cm from the left-hand and bottom edges. If one desires to locate a specific individual nucleoside trialcohol on the map, cochromatograph an excess of the respective radioactive reference compound with labeled digest or use 2-5 nmol of UV-absorbing marker. Be careful not to damage the fragile thin layer while spotting.

Develop in the first dimension with solvent A to 17 cm from the origin. Begin chromatography no later than 2-3 min after pouring the solvent into the tank.

Solvent A is acetonitrile/4 N ammonia (3.4:1, v/v), freshly prepared.

Thoroughly dry sheet after development. Attach a Whatman 1 paper wick to the original right-hand side of the sheet by stapling. Develop in the second dimension with solvent B to 4-5 cm on the wick, again without prior tank saturation.

Solvent B is t-amylalcohol/methyl ethyl ketone/acetonitrile/ethyl acetate/water/formic acid, specific gravity 1.2 (4:2:1.5:2:1.5:0.18, v/v), freshly prepared.

Dry the sheet after chromatography and remove the wick.

For the fluorographic detection and quantitative analysis of the ^{3}H-labeled nucleoside trialcohols, proceed as described in Section II.C.

Location of nucleoside trialcohols on the map.[1]   A representative separation of nucleoside trialcohols obtained by digestion and

---

[1]*Abbreviations*: m^{1}A, 1-methyladenosine; m^{2}A, 2-methyladenosine; m^{6}A, N^6-methyladenosine; m$^6_2$A, N^6,N^6-dimethyladenosine; i^{6}A, N^6-isopentenyladenosine; t^{6}A, N-[N-(9-β-D-ribofuranosylpurin-6-yl)carbamoyl]threonine; mt^{6}A,

labeling of mouse liver tRNA is shown in Fig. 5-1. The compounds indicated may be assayed reproducibly by the described procedure. In addition to these compounds, several other modified nucleosides occurring in tRNA may be determined, such as m^2A (Randerath, E., Yu, & Randerath, 1972), m_2^6A, i^6A (Randerath, E., et al., 1981), V (Randerath, E., Yu & Randerath, 1972), 5-hydroxyuridine (Murao et al., 1978), and mam^5s^2U (K. Randerath, unpublished experiments, 1972), as well as U* (see Section III.A above).

Calculation of Base Composition.   Calculate the base composition according to Eq. (2):

$$f_i = \left(\frac{cpm}{\sum\limits_{i=1}^{N}} cpm_i \right) \times 100 \quad [\%] \tag{2}$$

where f_i is the base composition value for an individual nucleoside expressed as the percentage of the total, N is the number of radioactive nucleoside derivatives, and cpm is the count rate of an individual radioactive nucleoside derivative.

Correct the amounts of certain labile constituents for known losses (Randerath, E., Yu, & Randerath, 1972; Randerath, E., et al., 1979; Randerath, K., & Randerath, 1973). Recoveries under the conditions described are 67% for m^7G, 85% for m^3C, 95% for hU, and 57% for ac^4C. Twelve to 15% of m^1A is recovered as m^6A', the

---

N-[N-(9-β-D-ribofuranosylpurin-6-yl)N-methylcarbamoyl]threonine; ms^2i^6A, 2-methylthio-N^6-isopentenyladenosine; m^5U, 5-methyluridine; ψ, pseudouridine; hU, dihydrouridine; s^4U, 4-thiouridine; $4abu^3U$, 3-(3-amino-3-carboxypropyl)-uridine; V, uridine-5-oxyacetic acid; mam^5s^2U, 5-methylaminomethyl-2-thiour-idine; U*, an unknown hypermodified uridine derivative occurring in the wobble position of mammalian mitochondrial tRNA (Randerath, E., Agrawal, & Randerath, 1981); m^3C, 3-methylcytidine; m^5C, 5-methylcytidine; ac^4C, N^4-acetylcytidine; m^1G, 1-methylguanosine; m^2G, N^2-methylguanosine; m_2^2G, N^2,N^2-dimethylguanosine; I, inosine; Q, 7-(4,5-*cis*-dihydroxy-1-cyclopenten-3-yl-aminomethyl)-7-deazaguanosine; Y, wybutosine, the fluorescent nucleoside present in tRNA^{Phe} of eukaryotic organisms. Nm, 2'-*o*-methylated nucleoside. Mn (probably a C derivative that is 2'-O-methylated), m^1G* (a m^1G derivative), and L (an unknown hypermodified nucleoside) occurring in rat liver and hepatoma cytoplasmic tRNA$^{Leu}_{MmAA}$ (Randerath, E., et al., 1980). PEI-cellulose, anion-exchange material obtained by treating cellulose with polyethyleneimine (Randerath, K., 1962a, 1962b).

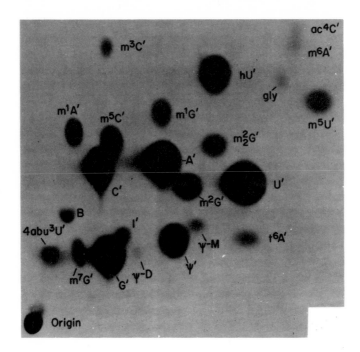

Figure 5-1. Fluorogram of a cellulose thin-layer map obtained by digestion to nucleosides and ^{3}H-postlabeling of mouse-liver tRNA. A′, C′, U′, G′, etc., ^{3}H-labeled trialcohols of adenosine, cytidine, uridine, guanosine, etc. For abbreviations of modified nucleosides, see footnote 1. ψ-D: a trace of a decomposition product of pseudouridine. ψ-M: monoaldehyde of pseudouridine (see text). Gly: glycerol. Spots marked B: background label (not from RNA). Reproduced by permission, from Lu.-J. W. et al., *Biochem. Biophys. Res. Comm.* 68:1094–1101 (1976).

rest as m^{1}A′. Ninety-five percent of ψ is recovered as ψ' and 5% as radioactive decomposition product ψ-D.

Calculate the base constituency of polynucleotides of known chain length from Eq. (3):

$$f_i = \left(\frac{cpm}{\sum\limits_{i=1}^{N} cpm_i}\right) \times \text{chain length} \quad [\text{mol}] \tag{3}$$

(ii)   Scaled-down Version 1 of Standard Procedure

In this scaled-down version of the standard procedure, all concentrations of buffers, enzymes, and reagents are maintained (10 mM MgCl$_2$,

30 mM bicene-Na, 0.20 μg/μl of RNase A, 0.20 μg/μl of snake venom phosphodiesterase, and 0.16 μg/μl of alkaline phosphatase for the enzymatic digestion; 1.4 mM NaIO$_4$, 15 mM potassium phosphate, and 7 mM borohydride for the labeling reaction), while volumes are scaled down. With RNA amounts of 2.5 μg or more, the RNA concentration for the enzymatic digestion (1 μg/μl) and the nucleoside concentration for the subsequent labeling reaction (0.6–0.7 mM) can also be maintained. For RNA amounts less than 2.5 μg, the RNA and nucleoside concentration will be lower. Whereas for the standard conditions the quantitative aspects of the procedure have been studied systematically, this is not the case for low concentrations ($<$2.5 μg) of RNA. However, our studies on a number of cytoplasmic (Randerath, E., et al., 1979, 1981a) and mitochondrial tRNAs (E. Randerath, H. P. Agrawal, & K. Randerath, unpublished experiments) indicate that quantitative data for most nucleosides can still be obtained with amounts of 0.5–2.5 μg of tRNA.

For 0.5–2.5 μg of RNA, use the following procedure: Add 2.5 μl of enzyme/buffer mixture (prepared as described for the standard procedure) to the dried residue of the RNA in a 25 × 6 mm siliconized glass tube. Incubate the reaction at 37°C for 6 hr. Dry sample in a desiccator *in vacuo* in the presence of P$_2$O$_5$ and KOH or by lyophilization.

For ^{3}H-labeling, take the residue up in 9 μl of water and 1.5 μl of freshly prepared 10 mM NaIO$_4$. Incubate the reaction mixture at 23°C for 2 hr in the dark. Cool sample on ice, and immediately add 0.5 μl of 0.33 M potassium phosphate, pH 6.8, followed by 1–2 μl of borohydride solution A (see standard procedure). Use (^{3}H)KBH$_4$ of a specific activity of 5–6 Ci/mmol. Incubate the reaction mixture at 23°C for 2 hr in the dark. Add 6 μl of 5 N acetic acid to decompose excess borohydride. Keep the solution in the open tube for 5–10 min at room temperature, and then evaporate it either in a stream of air or in the desiccator. Take the dry residue up to 10 μl of 0.1 N HCOOH and store the solution at −18°C. For chromatography, use 2–5 μl of the labeled solution per chromatogram depending on the amount of RNA digested and labeled. For further analysis, proceed as described above.

(iii) *Scaled-down Version 2 of Standard Procedure*

Add 4 μl of enzyme solution B to the dried residue of 2–6 pmol of a ^{3}H-labeled oligonucleotide 3′-dialcohol (see Section VII below).

Enzyme solution B contains 0.10 μg of RNase T$_2$, 0.20 μg of snake venom phosphodiesterase, 0.10 µg of alkaline phosphatase, 10 nmol of $MgCl_2$, and 20 nmol of bicine-Na, pH 7.7, per µl.

Incubate the solution at 37°C for 6 hr. Labeling steps are analogous to those described above. Add to the sample: 1 µl of 0.7 mM $NaIO_4$ (sufficient for oligonucleotides of chain length <10) for oxidation; 0.5 µl of 0.1 M potassium phosphate, pH 6.8, and 1 µl of borohydride solution B for reduction; 3 µl of 1 N acetic acid for decomposing residual borohydride.

Prepare borohydride solution B (25–30 Ci/mmol) as follows: Dissolve $(^3H)NaBH_4$ (specific activity >20 Ci/mmol) in 0.1 N KOH (CO_2-free) at a concentration of about 25 mM. If necessary, adjust the specific activity by addition of unlabeled KBH_4. Divide the solution into 10-µl portions, lyophilize, and store dry residues at −70 to −80°C. Dissolve individual residues in 10 µl of 0.1 N KOH each immediately before use.

Dry the sample and dissolve in 4 µl of water. Chromatograph 1–2 µl for base analysis in the presence of the four major nucleoside trialcohols serving as UV markers.

E.   Comments

Purity of Sample.   The purity of the sample to be labeled is important. Thus, a sample should not contain significant amounts of oxidizable or reducible material which would consume the reagents. For example, tris buffer must not be present because it reacts with periodate.

The ^3H derivative method is directly applicable to RNAs extracted from polyacrylamide gels (Chia, Randerath, & Randerath, 1973).

Improper Labeling Conditions.   If the conditions specified above are not closely adhered to, nucleoside monoaldehydes may form or losses of labile constituents may occur.

Formation of Monoaldehydes (Randerath, K., & Randerath, 1971). If the borohydride concentration during labeling is too low, nucleoside dialdehydes are reduced incompletely, resulting in the formation of ^3H-labeled monoaldehydes. These compounds are located on the two-dimensional maps to the right of the completely reduced tri-alcohols. A trace of ψ-monoaldehyde ($<5\%$ of ψ' radioactivity), which forms most readily, is acceptable. In this case, multiply the radioactivity of ψ-monoaldehyde by 2 and add this number to the radioactivity of ψ'. Insufficient excess of borohydride leads to the appearance of monoaldehyde derivatives of additional nucleosides. Such a chromatogram is not suitable for quantitative evaluation. The borohydride concentration may be too low for the following reasons: (i) Partial decomposition of borohydride in the reaction mixture if the pH is less than 7. (ii) Presence in the sample of impurities, which are reduced by borohydride. (iii) Partial decomposition of the borohydride stock solution due to prolonged storage and/or repeated thawing and freezing. The higher the specific activity of the borohydride, the less stable it is. (iv) Variations in the amount of active borohydride in commercial batches, as has been noted on several occasions.

Loss of Labile Constituents. If the borohydride concentration or the pH of the reaction mixture is too high, losses of labile constituents, particularly of hU, occur. Losses of alkali-labile constituents, for example m^7G, occur also if the digestion of the RNA is carried out for longer than the time specified (Randerath, E., Yu, & Randerath, 1972).

Test for Reducing Capacity of Borohydride. The reducing capacity of a borohydride solution may be assessed by labeling and analyzing a periodate-oxidized digest derived from a reference sample at 3 or 4 borohydride concentrations. An appropriate adjustment of the amount of borohydride to be added may then be applied in subsequent experiments.

Comparison of RNA Preparations. If two RNA preparations to be compared exhibit only small base composition differences, it is important that experiments be conducted in parallel (Randerath, E., et al., 1974). In particular, the labeling reaction should be carried out in this way using the same borohydride preparation. It is also essential to cut spots from the chromatograms in an identical manner. Several replicate chromatographic analyses (at least four) are carried out, and standard deviations are determined. Statistically significant differences between two samples are estimated on the basis of Student's t test (Randerath, K., & Randerath, 1973; Randerath, E., et al., 1974).

IV. RADIOIODINATION METHOD FOR RIBONUCLEOSIDE ANALYSIS

A. General Aspects

Although the sensitivity of the ^{3}H derivative method for base analysis of RNA is much greater than that of spectrophotometric methods, it is limited because the maximal specific activity of commercially available borotritide (40–60 Ci/mmol) is relatively low compared to that of ^{125}I (~2,000 Ci/mmol) or ^{32}P (~8,000 Ci/mmol) and, in addition, borotritide of this specific activity is rather unstable. It appeared therefore desirable to develop a new reagent that would react with nucleoside dialdehydes in a similar way as borotritide but would contain an isotope of higher specific activity than that of ^{3}H. To this end, a new ^{125}I-labeled reagent, 3-([3-^{125}I]iodo-4-hydroxy-phenyl)propionyl carbohydrazide, was developed in our laboratory (Randerath, K., 1981). This reagent was shown to convert nucleoside dialdehydes quantitatively to ^{125}I-labeled morpholine derivatives which could be separated chromatographically and quantified by liquid scintillation counting (Randerath, K., 1981). Thus, base analysis of RNA entailing the use of this reagent appears feasible and promising, particularly in view of its facile synthesis. The preparation of the reagent and its use for the analysis of the four major constituents of RNA will be described here. The application of the iodination reaction to the analysis of modified constituents is currently under investigation in our laboratory.

B. Outline of Method

The radioiodination reagent, 3-([3-^{125}I]iodo-4-hydroxyphenyl)-propionyl carbohydrazide (*III*, Fig. 5-2), is prepared as described by Randerath, K. (1981) by iodinating 3-(4-hydroxyphenyl)propionic acid *N*-hydroxysuccinimide ester (*I*) with (^{125}I)NaI in the presence of chloramine T, followed by reduction of the latter with sodium arsenite and treatment of the radioiodinated ester (*II*) with an excess of carbohydrazide. Treatment of periodate-oxidized nucleosides (see Section III) with this reagent quantitatively converts these compounds to ^{125}I-labeled morpholine derivatives (*V*, Fig. 5-3), which may then be separated by thin-layer chromatography and quantified by liquid scintillation counting (Randerath, K., 1981).

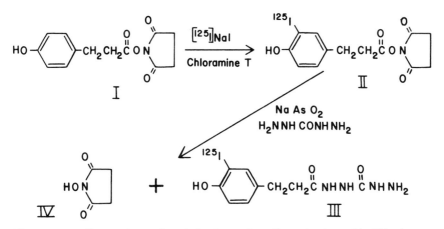

Figure 5.2. Conversion of 3-(4-hydroxyphenyl)propionic acid N-hydroxysuccinimide ester (I) to 3-([3-^{125}I]iodo-4-hydroxyphenyl)propionyl carbohydrazide (III) and N-hydroxysuccinimide (IV) via the intermediate formation of ^{125}I-labeled ester (II). Reproduced by permission, from Randerath, K., *Anal. Biochem.* 115:391–397 (1981).

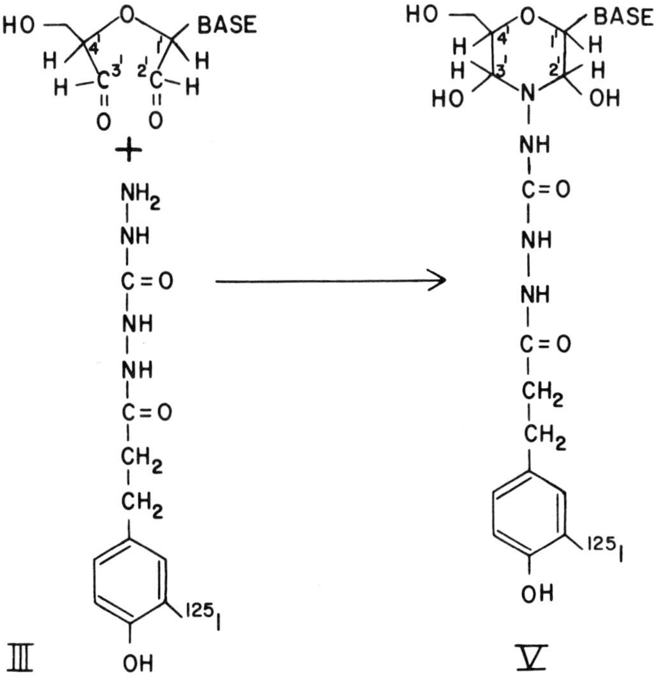

Figure 5-3. Reaction of a ribonucleoside dialdehyde with 3-([3-^{125}I]iodo-4-hydroxyphenyl)propionyl carbohydrazide (III) to give a ^{125}I-labeled morpholine derivative (V). Reproduced by permission, from Randerath, K. *Anal. Biochem.* 115:391–397 (1981).

189

C. Materials

(^{125}I)NaI (code NEZ-033L, carrier-free, 17 Ci/mg) from New England Nuclear Corp. (Boston, Massachusetts).

3-(4-hydroxyphenyl)propionic acid *N*-hydroxysuccinimide ester (code 57777) from Calbiochem-Behring Corp. (San Diego, California).

Carbohydrazide (code Cl, 100-6) from Aldrich Chemical Company (Milwaukee, Wisconsin). Chloramine T (code 1022) from Eastman-Kodak Co.

Sodium periodate (code S-398) and sodium arsenite (code S-225) from Fisher Scientific Co. (Pittsburgh, Pennsylvania).

Silica gel sheets with fluorescent indicator (code 5775) from EM Laboratories, Inc. (Elsmford, New York).

Benzene is dried over Davison Type 3A Grade 564 molecular sieve beads (Fisher Scientific Co.).

The scintillation fluid contains 0.3% Omnifluor (New England Nuclear) (w/v) and 25% Triton X-100 (v/v) in xylene.

D. Procedure

Preparation of the Reagent

The radioiodination reagent is prepared as described by Randerath, K. (1981). The procedure should be carried out in a well-ventilated hood, and it is recommended that the safety rules outlined by Prensky (1976) for handling ^{125}I-labeled compounds of high specific activity should be followed. The procedure is as follows:

(i) Prepare a 0.008 M solution of 3-(4-hydroxyphenyl)propionic acid *N*-hydroxysuccinimide ester in dry benzene. (This solution can be kept in a stoppered tube at −18°C for several weeks.)

(ii) Place a mixture of 2 μl of this solution and 10 μl of dry benzene at the bottom of a 15-ml glass tube. Evaporate the solution to dryness by briefly warming the outside of the tube with a warm-air blower.

(iii) Add 25 μl of 0.002 M (^{125}I)NaI solution (pH 9–9.5) of the desired specific activity to the dried residue.

(iv) Add 200 μl of freshly prepared 0.02 M chloramine T solution in 0.2 M sodium phosphate (pH 7.7). Mix thoroughly (vortex) for 15 sec.

(v) Add 200 μl of 0.06 M NaAsO$_2$ solution in water. Mix briefly.

(vi) Add 200 μl of 0.15 M KI solution in water. Mix briefly.

(vii) Add 200 μl of 0.2 M carbohydrazide solution in water.

Mix. Steps iv to vii should take no more than 45 sec; so prepare and measure these four solutions in advance.

(viii) Evaporate the solution in a current of dry air.

(ix) Add 5 ml of dry benzene to the dry residue.

(x) Stopper the tube and let sit at room temperature for 1-2 hr with gentle agitation.

(xi) Remove supernatant benzene phase carefully with a Pasteur pipet and transfer it to a fresh glass tube.

(xii) Repeat steps ix to xi to obtain a second extract of the reagent.

(xiii) Combine both extracts.

(xiv) Count 1-5 μl of benzene phases, if necessary after appropriate dilution with benzene.

(xv) Calculate yield of reagent.

(xvi) Place 0.5-2 μl of reagent solution at the origin of a silica gel thin layer (not activated by heating). Develop with ethyl acetate/toluene (1:1, v/v) to 15-18 cm. (Do not saturate chamber with solvent vapors.)

(xvii) Prepare an autoradiogram of the chromatogram at $-70°$C using Kodak XAR film and an intensifying screen (Section II above).

(xviii) Store the benzene solution of 3-([3-^{125}I]iodo-4-hydroxyphenyl)propionyl carbohydrazide in a stoppered tube at 0-4°C.

Periodate Oxidation of Nucleosides

Nucleosides are oxidized as described in Randerath, K. (1981). Perform the oxidation of 1-10 mM nucleoside concentration in the presence of an equimolar amount or a slight excess of $NaIO_4$. Allow the reaction to proceed at 23°C for 30 min in the dark. Destroy excess $NaIO_4$ by adding a 4-fold molar excess of $NaAsO_2$ over original $NaIO_4$, and incubate at 23°C for an additional 45-60 min. After appropriate dilution, this solution can be used for the reaction with the radioiodination reagent.

Reaction of Nucleoside Dialdehydes
with 3-([3-^{125}I]Iodo-4-hydroxyphenyl)propionyl Carbohydrazide and Chromatography

The iodination reaction is carried out as described by Randerath, K. (1981). Dry a 25-μl portion of the benzene solution of the reagent (20-30 pmol, 0.2-0.3 μCi), take the residue up in 30 μl of nucleoside dialdehyde solution (0.5-1 mM), and keep the solution at 38°C for 70 min. Spot 3 μl of the solution in 1-μl portions on each of two silica gel thin-layer plates and develop in (a) ethyl acetate/methyl ethyl ketone/methanol/water (5:5:1:0.9, v/v) and (b) ethyl acetate/ methyl ethyl ketone/methanol/water/90% formic acid (5:5:1:0.8: 0.1, v/v) to 18 cm above the origin, without tank saturation. Detect the reaction products by screen-intensified autoradiography at $-70°C$ for at least 3 hr. Store the reaction mixture at $-18°C$.

For the separation of the ^{125}I-labeled morpholine derivatives in the above chromatographic systems, see Randerath, K. (1981).

Extraction of Spots from Chromatograms for Quantitative Analysis

Excise spots located by autoradiography from chromatogram and extract them with 1 ml of 0.5 M LiCl solution. Transfer 500-μl aliquots to plastic vials, add 10 ml of scintillation fluid, and determine the radioactivity in a liquid scintillation counter.

E. Comments

The novel ^{125}I derivative method described here is quantitative. Thus, the same base composition results were obtained with the ^{125}I and the ^{3}H method when the same nucleoside mixture was analyzed by both methods (Randerath, K., 1981).

It should be noted that the iodinated nucleoside morpholine derivatives are stable at neutral and weakly acidic pH, but unstable at alkaline pH in the presence of ammonia or primary amines. If chromatography of the derivatives at alkaline pH is desired, the morpholine derivatives can be stabilized prior to chromatography by reduction with cyanoborohydride (Randerath, K., 1981).

The reagent is also useful for the analysis of other carbonyl compounds, such as simple aldehydes and ketones, and periodate-oxidized glycoproteins. Its utility for 3′-terminal RNA labeling is under investigation.

V. [32]P DERIVATIVE METHOD FOR BASE ANALYSIS OF NORMAL AND MUTAGEN/CARCINOGEN-MODIFIED DNA

A. General Aspects

The identification of DNA-reactive chemicals in the human environment has become an important task, as damage to DNA by chemicals is likely to be a major cause of cancer (Ames, 1979; Miller, J. A.,

1970; Miller, E. C., 1978) and may also play a role in other diseases (Ames, 1979). Until recently, no sensitive and generally applicable test has been available to detect directly the presence of chemically altered bases in DNA. So far, most assays were based on the binding of radioactively labeled or fluorescent test compounds to DNA and are therefore not generally applicable to screening large numbers of chemicals (Hollstein et al., 1979). Recently, our laboratory has initiated the development of a rapid ultrasensitive assay for the direct detection of covalent binding of chemicals to DNA, which involves base composition analysis of the modified DNA by a ^{32}P-postlabeling method (Randerath, K., et al., 1981a; Reddy et al., 1981). The lack of requirement for radioactive DNA-modifying agents represents a distinct advantage of this assay because (i) the vast majority of chemicals is not available in radioactive form; (ii) any chemical can be screened; (iii) the specific activity obtainable by ^{32}P-labeling ($\sim 8,000$ Ci/mmol) is much greater than the specific activity of available radioactive mutagens and carcinogens; and (iv) the sensitivity of detection of ^{32}P exceeds that of ^{3}H or ^{14}C.

The development of this assay involves three stages: (i) The development of a quantitative ^{32}P derivative method for base composition analysis of normal untreated DNA (Reddy, Gupta, & Randerath, 1981), (ii) extension of the method to make it applicable to DNA modified by mutagens/carcinogens *in vitro* (Randerath, K., Reddy, & Gupta, 1981), and (iii) further refinement of the method to make it applicable to DNA modified by mutagens/carcinogens *in vivo*. While work concerning stage (i) has been completed and most aspects of stage (ii) have been dealt with, work on stage (iii) began only recently. This part of the project deals with some of the more difficult aspects of the assay, including techniques for optimizing its sensitivity, an investigation of radioactive background components which may interfere with the analysis of low levels of certain labeled DNA adducts, the development of chromatographic systems suitable for the separation of a multitude of adducts with very different properties, as determined by the type of DNA-modifying chemical, a study of *in vivo* effects on DNA of a large number of chemicals, and the quantitation of the most important adducts. As standard procedures for stage (iii) have not as yet been developed, we will deal here only with stages (i) and (ii).

B. Outline of Method

Normal DNA or DNA modified *in vitro* by mutagens/carcinogens is degraded enzymatically by incubation with micrococcal endonuclease and spleen exonuclease to deoxynucleoside 3′-monophosphates. The digests are treated with (γ-^{32}P)ATP and T$_4$ polynucleotide kinase to convert the monophosphates to (5′-^{32}P)-labeled deoxynucleoside 3′,5′-bisphosphates. These compounds are then separated by two-dimensional polyethyleneimine-cellulose thin-layer chromatography. The labeled compounds are detected by autoradiography. As can be seen in Fig. 5-4, a characteristic chromatographic pattern of adducts is obtained for each DNA-modifying chemical. For quantitative evaluation, radioactive spots are cut out and assayed by Cerenkov counting. While the conditions worked out for the analysis of normal DNA yield quantitative base-composition data, labeling of DNA adducts, particularly those with bulky residues, may not always be quantitative under the conditions specified here. For this purpose, conditions may need modification (see also Comments below).

C. Materials

Micrococcal endonuclease (Grade VI, 0.21 U/μg) from Sigma Chemical Co. (Saint Louis, Missouri) and spleen exonuclease (0.002 U/μg) from Boehringer Mannheim Biochemicals (Indianapolis, Indiana) are dialyzed against water at 4°C for 15 hr.

T$_4$ polynucleotide kinase from P-L Biochemicals (Milwaukee, Wisconsin).

Potato apyrase (code A7646) from Sigma Chemical Co.

(γ-^{32}P)ATP from Amersham Corp. (Arlington Heights, Illinois).

Commercial PEI-cellulose sheets without indicator (20 × 20 cm; Brinkmann Instruments) are used for the analysis of normal DNA.

Figure 5-4. ^{32}P maps of digests of control calf-thymus DNA (a) and calf-thymus DNA treated with the indicated electrophilic chemicals (b to i). For conditions for the alkylation of DNA with the various chemicals specified, see Randerath, K., Reddy, and Gupta (1981). The ^{32}P-labeled digests were obtained as described in the text. (γ-^{32}P)ATP had a specific activity of 100 Ci/mmol. Chromatography of the labeled digests (total radioactivity, 1.2 μCi; 3.5 pmol (0.35 μCi) of ^{32}P-deoxynucleoside 3′,5′-bisphosphates) on anion-exchange thin-layer sheets of PEI-cellulose in formate (bottom to top) and sulfate (left to right) solutions. Autoradiography was at −70°C for 7–9 hr, except for samples a and i, which were exposed for 12 and 16 hr, respectively. Spots of dpGp (G), dpAp (A), dpTp (T), dpCp (C), and dpm^{5}Cp (m^{5}C) have been indicated for control (a), the formaldehyde (d), and β-propiolactone (e) samples. Arrows indicate background spots, which were also present in reactions without DNA (not shown). The intensity, but not the location, of these spots varied with the batch of (γ-^{32}P)ATP. Taken from Randerath, K., Reddy, and Gupta (1981).

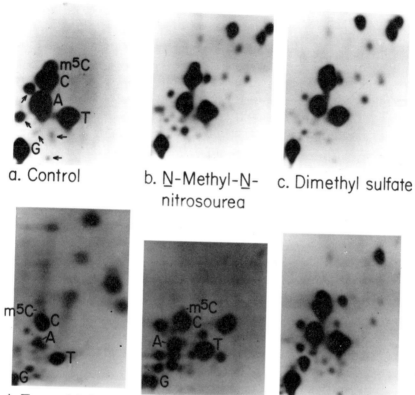

a. Control b. N-Methyl-N-nitrosourea c. Dimethyl sulfate

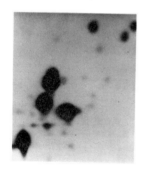

d. Formaldehyde e. β-Propiolactone f. Propylene oxide

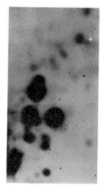
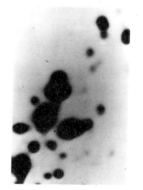

g. Streptozotocin h. Nitrogen mustard i. 1,3-Bis(2-chloro-ethyl)-1-nitrosourea

For carcinogen-modified DNA, PEI-cellulose sheets are prepared in the laboratory as described (Randerath, K., & Randerath, 1966, 1967). A 0.5% neutralized solution of PEI-1000 (Dow Chemical Co., Freeport, Texas) may be substituted for 1% Polymin P (BASF) solution. The size of these sheets is 25×20 cm. Layers are prewashed by ascending development in water.

D. Procedures

Procedure for the Quantitative Base Analysis of Normal DNA

Digestion and ^{32}P-Labeling. For quantitative digestion of DNA to deoxynucleoside 3'-monophosphates, incubate 1 μg of DNA ($\sim$3 nmol of DNA-P) at 37°C for 2 hr with 1 μg each of micrococcal nuclease (0.10–0.15 units) and spleen phosphodiesterase in 10 μl of a solution containing 20 mM sodium succinate and 8 mM $CaCl_2$, pH 6.0. Treat a control without DNA in the same way as the sample. Dilute the digest 10-fold. Mix a 2-μl aliquot of the diluted digest containing approximately 60 pmol of deoxynucleoside 3'-monophosphates with a solution of (γ-^{32}P)ATP (60 μM, 10–20 Ci/mmol) and T_4 polynucleotide kinase (0.12 U/μl) in 10 μl of a solution containing 40 mM bicine-NaOH, 10 mM $MgCl_2$, 10 mM dithiothreitol, and 0.1 mM spermidine, pH 9.0. Incubate the reaction mixture at 37°C for 1 hr. Incubate the ^{32}P-labeled mixture (10 μl) with 2.5 μl of a solution of potato apyrase (5 mU/μl) at 37°C for 1 hr to convert excess (γ-^{32}P)ATP and ^{32}P-labeled impurities to (^{32}P)P_i before chromatography.

Chromatography and Quantitative Analysis. Preequilibrate four PEI-cellulose sheets (20×20 cm) by soaking in 500 ml of 0.1 M ammonium formate, pH 3.5, for 30 min and drying, first in a current of cool air, then warm air (5 min each). Apply aliquots (2 μl) of the reaction mixture to each of the four sheets. Develop overnight (at least 15 hr) in 0.8 M ammonium formate, pH 3.5, to about 25 cm on a Whatman 1 wick which had been attached to the top of the sheet by stapling and then folded back. Continue development for 2–3 hr in 4 M ammonium formate, pH 3.5, without intermediate drying. Dry sheets as above, soak them in 250 ml of methanol for 10 min, then dry them briefly in cool air. For the second dimension, develop with water to the origin, followed by 0.3 M ammonium sulfate to 4–5 cm on a Whatman 1 wick attached to the top of the sheet. Locate individual nucleotide spots by autoradiography, cut them from the chromatogram, and determine their radioactivity by Cerenkov counting. Calculate base ratios, after subtracting blank values, from the four replicate analyses.

For a map of the four major p*N*p compounds and pm⁵Cp obtained with the above chromatographic system, see Reddy, Gupta, and Randerath (1981).

Procedure for Base Analysis of Mutagen/Carcinogen-treated DNA

Digestion and ^{32}P-Labeling. Convert treated and control DNA quantitatively to deoxynucleoside 3'-monophosphates by incubating 1 μg of DNA (∼3 nmol of DNA-P) at 37°C for 2 hr with 2 μg each of micrococcal endonuclease and spleen exonuclease in 10 μl of 20 mM sodium succinate, 8 mM CaCl$_2$, pH 6.0. For ^{32}P-labeling, dilute the digest 1:10 and add a 1-μl aliquot containing approximately 30 pmol of deoxynucleoside 3'-monophosphates to 10 μl of 10 μM (γ-^{32}P)ATP (100 Ci/mmol), 40 mM bicine-NaOH, pH 9.0, 10 mM MgCl$_2$, 10 mM dithiothreitol, 0.1 mM spermidine, T$_4$ polynucleotide kinase (0.12 U/μl). Incubate at 37°C for 1 hr. Incubate the ^{32}P-labeled reaction mixture further with 3 μl of apyrase (5 mU/μl) at 37°C for 30 min to convert untreated (γ-^{32}P)ATP to (^{32}P)P$_i$.

Mapping of Labeled Digests (Randerath, K., Reddy, & Gupta, 1981). Add 3 μl of 0.2 M sodium tungstate to the reaction mixture to complex P$_i$ so that most of it will be retained at the origin of the chromatogram. Apply 1 μl of a mixture of authentic dpGp, dpAp, dpCp, and dpTp (2 μg each) and 2 μl of the reaction mixture (3.5 pmol of ^{32}P-deoxynucleoside 3',5'-bisphosphates) to a PEI-cellulose sheet (25 × 20 cm) at 1.5 cm from the lower edge and 2.5 cm from the left-hand edge.

Develop the chromatogram with water to the origin, then with 4 M lithium formate, 7.5 M urea, pH 4.0 (prepared by titration of a formic acid/urea mixture with lithium hydroxide) to 5 cm above the origin. Soak the sheet in 300 ml of methanol for 10 min, dry in a current of cool air, and cut 1 cm above the origin. Discard the lower part, containing mainly (^{32}P)P$_i$ (retained at the origin as a tungstate complex). Develop the upper portion with 0.04 M ammonium formate, pH 3.5 (prepared by titration of formic acid with ammonium hydroxide) to 3 cm from the cut edge, then, without intermediate drying, with 0.7 M ammonium formate, pH 3.5, to 6 cm on a Whatman 1 wick which had been attached 21–22 cm from the bottom edge by stapling and folded back. After chromatography, dry the layer in a current of cool air, followed by warm air (5 min each). Soak the sheet in 300 ml of 0.01 M tris base in water for 10 min, then in 300 ml of water for 5 min and dry.

Develop in the second dimension with 0.01 M tris-HCl, pH 8.0, to the origin, then with 0.1 M ammonium sulfate, 0.01 M tris-HCl, pH 8.0, to 7 cm on a Whatman 1 wick attached to the top of the sheet. Locate spots by screen-intensified autoradiography (Section II.E). To distinguish spots representing DNA adducts from radioactive background spots, compare sample and control maps (see also Fig. 5-4).

Cut spots out and determine their radioactivity by Cerenkov assay. For quantitative evaluation, an improved resolution of dpm ^{5}Cp and dpCp may be achieved by developing the chromatograms in the first dimension to about 25 cm rather than 6 cm on the wick.

E. Comments

Various conditions for the enzymatic digestion of DNA to deoxyribonucleoside 3'-monophosphates have been studied (Reddy, Gupta, & Randerath, 1981). The conditions described above for the digestion of normal and alkylated DNA were found to convert DNA quantitatively to its monomeric constituents. As is evident from Fig. 5-4, DNA alkylated with various reagents also yielded monomeric constituents under the above digestion conditions. Moreover, preliminary experiments in our laboratory (K. Randerath, R. C. Gupta, & M. V. Reddy, unpublished experiments) have shown that DNA containing bulky substituents, e.g., derived from benzo(a)-pyrene or 2-acetylaminofluorene, can also be degraded under the above conditions, thus yielding labeled adducts when the digest is labeled by the (γ-^{32}P)ATP/polynucleotide kinase reaction.[2]

While the assay conditions for normal DNA described above result in quantitative labeling of the deoxyribonucleoside 3'-monophosphates, adducts containing bulky substituents may not always be labeled quantitatively under the above assay conditions, as hyper-

---

[2] Conditions for labeling, separation, and ultrasensitive detection (1 adduct in $>10^7$ normal nucleotides) of aromatic carcinogen-DNA adducts were recently published (Gupta, Reddy, & Randerath, 1982).

modified ribonucleoside 3′-monophosphates, such as ms^{2}i^{6}A, were found to react more slowly with (γ-^{32}P)ATP/polynucleotide kinase than the four major ribonucleoside 3′-monophosphates (H. P. Agrawal, E. Randerath, & K. Randerath, unpublished experiments). Recoveries of such adducts still have to be determined under the standard reaction conditions, so that correction factors can be applied if necessary for quantitative analysis. An analogous approach was used in the development of the ^{3}H derivative method for base composition analysis of RNA (Section III). Adducts of DNA treated with either N-methyl-N-nitrosourea or 1,3-bis(2-chloroethyl)-1-nitrosourea were found to be labeled quantitatively under the above conditions [(γ-^{32}P)ATP/substrate ratio of 3:1, 0.12 U/μl poly-nucleotide kinase, and 1 hr incubation at 37°C] (K. Randerath, M. V. Reddy, & R. C. Gupta, unpublished experiments).

The overall sensitivity of the assay is influenced by background radioactivity. Chromatograms obtained from control DNA always contain a number of spots that are not derived from the normal DNA constituents. Some of these spots are present in certain batches of commercial (γ-^{32}P)ATP, others are formed during the labeling reaction by as yet unknown mechanism(s). This point is under current investigation. So far we have reduced the radioactive back-ground by several means, including the purification of (γ-^{32}P)ATP, the use of a relatively low ATP concentration, apyrase treatment to convert excess (γ-^{32}P)ATP and some of its contaminants to (^{32}P)P$_i$, and addition of tungstate to the labeled digest to convert (^{32}P)P$_i$ into a complex that is retained at the origin of the chromato-gram. The method currently detects a single adduct in about 10^5 nucleotides. Work is now under way to scale the assay down further by one to two orders of magnitude. As preliminary experiments (K. Randerath, M. V. Reddy, & R. C. Gupta, unpublished) have shown, sensitivity is greatly increased if the normal nucleotides are removed, e.g., by HPLC, before the final analysis of the adducts. Thus, reversed-phase HPLC appears to be a useful tool to purify hydrophobic nucleotide adducts. Since most carcinogens lead to the formation of such adducts, this technique should be widely applicable in conjunction with the assay. In conclusion, the approach outlined in this section makes possible the detection of a large number of carcinogen-DNA adducts without the use of radioactive carcinogens. We expect postlabeling methods to become powerful tools in investi-gations of the interactions of chemicals with nucleic acids. The methodology needs further development to extend its scope and to increase its sensitivity.

VI. ^{3}H DERIVATIVE METHOD FOR SEQUENCE ANALYSIS OF RNA

A. General Aspects

Since the ^{3}H-sequencing method (Sivarajan et al., 1974; Randerath, K., Gupta, & Randerath, 1980) requires considerably more RNA (e.g., 80–120 µg of tRNA) than the more sensitive ^{32}P methods, it will only be outlined here. For experimental details—in particular, procedures for the separation of complex oligonucleotide mixtures on PEI-cellulose and for the isolation of oligonucleotides from these layers—the reader is referred to Randerath, K., Gupta, and Randerath (1980).

B. Outline of Method

A complete RNase T_1 or A digest of the RNA is separated by two-dimensional PEI-cellulose anion-exchange thin-layer chromatography (Randerath, K., 1962; Randerath, K., & Randerath, 1964a; Randerath, K., Gupta, & Randerath, 1980). Oligonucleotides are located under short-wave UV light and eluted (Randerath, E., et al., 1979; Randerath, K., Gupta, & Randerath, 1980). Their base compositions and molar ratios are determined by ^{3}H-postlabeling techniques. They are then sequenced by a ^{3}H derivative method [formerly called procedure II (Sivarajan et al., 1974)], which entails the following steps: (i) controlled digestion of the oligonucleotide with snake venom phosphodiesterase/alkaline phosphatase; (ii) periodate oxidation of the 3′ ends of the partial digestion products; (iii) borohydride reduction of oligonucleotide 3′-dialdehydes obtained in step (ii) to 3′-terminally ^{3}H-labeled oligonucleotide 3′-dialcohols [see Eq. (1), Section III.B; R = mono- or oligonucleotide residue]; (iv) deduction of the sequence by separation of the ^{3}H-labeled oligonucleotide 3′-dialcohols according to chain length, *in situ* enzymatic liberation of ^{3}H-labeled 3′-terminal trialcohols, contact transfer, and identification of the trialcohols by two-dimensional silica gel thin-layer chromatography and scintillation counting. The 5′ termini of the oligonucleotides are determined separately as ^{3}H-labeled nucleoside trialcohols after venom phosphodiesterase treatment of the oligonucleotides and labeling (Randerath, K., Randerath, et al., 1974).

2′-*O*-Methylated nucleosides are also determined separately (Gupta, Randerath, & Randerath, 1976c). The presence of a 2′-*O*-

methylated nucleoside in an oligonucleotide can be inferred from a gap in the series of oligonucleotide 3'-dialcohol intermediates generated during the ^{3}H-sequencing procedure. Such oligonucleotides are subjected to RNase T_2 digestion resulting in the formation of a dinucleotide containing the 2'-*O*-methylated nucleoside (*N*mp*N*p) and nucleoside 3'-monophosphates. After dephosphorylation and inactivation of phosphatase, *N*mp*N* is converted to (^{32}P)p*N*mp*N* by the (γ-^{32}P)ATP/polynucleotide kinase reaction. (^{32}P)p*N*m is subsequently released by venom phosphodiesterase treatment and chromatographed together with appropriate markers (Gupta, Randerath, & Randerath, 1976b).

To obtain necessary overlaps, partial fragments are analyzed by methods similar to those by which the fragments in complete digests or the positions of A, G, and pyrimidines in the RNA or in partial fragments are identified by the gel readout technique.

VII.   COMBINED ^{3}H/^{32}P DERIVATIVE METHOD FOR SEQUENCE ANALYSIS OF RNA

A.   General Aspects

In the method outlined in Section VI, fragments in complete digests are detected on chromatograms by UV absorption; however, in the present method (Gupta, Randerath, & Randerath, 1976a; Randerath, K., Gupta, & Randerath, 1980) such fragments are 3'-terminally labeled with ^{3}H before chromatography and detected by fluorography. The present method is therefore over 10 times more sensitive than the former. Although the combined ^{3}H/^{32}P method is more time-consuming and less sensitive than the thin-layer readout method (Section IX), it does have some utility as a complementary technique to the latter. For example, it may be used to locate and identify modified nucleotides. Only a brief outline of the method will be given here. For experimental details, see Randerath, K., Gupta, and Randerath (1980).

B.   Outline of Method

Oligonucleotides in a complete RNase T_1/alkaline phosphatase or RNase A/alkaline phosphatase digest of the RNA are converted to

3′-terminally ^{3}H-labeled oligonucleotide 3′-dialcohols [Eq. (1), Section III.B; R = mono- or oligonucleotide residue] and mapped on PEI-cellulose (Randerath, K., & Randerath, 1973; Gupta, Roe, & Randerath, 1979). The labeled derivatives are located by fluorography and eluted. Base compositions of the oligonucleotide 3′-dialcohols are determined by the ^{3}H derivative method. Molar ratios of the oligonucleotides are calculated from the count rates of the ^{3}H-labeled oligonucleotide 3′-dialcohols.

The sequences of the oligonucleotide 3′-dialcohols are deduced by a readout procedure entailing partial digestion of these compounds with specific endonucleases (RNases T_1, U_2, A, Phy$_1$) and resolution of the products by size on PEI-cellulose (Gupta & Randerath, 1977a, 1977b). These partial digests are chromatographed alongside a controlled nuclease S_1/alkaline phosphatase digest, which contains all possible cleavage products. The sequence of the four major constituents can then be read directly from the chromatogram.

The positions of modified constituents can usually be inferred from base composition data. If necessary, internal modified constituents in the oligonucleotide chain can be determined as follows (Gupta, Randerath, & Randerath, 1976a, 1976b): The oligonucleotide 3′-dialcohol is partially digested with nuclease S_1/alkaline phosphatase to give two sets of labeled and unlabeled fragments, respectively. The undesired nonradioactive fragments are oxidized with periodate. Subsequent thin-layer chromatography on PEI-cellulose separates the labeled fragments by size, while the oxidized nonradioactive fragments are retained at the origin. The labeled fragments are located by fluorography, cut out, and eluted. The sequence of the parent compound is then deduced by enzymatic incorporation of ^{32}P-label into the 5′ termini of the eluted fragments, enzymatic release of ^{32}P-labeled nucleoside 5′-monophosphates, two-dimensional thin-layer chromatography, and autoradiography.

VIII. ^{32}P GEL READOUT METHOD ENTAILING PARTIAL ENZYMATIC DEGRADATION OF END-LABELED RNA

A.   General Aspects

In principle, the enzymatic gel readout method is similar to the method of Maxam and Gilbert for sequencing DNA (Maxam & Gilbert, 1977) and the method of Peattie for sequencing RNA (Peattie, 1979). However, unlike these methods, it utilizes enzymes of different base specificities rather than chemical reagents for partial cleavages of ^{32}P-end-labeled RNA. This method is fast and highly sensitive, requiring only a few μg of a short RNA (chain length $\leqslant$150). It affords the identification of A, G, and pyrimidine residues in the polynucleotide chain, but often it cannot distinguish clearly between U and C residues due to the lack of highly U- or C-specific endoribonucleases. This latter problem has become particularly apparent with RNAs rich in C—A or U—A bonds, such as mammalian mitochondrial tRNAs (H. P. Agrawal, E. Randerath, & K. Randerath, unpublished experiments), as the currently available pyrimidine-specific enzymes exhibit insufficient specificity for these bonds. The gel readout method is not capable of identifying modified nucleotides and therefore can only be used in conjunction with other methods (see also Section VII). Certain RNA regions exhibiting tight secondary structure due to the presence of several adjacent G·C base pairs may give rise to gaps or band compression in the gel patterns and thus make reading of the sequence in these regions difficult. This, on the other hand, may be exploited to locate areas of tight secondary structure.

A number of papers have dealt with the development and use of this method. Its basic underlying principle, i.e., the location of defined positions in the chain relative to the labeled end of the nucleic acid, was first established for DNA sequence analysis (Maxam & Gilbert, 1977). An enzymatic method for the sequence analysis of end-labeled ribooligonucleotides (Gupta & Randerath, K., 1977a) and end-labeled RNA (Donis-Keller, Maxam, & Gilbert, 1977; Simoncsits et al., 1977; Lockard et al., 1978) based on the same principle was subsequently described.

The procedures to be described here are those used in our laboratory (Randerath, E., et al., 1979; Gupta, Roe, & Randerath, 1980; Randerath, K., Gupta, & Randerath, 1980). Most conditions are similar to those described by others (Donis-Keller, Maxam, & Gilbert, 1977; Simoncsits et al., 1977; Silberklang et al., 1977; Lockard et al., 1978; Donis-Keller, 1980; Boguski, Hieter, & Levy, 1980). Conditions for the use of RNase Phy$_1$ are those first developed in our laboratory (Gupta & Randerath, 1977b).

B. Outline of Method

The RNA is labeled with ^{32}P either at its 5′ terminus by the (γ-^{32}P)ATP/T$_4$ polynucleotide kinase reaction (Szekely & Sanger, 1969; Szekely, 1971; Richardson, 1971; Simsek et al., 1973) or at its 3′ terminus by (5′-^{32}P)pCp/T$_4$ RNA ligase (England & Uhlenbeck,

1978; Bruce & Uhlenbeck, 1978; England, Bruce, & Uhlenbeck, 1980). Terminal labeling establishes a unique reference point at one end of the RNA. Subsequently, the RNA is digested with base-specific endoribonucleases, such as RNase T_1 and U_2, cleaving at G and A residues, respectively. These reactions generate nested sets of labeled fragments, each extending from the labeled terminus to an internal G or A residue. A limited alkaline hydrolyzate of the RNA is also prepared, which contains another set of nested fragments, extending from the labeled end to random bases in the RNA. Similar fragments may also be obtained when water, formamide, or RNase T_2 is used for partial digestion of the RNA. The partial enzymatic digests and the limited hydrolyzate are electrophoresed on a polyacrylamide gel, which separates the partial fragments by size. The end-labeled fragments are detected by autoradiography. The above samples will produce three parallel band patterns on the autoradiogram, i.e., one from cleavage at every G, another from cleavage at every A, and the third from cleavage at every base. Since these cleavage products are ordered by size, the position of each A and G in the polynucleotide chain can be read directly from the autoradiogram, while the positions of pyrimidines can be inferred. Cleavages at pyrimidines may be obtained with RNase A or with an extracellular RNase from *B. cereus* (Lockard et al., 1978). Of these two enzymes, the former is more sequence-specific and thus yields a less-even distribution of radioactivity among the partial fragments. Several enzymes exhibit a certain preference for U—N or C—N bonds, respectively. These include RNase Phy_1 (Gupta & Randerath, 1977b; Randerath, E., et al., 1979; Randerath, K., Gupta, & Randerath, 1980) and RNase Phy M (Donis-Keller, 1980), both extracellular enzymes of the slime mold *P. polycephalum* (Braun & Behrens, 1969; Pilly et al., 1978) which exhibit a certain preference for U—N bonds, and a chicken-liver ribonuclease with a preference for C—N bonds (Levy & Karpetsky, 1980; Boguski, Hieter, & Levy, 1980). The presence of U or C residues can also be inferred from the known specificities of endonucleases from *S. aureus* and *N. crassa* (Krupp & Gross, 1979). Another possibility for distinguishing between U and C residues is through the use of mobility-shift procedures on anion-exchange thin layers (Gupta & Randerath, 1977a) and on polyacrylamide gels (Lockard et al., 1978).

C.   Materials

RNase T_1 (No. 556785) and RNase U_2 (No. 556877) from Calbiochem-Behring Corp. (San Diego, California). RNase Phy_1 is prepared as described by Pilly et al. (1978). One unit of enzymatic activity is defined as the amount of enzyme that solubilizes 35 μg of RNA per min at pH 4.5 and 40°C (Pilly et al., 1978). The enzyme is stored in the presence of 40% glycerol at −18°C. This enzyme is commercially available from P-L Biochemicals Inc. (Milwaukee, Wisconsin; No. 0924) and from Enzo Biochem. Research Products (New York, N.Y.; No. ERN-Phy$_1$). *B. cereus* ribonuclease (No. 0993) and ribonuclease Phy M (No. 0994) from P-L Biochemicals Inc. Prepare stock solutions of *B. cereus* RNase and RNase Phy M by adding 20 μl of water and 20 μl of 50% glycerol, respectively, to 200 U of enzyme. Store both enzyme solutions at −20°C.

A RNA sequencing enzyme kit is also available from P-L Biochemicals Inc. (No. 0997), containing, in addition to *B. cereus* RNase and RNase Phy M, RNase T$_1$ and RNase U$_2$. (RNase T$_1$ and U$_2$ used by us were from Calbiochem-Behring Corp. Conditions recommended by P-L Biochemicals Inc. for RNase T$_1$ and U$_2$ digestions are somewhat different from the ones described here.)

Chicken-liver ribonuclease was obtained from Dr. C. C. Levy as a fraction eluted from a CM-Sephadex column (Levy & Karpetsky, 1980). To remove salt, the preparation was dialyzed against water.

T$_4$ polynucleotide kinase from P-L Biochemicals or New England Nuclear Corp. (Boston, Massachusetts).

Yeast transfer RNA (Type I) from Sigma Chemical Co. (Saint Louis, Missouri).

Ultrapure urea from Schwarz/Mann (Orangeburg, N.Y.); acrylamide and methylene bisacrylamide from BioRad (Richmond, California).

A surface temperature thermometer, Model 309C, PTC Instruments, from Fisher Scientific Co. (Pittsburgh, Pennsylvania).

Cronex 4 X-ray film from DuPont (Wilmington, Delaware).

For additional materials, see Sections III.C and V.C.

(γ-^{32}P)ATP may also be obtained from ICN Pharmaceutical Inc. (Irvine, California) or New England Nuclear Corp.

D.   Procedures

Preparation of 5′-terminally ^{32}P-labeled RNA

If the RNA has 5′-terminal phosphate, this needs to be removed first. For this purpose, treat RNA with phosphatase as follows. Incubate 5 μg of RNA (chain length <150) in 5 μl of a solution containing 15 mM tris-HCl, pH 8.0, 0.1 mM EDTA, and 0.2 μg/μl of calf intestinal alkaline phosphatase. Keep at 55°C for 25 min.

Before polynucleotide kinase labeling, inactivate or remove the phosphatase. To inactivate the phosphatase, add 1 μl of 45 mM nitrilotriacetic acid, incubate at 23°C for 20 min, and then heat sample to 100°C for 3 min. Dry sample in a stream of air or *in vacuo.* Alternatively, the phosphatase may be removed by phenol extraction. For this purpose, add 200 μl of water, 2 μl of 100 mM EDTA, and 200 μl of dist. phenol saturated with water. After thorough mixing, spin sample at 12,000 rpm for 10 min. Reextract the upper aqueous phase with another portion of 200 μl of phenol. Precipitate the RNA from the aqueous phase by the addition of 20 μl of 20% potassium acetate, pH 5.0, and 660 μl of acetonitrile/ethanol (4:1). Keep the sample on ice for at least 1$\frac{1}{2}$ hr. Spin at 11,000 rpm for 10 min and discard the supernatant solution. Wash the precipitate with ice-cold abs. ethanol. Recentrifuge and discard ethanol. Dry precipitate in vacuo.

To label the 5' terminus with ^{32}P, incubate the dried residue of the RNA in 5 μl of a mixture of 50 mM tris-HCl, pH 8.7, 30 mM MgCl$_2$, 25 mM dithiothreitol, 60 μM (γ-^{32}P)ATP (800–1,500 Ci/mmol), and 5 U of polynucleotide kinase at 37°C for 25 min. Dry sample in a stream of air.

To purify the end-labeled RNA, dissolve the dried residue in 5 μl of 90 mM tris, 90 mM boric acid, 4 mM EDTA, pH 8.3 (TBE buffer), containing 1 μg/μl of xylene cyanole (XC) and bromophenol blue (BP), and saturate solution with urea. Electrophorese sample on a 12% polyacrylamide/8 M urea gel (acrylamide/methylene bisacrylamide, 20:1) at pH 8.3 (TBE buffer) (Peacock & Dingman, 1967). Example for electrophoresis conditions: Gel dimensions, 50 × 20 × 0.06 cm; electrophorese at 800–1,000 V, 35–40°C, for 11 hr until XC is at approximately 45 cm. Locate the RNA by autoradiography.

Extract the RNA from the gel by homogenization in the presence of phenol (Chia, Randerath, & Randerath, 1973). For this purpose, add 4 ml of 0.05 M NaCl, 0.02 M sodium acetate, pH 5.1, containing 10–20 μg of carrier yeast tRNA, and 4 ml of dist. phenol. Precipitate the RNA from the aqueous phase by adding 0.1 volume of 20% potassium acetate, pH 5.0, and 3.3 volumes of acetonitrile/ethanol (4:1).

Preparation of 3'-Terminally ^{32}P-Labeled RNA

For a description of the conditions for 3'-terminal labeling of RNA by the (5'-^{32}P)pCp/T$_4$ RNA ligase reaction, see Chapter 7 or references quoted above.

Preparation of Partial Digests

The conditions described below will yield an amount of digest sufficient for five sample applications to the subsequent gel. The preparations of the digests may be scaled down depending on the amount of RNA available. The conditions given may not be optimal in terms of the enzyme/substrate ratio. Optimal conditions should be established by serial dilution of enzyme into the reaction mixtures containing all ingredients except the enzyme (Lockard et al., 1978).

RNase T_1 Digest. To 0.3–1.0×10^6 dpm of the labeled RNA add yeast tRNA carrier to give a total of 5 μg of RNA, then dry the sample. Add 20 μl of buffer A and 2 μl of a RNase T_1 solution containing 0.01 μg/μl.

Buffer A is 20 mM sodium citrate, 7 M urea, 1 mM EDTA, pH 5.0, containing 0.2–0.5 μg/μl each of the tracking dyes XC and BP.

Keep at 50°C for 15 min. Additional carrier RNA (20–30 μg) in 2 μl of water may be added to avoid overdigestion during further handling of the sample. (This may also be done with the other partial enzymatic digests described below.) Saturate sample with solid urea. This will result in a better resolution on the gel (see below). Freeze sample immediately in Dry Ice/acetone.

RNase U_2 Digest. Use the same conditions as for the RNase T_1 digest but, instead of using RNase T_1 solution, add 1 μl of a solution containing 25 mU of RNase U_2. Incubate at 50°C for 10 min. Saturate sample with solid urea and freeze it in Dry Ice/acetone.

RNase A Digest. Digest the same amount of RNA as above. Add 20 μl of buffer B and 2 μl of a RNase A solution containing 0.001 μg/μl.

Buffer B is 30 mM tris-HCl, 70% DMSO, pH 7.8, containing tracking dyes (see above).

Keep at 37°C for 20 min. Saturate sample with solid urea before freezing it.

B. cereus RNase Digest. To 0.3–1.0×10^6 dpm of labeled RNA add yeast carrier tRNA to give a total of 8 μg RNA, then dry the sample. Add 12 μl of 25 mM sodium citrate, pH 5.0, 1 mM EDTA, and 1 μl of a diluted solution of *B. cereus* RNase (3.33 U/μl). Digest the RNA at 50°C and withdraw 3-μl aliquots at 5, 10, 20, and 30 min. Combine the aliquots as they are withdrawn; keep combined digest on Dry Ice. After completion of the reaction, add 1 μl of tracking dye solution, containing 4–10 μg each of XC and BP, and saturate sample with solid urea. Freeze sample immediately.

RNase Phy₁ Digest. To two portions of $0.3–1.0 \times 10^6$ dpm of labeled RNA add yeast tRNA carrier to give 1 μg (sample 1) and 1.5 μg (sample 2), respectively, of total RNA, then dry the samples. Take each residue up in 20 μl of buffer A and dry again. Add 20 μl of RNase Phy₁ containing 3.4 mU/μl to each dried sample. Incubate sample 1 and 2 at 50°C and 37°C, respectively, for 10 min. The two samples may be combined or may be kept separately for subsequent gel electrophoresis. Saturate samples with solid urea, and freeze them immediately.

RNase Phy₁ digests may also be prepared in the absence of urea. For experimental details, see Randerath, K., Gupta, and Randerath (1980). Such digests will display gaps in the gel patterns corresponding to regions of tight secondary structure.

RNase Phy M Digest. Prepare two samples, each with the same amount of RNA as for the *B. cereus* RNase digest. To each dried residue add 12 μl of 25 mM sodium citrate, pH 5.0, 1 mM EDTA, 7 M urea. To sample 1 add 1 μl of a RNase Phy M solution containing 2 U; to sample 2 add 1 μl of a solution containing 1 U. Digest the RNA at 50°C for 15 min. Add 0.5 μl of the above tracking dye solution, saturate samples with solid urea, and freeze them immediately.

Chicken-liver Ribonuclease Digest. To $0.3–1.0 \times 10^6$ dpm of the labeled RNA add yeast tRNA carrier to give a total of 2 μg of RNA, then dry the sample. Take residue up in 10 μl of 20 mM sodium phosphate, pH 6.5, and add 1–10 mU of chicken-liver ribonuclease in a volume of 10 μl. Incubate at 37°C for 5 min. Add 2 μl of the above tracking dye solution, saturate sample with solid urea, and freeze it immediately.

No-enzyme Control(s). Treat labeled RNA in the presence of buffer but omit enzyme from the reaction mixture. Gel electrophoresis of control samples shows labeled fragments arising from nicks in the RNA.

Alkaline Hydrolyzate. Use the same amounts of labeled and carrier RNA as for the RNase T_1 digest. To the dried residue add 20 μl of 50 mM sodium bicarbonate, pH 9.0 (NaOH), 1 mM EDTA, containing 0.2-0.5 μg/μl each of XC and BP, and incubate at 90°C for 12 min. Saturate sample with solid urea and freeze it.

Gel Electrophoresis and Autoradiography. Separate partial digests on 12 and 20% polyacrylamide/8 M urea gels (acrylamide/methylene bisacrylamide, 30:1 and 20:1, respectively) at pH 8.3 (TBE buffer). Gel dimensions are 50 × 30 × 0.06 cm. The 12% gel is for the separation of long and intermediate fragments, the 20% gel is for short fragments. Apply 4-5 μl of each digest. Adjust the voltage during electrophoresis so that a temperature of 40-45°C is maintained. Use a surface-temperature thermometer to measure the gel temperature. Determine the electrophoresis time according to the migration of the tracking dyes and the desired fragment sizes. On 12% gels, XC and BP travel approximately with fragments of chain lengths 40 and 12, respectively. On 20% gels, the corresponding chain lengths are 23 and 7, respectively. For the detection of the radioactive fragments, expose Cronex 4 X-ray film to the gel. For screen-intensified autoradiography, see Section II.E.

Reading of the Sequence. The sequence is read from the autoradiogram as explained above in Outline of Method. The main cleavage rules for RNase Phy_1 under partial digestion conditions (Gupta & Randerath, 1977b) may be summarized as follows:

(i) C—*N* is resistant with the exception of C—A, which may be digested under certain conditions.
(ii) U—*N* is always cleaved.
(iii) *N*—C is cleaved with the exception of C—C.
(iv) *N*—U is resistant with the exception of U—U.

In cases where RNase Phy_1 does not cleave next to U derivatives, e.g., T—ψ, these may be distinguished from C by an alternative procedure (Section VII). RNase Phy M uniformly cleaves A—*N* and U—*N* bonds in RNA when used in 7 M urea, pH 5.0, at 50°C (Donis-Keller, 1980). In conjunction with the *B. cereus* RNase, which cleaves C—*N* and U—*N* bonds (Lockard et al., 1978), RNase Phy M was thus reported to allow the distinction of the pyrimidines (Donis-Keller, 1980). However, commercial preparations of this enzyme were found by us to cleave C—A bonds also (unpublished experiments).

Examples illustrating the reading of a RNA sequence from the cleavage patterns of the partial digests, particularly with regard to the RNase Phy_1 cleavage pattern, may be found in Gupta and Randerath (1977b); Randerath, E., et al. (1979); Gupta, Roe, and Randerath (1980); Randerath, K., Gupta, & Randerath (1980).

E. Comments

Gaps or compression effects are frequently observed on the gels. The appearance of gaps is due to regions of tight secondary structure that are resistant to partial enzymatic cleavage, even in 7 M urea at 50°C. In such cases, partial degradation of RNA by chemical rather than enzymatic means, followed by reading the sequence on gels or thin layers, is recommended. Band compression reflects a lack of resolution by size of neighboring fragments, due to the presence of regions of tight secondary structure in these fragments. Polyacrylamide gels containing high concentrations of formamide (~20 M) may conceivably enable the resolution of such fragments, as such gels have been shown to be superior to urea gels in resolving DNA fragments exhibiting tight secondary structure (Frank, Müller, & Wolff, 1981).

IX. ^{32}P THIN–LAYER READOUT METHOD
BASED ON SINGLE-HIT CHEMICAL CLEAVAGES OF RNA

A.   General Aspects

The ^{32}P thin-layer readout method (Gupta & Randerath, 1979) is based on random single-hit chemical cleavages of RNA (Stanley & Vassilenko, 1978). It allows one to identify and display directly the positions in the nucleic acid chain of modified nucleosides along with those of the four major nucleosides. Modified constituents amenable to this analysis include m^5C, m^3C, ac^4C, m^1A, m^2A, m^6A, i^6A, t^6A, mt^6A, ms^2i^6A, ψ, hU (D), m^5U (T), s^4U, V, mam^5s^2U, m^1G, m^2G, m_2^2G, m^7G, I, and Q (Gupta & Randerath, 1979; Randerath, E., et al., 1981; Randerath, E., Agrawal, & Randerath, 1981; K. Randerath, H. P. Agrawal, and E. Randerath, unpublished experiments). This method also permits the detection and chromatographic characterization of unknown modified constituents. Thus, using this method, we detected two unknown hypermodified nucleosides in rat liver and Morris hepatoma cytoplasmic tRNA$_{MmAA}^{Leu}$ (Randerath, E., et al., 1980), as well as a hypermodified uridine derivative, U*, in rat liver mitochondrial tRNA$_{U*UU}^{Lys}$ (Randerath, E., Agrawal, & Randerath, 1981; see also Fig. 5-5 below). A problem that is frequently encountered with the enzymatic gel readout method (Section VIII above), i.e., the resistance of highly base-paired regions of RNA to enzymatic cleavage, does not occur in this method. This is also the case for another problem of the gel readout method: the lack of distinction between C and U residues. Additional advantages of the thin-layer readout method are its high sensitivity, requiring as little as 0.5 μg of a tRNA for complete sequence analysis, and its speed when compared with other sequencing methods. In contrast to other sequencing methods, the thin-layer readout method, except for the determination of the 3' terminus, normally does not require any complementary procedures to derive a complete RNA sequence. It thus is the method of choice for sequencing low-molecular weight RNAs (chain length <150), in particular RNAs containing modified constituents. If only small amounts (0.5–3 μg) of such RNAs are available, this is the only method by which complete sequences can be determined. For example, the elucidation of the sequences of a number of mammalian mitochondrial tRNAs in our laboratory (Randerath, K., et al., 1981b; Agrawal, Randerath, & Randerath, 1981a; Agrawal et al., 1981b; Randerath, E., Agrawal, & Randerath, 1981; and our unpublished experiments) was made possible by this method.

B. Outline of Method

The method is in part based on the observation by Stanley and Vassilenko (1978) that under sufficiently mild conditions of chemical degradation, RNA cleavage products arise mostly from random single "hits," leading to the formation of two sets of fragments. The first set consists of chains extending from the phosphorylated 5'-terminus of the RNA to internal residues carrying 3'-terminal phosphate groups. The second set contains chains extending from internal positions carrying a free 5'-hydroxyl group to the 3'-terminus of the RNA. To generate the two sets of fragments, RNA is partially hydrolyzed by brief heating in water. The 5'-hydroxyl groups of the second set of fragments are then labeled with ^{32}P by the (γ-^{32}P)ATP/

polynucleotide kinase reaction. The fragments are separated by size on a denaturing polyacrylamide gel, giving rise to a "ladder" of radioactive bands, each differing from its neighbor by the addition or removal of a single 5′-terminal nucleotide. The fragments are then contact-transferred to a PEI-cellulose anion-exchange thin-layer sheet ("print" step). The fragments are digested *in situ* with RNase T$_2$ (Randerath, K., & Randerath, 1964b; Randerath, K., Randerath, et al., 1974; Sivarajan et al., 1974), and the 5′-^{32}P-labeled 5′-terminal nucleoside 3′,5′-bisphosphates thus released are resolved by thin-layer chromatography on PEI-cellulose. Two complementary systems (ammonium formate and ammonium sulfate) have been developed for this purpose so that two different readouts are obtained for the same sequence. After autoradiography, the nucleotide sequence can be read directly from the spot patterns on the X-ray films.

2′-*O*-Methylated nucleosides give rise to a gap in the "ladder" on the polyacrylamide gel and the appearance of dinucleotides [(^{32}P)pNmp$N$p] on the readouts. (^{32}P)pNm can be released enzymatically from such compounds and further analyzed by chromatography in several thin-layer systems (Randerath, E., et al., 1980). Similarly, unknown (^{32}P)pNp compounds can be converted enzymatically to (^{32}P)pN for further characterization by chromatography.

An example illustrating the thin-layer readout method is given in Fig. 5.5, exemplifying the sequence of rat-liver mitochondrial tRNA$_{\text{U*UU}}^{\text{Lys}}$ (Randerath, E., Agrawal, & Randerath, 1981). The sequence is read from left to right in the 5′ → 3′ direction. Note that several modified constituents are displayed, i.e., m^{1}A9, m^{2}G10, m^{2}G19, ψ20, U*27, and t^{6}A30.

C. Materials

Nuclease P$_1$ from Yamasa Shoyu Co. (Tokyo).

T$_4$ polynucleotide kinase from New England Nuclear (Boston, Massachusetts); preparations from other suppliers (e.g., P-L Biochemicals Inc. or Bethesda Research Laboratories) have also been used.

Macherey and Nagal PEI-cellulose sheets (40 × 20 cm and 20 × 20 cm) without indicator from Brinkmann Instruments (Westbury, New York). For purification, predevelop the 20 × 20 cm sheets in water (Randerath, K., & Randerath, 1967). After the front has reached the top of the sheets, move the cover of the chromatographic tank back 3–5 cm to allow water to evaporate from the upper portion of the sheets. After 8–15 hr, allow sheets to dry at

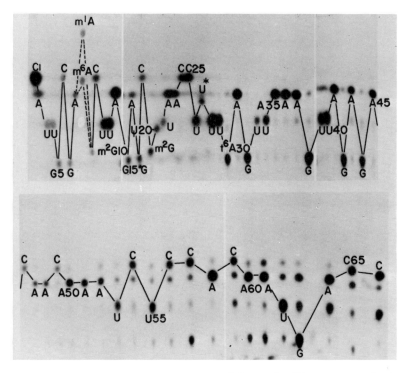

Figure 5-5.   PEI-cellulose thin-layer readout of the nucleotide sequence of rat-liver mitochondrial tRNA$^{Lys}_{U*UU}$. The solvent was 1.2 M ammonium formate, pH 3.5. The autoradiograms display $(5'-{}^{32}P)$nucleoside $3',5'$-bisphosphates derived from single-hit cleavages of the RNA. The presence of m^6A in position 10 is due to partial conversion of m^1A to m^6A during the experimental manipulations. Extra spots in the $3'$-terminal area are due to overdigestion of the RNA. For abbreviations of modified nucleotides, see footnote 1. Reproduced by permission from Randerath, E., et al., *Biochem. Biophys. Res. Comm.* 103:739–744 (1981).

room temperature for several hours before use. If not used the same day, store sheets at −20°C. Use the 40 × 20 cm sheets for the "print" step without predevelopment. For the preparation of PEI-cellulose sheets in the laboratory, see Section V.C; for additional materials, see Sections III.C, V.C, and VIII.C.

TLC scraper, 2 mm (No. 7600) from Alltech Associates (Deerfield, Illinois).

Button-type permanent Alnico magnets ($\frac{1}{2}$-inch diameter × $\frac{3}{8}$-inch) (General Hardware Manufacturing Co., New York, N.Y.) from local stationery stores. Prepare magnetic bars from the magnets as follows (Gupta & Randerath, 1977a): Align a series of magnets on the sticky side of a piece of 1-inch wide surgical tape so that opposite poles of adjacent magnets face each other. Bend excess tape upwards so that it will adhere to the sides of the row of magnets. Fasten one piece each of $\frac{1}{2}$-inch-wide double-coated tape to the top and the bottom of the assembly. Enclose the bar in a piece of 4-inch-wide Parafilm as follows: Attach the Parafilm to the double-coated tape on the bottom of the assembly (align edges of Parafilm and bar), and wrap Parafilm tightly around the assembly once. Fasten another piece of double-coated tape to the top of the assembly, and continue wrapping the Parafilm one more turn. Fold Parafilm protruding from the ends of the bar toward the bottom. Color code each magnetic bar appropriately for correct alignment of two bars, so that optimal contact between donor strip and acceptor sheet (see below) is achieved during chromatography. The magnetic bars can be reused many times before the Parafilm wrapping needs to be replaced.

D.   Procedure

The procedure described here is a slightly modified version of the original procedure (Gupta & Randerath, 1979; Randerath, K., Gupta, & Randerath, 1980).

Controlled Hydrolysis of RNA

Place a sealed siliconized glass tube, containing 0.5–2 μg of a low-molecular weight RNA (e.g., tRNA) in 10 μl of water, in a boiling water bath for 35–45 sec. Alternatively, hydrolyze sample at 80°C for 4–6 min. Dry the sample in a desiccator *in vacuo* over P_2O_5 and KOH.

^{32}P-Labeling of 5′ Ends

Dissolve the residue in 5 μl of 80 mM tris-HCl, pH 8.7, 15 mM $MgCl_2$, 10 mM dithiothreitol, 25 μM (γ-^{32}P)ATP (1,500–3,000 Ci/mmol). Add 2 μl of a solution of polynucleotide kinase containing 10 U and keep the solution at 37°C for 25 min. Add 3 μl of a solution of apyrase containing 7 mU and incubate the mixture at 37°C for 30 min. Dry sample as above.

Gel Electrophoresis

Dissolve the residue in 15 μl of 90 mM tris base, 90 mM boric acid, 4 mM EDTA, pH 8.3 (TBE buffer), add 5 μl of a solution of xylene cyanole containing 10 $\mu g/\mu l$, then saturate the solution with solid urea. Fractionate 4–5 μl aliquots of the sample on a 12% (for long and intermediate fragments) and a 20% (for short fragments) poly-acrylamide gel. Gels are $50 \times 30 \times 0.06$ cm, with sample wells of $1 \times 1 \times 0.06$ cm.

Prepare the 12% gel from about 140 ml of a solution containing, per 100 ml, 11.61 g of acrylamide, 0.39 g of methylene bisacryl-amide, 42 g of urea, 40 mg of ammonium persulfate, and 40 μl of N,N,N',N'-tetramethylethylenediamine (TEMED) dissolved in TBE buffer. Ingredients for the 20% polyacrylamide gel are 19.05 g of acrylamide, 0.95 g of methylene bisacrylamide, 48 g of urea, 40 mg of ammonium persulfate, and 40 μl of TEMED. Always add TEMED last, immediately before pouring the solution into the gel mold. Allow the gels to age for 5–6 hr.

Preelectrophorese gels overnight at 700 V (12% gel) and 1,000 V (20% gel) using TBE as the electrophoresis buffer. Load 4–5 μl of sample per well. On the 12% gel, load sample at 0, 2.5–3, and 5.5 hr and electrophorese at about 1,100–1,200 V. Adjust voltage so that a gel temperature of 44–46°C is maintained (surface-temperature thermometer). Terminate electrophoresis when the XC marker has reached 46, 35, and 24 cm in the three sample lanes, respectively (total electrophoresis time, about 10–11 hr). On the 20% gel, load sample at 0 and 3.5 hr. Electrophorese at about 1,500 V (44–46°C), until the XC marker has reached 27 and 17 cm, respectively (about 9.5 hr). For the migration rate of the XC marker, see also Section VIII.D, Gel Electrophoresis and Autoradiography.

Print Step

Soak a PEI-cellulose sheet (40×20 cm) in 500 ml of deionized water for 10 min. Remove one glass plate from the gel by gentle prying with a spatula at one corner of the assembly. Drain the PEI-cellulose sheet to be placed on the 12% gel, but do not drain the sheet for the 20% gel. Allow the sheet to make contact at the bottom of the gel first, then lower sheet gradually until it makes complete contact with the gel. Putting the PEI-cellulose sheet down this way minimizes trapping of air bubbles. Extrude any trapped air and water by gently stroking the plastic backing of the thin layer. Soak up expelled water with a paper towel. Cover the gel/PEI-cellulose assembly with Saran wrap to prevent evaporation. Place a piece of aluminum foil on top, followed by a glass plate and four glass thin-layer chromatography tanks (or similar weight).

After 15–20 hr, remove the PEI-cellulose sheets. Dry and then soak them in 500 ml of MeOH for 10 min with agitation to remove urea and buffer. Mark sheets with ^{99}Tc-ink and autoradiograph them in the presence of an intensifying screen. Expose at $-70°C$, $-20°C$, or room temperature for 15 min to a few hours, depending on the amount of radioactivity to be detected. See also Section II.E for screen-intensified autoradiography. If desired, an autoradiogram can also be made of the polyacrylamide gel before the print step; see Section II.E. Examples depicting polyacrylamide gel and PEI-cellulose print "ladders" can be found in Gupta and Randerath, (1979), Randerath, K., Gupta, & Randerath (1980), and Gupta, Roe, and Randerath (1979, 1980).

5'-Terminal Analysis and Reading of the Sequence

This involves the following manipulations: (i) excision of the "ladder"; (ii) treatment of the "ladder" or of individual fragments *in situ* with RNase T$_2$ and preparation for contact transfer; (iii) contact transfer and chromatography of RNase T$_2$ digestion products; (iv) autoradiographic detection of radioactive digestion products and reading of the sequence.

(i) Mark portions of the "ladders" to be analyzed with a blue marker (e.g., Sharpie) on the film. Align film and "print" on a light box and mark these portions on the plastic backing of the "print" sheet. If individual bands are to be treated with RNase T$_2$, indicate them on the film with blue marker either by circling widely spaced bands or by drawing a line through individual closely adjacent bands so as to make the bands more clearly visible for subsequent marking on the backing of the "print." Cut the sheet midway between individual ladders of interest, leaving 8–10-mm-wide nonradioactive margins parallel to each side of the "ladders." Then excise individual portions selected for analysis. To obtain two half-"ladders," each suitable for an individual analysis, cut the strips lengthwise.

(ii) Prepare a solution of RNase T$_2$ (0.5 U/μl) in 0.1 M sodium acetate, pH 4.5, by adding 1 M sodium acetate to an aqueous stock solution of the enzyme.

Place strip to be treated on a glass plate and apply 20-μl portions of the enzyme solution evenly by streaking with a disposable capillary micropipette close to the nonradioactive margin. One 20-μl portion is sufficient for 5–6 cm. Cover the treated strip immediately with Teflon tape and another glass plate, to retard evaporation. Place the next strip on this glass plate and proceed as above. If individual bands are to be treated, wet strip immediately before treatment with methanol and scrape off the layer between the bands with a sharp spatula or with a 2-mm TLC scraper, then spot a small volume ($\sim 0.5\,\mu$l) of the enzyme solution on each band. Then proceed as above. Tape and wrap the assembly in Saran wrap, then keep at $37°$C for several hours or overnight.

Soak up to 10 strips in 200 ml of methanol for 7 min with agitation and dry them. Trim strips lengthwise so that a 4-mm-wide strip of the "ladder" and a 6-mm-wide strip of the nonradioactive margin remain. Discard the outer portions of the original strip.

(iii) For contact transfer and chromatography of digestion products in ammonium sulfate, use a commercial PEI-cellulose sheet (20 x 20 cm), prewashed with water. Place the layer of the donor strip on the acceptor sheet in such a way that its nonradioactive portion is below the origin line (about 2 cm from the bottom). Mark the corners of the donor strip with pencil on the acceptor sheet. Sandwich the donor strip and the acceptor sheet between two strong magnetic bars. For the preparation of such bars, see Section C above.[3] Develop the chromatogram at $4°$C with water to the origin line, then with 0.55 M ammonium sulfate to the top or to 2–4 cm on a Whatman 1 wick attached to the top of the sheet. (The latter results in a better separation of some modified nucleotides from the major ones but may entail losses of some fast-moving compounds.) Remove the magnetic bar and donor strip, then dry the layer in a current of warm air.

For chromatography in the ammonium formate system, use PEI-cellulose sheets prepared in the laboratory (length, 20 cm; width, 20–25 cm; predeveloped with water in the direction of subsequent chromatography). For preequilibration, soak several ladder strips in 100 ml of 0.1 M ammonium formate, pH 3.5, for 7 min with agitation. Dry the strips with warm air and attach them to acceptor sheets as described above. Develop the chromatogram at room temperature with water to the origin line, then with 4 M lithium formate, 7 M urea, pH 3.5, to 5.5 cm from the bottom of the acceptor sheet. The nucleotides will migrate onto the acceptor sheet in this solvent. After removal of the magnets and the donor strip, soak the wet sheet in 200 ml of methanol for 7 min with agitation, dry, and then soak it

---

[3]Magnetic bars prepared as described above ensure the quantitative transfer of the labeled mono- and dinucleotides to the acceptor sheets without streaking. Elongated spots are obtained if the magnets are too weak to cause sufficient contact between donor layer and acceptor layer. For instance, magnetic bars assembled from standard bulletin board magnets or strips of magnetic rubber are not suitable for contact transfer.

in 200 ml of 0.1 M ammonium formate, pH 3.5, for 7 min, again with agitation. Allow the sheet to drain before drying it thoroughly and evenly (5 min) with warm air. Attach a Whatman 1 wick to the top of the sheet and develop it in 1.2 M ammonium formate, pH 3.5, to 2-3 cm on the wick under conditions of tank saturation. Dry sheet in a current of warm air and cut the wick off.

(iv) Before autoradiography, the dry donor strips may be reattached with tape to the chromatogram in their original position to show undigested radioactive material on the autoradiograms. This helps align the released terminal nucleotides with the original bands on the "ladder."

For screen-intensified autoradiography, see Section II.E. Expose films at $-20°C$ or $-70°C$ for up to several hours or overnight as required. If weakly and strongly labeled nucleotides are present, repeat exposure for a different length of time.

Read the nucleotide sequence from the autoradiograms. For an example of a readout in ammonium formate, see Fig. 5-5. Other examples of thin-layer readouts may be found in Gupta and Randerath (1979); Gupta, Roe, and Randerath (1979, 1980); Randerath, K., Gupta, and Randerath (1980); Randerath, K., Agrawal, and Randerath (1981b); Randerath, E., et al. (1980, 1981); Randerath, E., Agrawal, and Randerath (1981); Agrawal et al. (1981); and Arawal, Randerath, and Randerath (1981).

For relative R_F values of nucleoside $3',5'$-bisphosphates and ribose-methylated dinucleoside triphosphates, consult Fig. 5-6. Results obtained in the two solvent systems complement each other so that most modified tRNA constituents can be identified on the basis of their relative R_F values.

Further Characterization of Unknown Nucleoside $3',5'$-bisphosphates and of Ribose-methylated Dinucleoside Triphosphates

Approaches to the further characterization of nucleotides from thin-layer chromatograms have been described (Randerath, K., Gupta, & Randerath, 1980; Randerath, E., et al., 1980). Nucleotides may be subjected to chemical or enzymatic treatments after elution or *in situ.* The reaction products are characterized chromatographically.

For the elution of nucleotides, use several replicates of ammonium formate chromatograms. Cut out replicate spots to be eluted, then soak cutouts in 50-100 ml of methanol for 10 min with agitation. Apply a mixture of the four major nucleoside $3',5'$-bisphosphates (5-10 nmol each) to the spots to serve as carriers. Elute

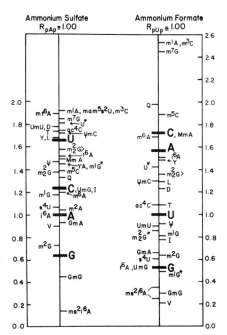

Figure 5-6. Diagram of relative R_F values of nucleoside $3',5'$-bisphosphates and ribose-methylated dinucleoside triphosphates in ammonium sulfate at $4°C$ ($R_{pAp} = 1.00$) and in ammonium formate, pH 3.5, at $23°C$ ($R_{pUp} = 1.00$). For abbreviations, see footnote 1. Updated (Randerath, E., et al., 1980, 1981; Randerath, Agrawal, & Randerath, 1981) version of original diagram of Gupta and Randerath (1979). Relative R_F values of m^2A, mt^6A, and mam^5s^2U in the ammonium sulfate system were kindly provided by Drs. E. J. Murgola and N. Prather of M. D. Anderson Hospital and Tumor Institute, Houston, Texas.

compounds with 2 M LiCl and remove salt as described (Randerath, E., et al., 1979; Randerath, K., Gupta, & Randerath, 1980).

To investigate the chemical stability of an eluted compound, divide the pooled eluates into several aliquots, each containing at least 500-700 dpm, and treat these, for example, with acid (HCl), alkali (piperidine), oxidizing and/or reducing agents, etc. Remove HCl or piperidine in a stream of air before subsequent chromatography. After addition of pNp markers (if desired), chromatograph treated sample(s) alongside untreated compound in the ammonium formate and ammonium sulfate systems described above. Prepare an autoradiogram to see whether treatment has resulted in alteration of mobility of original compound.

To remove $3'$-phosphomonoester groups from (^{32}P)pNp and (^{32}P)pNmpNp, respectively, treat an aliquot of the eluted compound with nuclease P_1: Dry the aliquot (1,500–1,700 dpm) in a stream of air, and take residue up in 20 μl of a solution containing 0.1 mM $ZnCl_2$, 20 mM sodium acetate, pH 5.0, and 0.1 μg/μl nuclease P_1. Keep at $50°C$ for $2\frac{1}{2}$ hr. Add 5 μl of a solution of the four major ribonucleoside $5'$-monophosphates (about 30 nmol of each) to a 5-μl aliquot of the reaction mixture, and chromatograph the sample by two-dimensional PEI-cellulose thin-layer chromatography

(Gupta, Randerath, & Randerath, 1976b). Other systems are also suitable, e.g., 0.4-0.6 M LiCl or 0.6-1 M ammonium formate, pH 3.5.

Alternatively, (^{32}P)pNp spots may be treated *in situ* with nuclease P_1 and contact-transferred to PEI-cellulose for chromatography (Randerath, K., Gupta, & Randerath, 1980). Conversion of (^{32}P)pNp to (^{32}P)pN under these conditions may not be quantitative, while complete 3'-dephosphorylation is usually obtained in solution.

(^{32}P)pNmp$N$p may not always be degraded to (^{32}P)pNm with nuclease P_1, but may yield (^{32}P)pNmp$N$ instead. This was observed with (^{32}P)pMmpAp derived from mammalian cytoplasmic tRNA$_{MmAA}^{Leu}$, where Mm is a hypermodified 2'-*O*-methylated nucleoside, probably a cytidine derivative (Randerath, E., et al., 1980). In this case, (^{32}P)pNmp$N$ may be converted to (^{32}P)pNm by snake venom phosphodiesterase treatment. The nucleoside 3'-adjacent to the 2'-*O*-methylated nucleoside may be identified by gel readout sequencing.

E.   Comments

The method described in this section allows the unambiguous identification and positioning of major and modified nucleotides in the ribopolynucleotide chain. *In situ* digestion of the "ladder" fragments constitutes an essential element of this sequencing strategy because it eliminates experimental errors that may arise if the fragments are extracted from excised gel bands before digestion, a procedure that may result in mistakes due to incorrect cutting of gel bands or the handling of a large number of samples.

The following points should be helpful in practical applications of the method:

(i) The RNA must be at least 90% pure since contaminating RNAs may give rise to additional nucleotide spots, which may make reading of the actual sequence difficult or impossible. The presence of strong background spots in the 5'-terminal portion of the readout indicates contamination by other RNA(s), while their presence in the 3'-terminal portion may have other reasons [see (ii) and (iii) below].

(ii) The RNA should be intact. Degraded RNA gives rise to additional nucleotide spots on the readouts, particularly in the area of the shorter fragments. Electrophoresis of the RNA preparation on a denaturing gel at 40-45°C is therefore recommended as a final purification step.

(iii) Overdigestion of the RNA should be avoided since secondary cleavages again will give rise to additional bands on the gels and additional nucleotide spots on the readouts, particularly in the area of the shorter fragments. Also, overdigestion may result in loss of 5'-terminal fragments. Under the conditions specified above, tRNA normally will not be overdigested. However, some variation of background on readouts of shorter fragments has been noted for various tRNAs digested under identical conditions (unpublished experiments). This indicates different susceptibilities of the RNAs to secondary cleavages due to individual structural features. Although, in general, the 3'-terminal sequence can be read without difficulties, milder digestion conditions may be required in special cases. Hydrolysis in water offers advantages over formamide digestion (Stanley & Vassilenko, 1978), because the sample is ready for the labeling reaction without additional steps, whereas formamide interferes with the polynucleotide kinase reaction and thus needs to be removed by ethanol precipitation of the RNA fragments or by freeze-drying.

(iv) Doublets of closely spaced bands are occasionally seen on the "ladders," particularly for shorter fragments on 20% gels. The same 5' termini are usually released from such doublets, which therefore may be due to conformational isomers.

(v) Treatment of the entire "ladder" with RNase T_2 ensures that every position of the sequence is represented on the final thin-layer readout. This technique, however, may lead to the release of terminal nucleotides from contaminating fragments migrating between the actual fragments on the gels. For the widely spaced 3'-terminal fragments on 20% gels, RNase T_2 digestion of individual bands of the "ladder" therefore is the method of choice, while bands of larger fragments may be too close for individual application of the enzyme. We recommend, however, that both techniques should be employed whenever possible; this will ensure a complete readout and the recognition of contaminating fractions. In addition, we suggest that the thin-layer readout should be aligned with the original "ladder" so as to obtain evidence that a particular spot on the former originated from the corresponding band on the gel.

(vi) Occasional gel band compression is due to tight secondary structure caused by several adjacent G·C base pairs, thus making the thin-layer readout hard to read in such an area. This problem can possibly be solved by separating the fragments on a denaturing polyacrylamide gel containing a high concentration of formamide (20 M) rather than urea (Frank, Müller, & Wolff, 1981).

(vii) Although polynucleotide kinase does not label all termini to the same extent, this has not been a problem in our experience.

Thus, longer film exposures will be needed for certain weakly labeled constituents. Dimethylsulfoxide may be added to the polynucleotide kinase reaction mixture to obtain more uniform labeling (Tanaka, Dyer, & Brownlee, 1980).

(viii) Fresh (γ-^{32}p)ATP is recommended for the labeling reaction. Aged (γ-^{32}P)ATP may contain radioactive contaminants, giving rise to background spots particularly in the 3'-terminal portion of the readout.

(ix) Apyrase treatment after the labeling reaction helps reduce radioactive background contributed by (γ-^{32}P)ATP, resulting in a clearer readout of the 3'-terminal portion of the RNA. At the dilution used, the commercial apyrase preparation was found not to contain interfering phosphatase and nuclease activities.

(x) To avoid streaking of the labeled digest during gel electrophoresis, it is important to saturate the sample with solid urea before application to the gel. Obviously, a clear readout can be obtained only if the fragments form sharp bands on the gel.

(xi) The transfer ("print") step involves the diffusion of the labeled polynucleotide fragments from a polyacrylamide gel into a PEI-cellulose thin layer, where the fragments bind to the positively charged anion-exchanger. The rate of diffusion depends on the amount of water present in the system. If too little water is present, diffusion will not proceed at a sufficient rate. On the other hand, if the space between gel and thin layer contains an excess of water, lateral diffusion will take place, resulting in irregularly diffuse bands on the "print." Some experience is needed to obtain optimal transfer. Whereas fragments up to 25–30 nucleotides long are being transferred almost quantitatively, the transfer of large fragments is less efficient, but sufficient is being transferred for terminal analysis of fragments up to 120 nucleotides long. (Larger fragments have not been investigated thus far.)

(xii) For chromatography in the ammonium formate, pH 3.5, system, we found it advantageous to use PEI-cellulose sheets prepared in the laboratory, rather than commercial sheets, because they give sharper separations at pH 3.5. It is also important to preequilibrate sheets evenly with formate and to saturate the tank atmosphere before beginning chromatography. This way the separation of pCp and pAp is improved.

(xiii) To deduce an unambiguous sequence, the two complementary thin-layer chromatographic systems described above should be used. Occasionally, it may be desirable to vary the gel electrophoretic conditions in terms of acrylamide/bisacrylamide content (8–22%) and temperature (22–50°C). Running the gels at even higher temperatures rarely improves the results.

(xiv) If two nucleotides are found for one position and artifacts due to points (i) to (iii) have been excluded, several other explanations may be considered: Some labeled termini are released, in addition to $(^{32}P)pNp$, as a cyclic nucleotide $(^{32}P)pN{>}p$, which migrates faster than the noncyclic compound. This has been found for m_2^2G (Gupta & Randerath, 1979) and may also occur with certain hypermodified constituents (e.g., ms^2i^6A). Similarly, the bond between such nucleotides and their 3'-adjacent neighbor may be incompletely cleaved, which leads to the appearance of a dinucleotide spot in addition to the regular mononucleotide spot. This was found to be the case for 5'-terminal $pm_2^2Gp\psi p$, for example. Some nucleotides are not completely stable under the experimental conditions and thus may give rise to additional spots, e.g., in addition to pm^1Ap, readouts also have pm^6Ap. The presence of two nucleotides, particularly of a modified constituent and its parent compound, may indicate a partial modification. Finally, two major constituents may be found for one position if the RNA analyzed is a mixture of two species differing in this position (Randerath, E., et al., 1980).

X. SEQUENCE ANALYSIS OF LARGE RNAs

Large RNAs are usually subjected to partial enzymatic digestions [e.g., see Brownlee (1972) for standard conditions], thereby exploiting the protection of certain sites in the RNA by secondary/tertiary structure against enzymatic cleavage.

Partial fragments may be separated by polyacrylamide gel electrophoresis (De Wachter & Fiers, 1972). If the fragments are terminally labeled with ^{32}P prior to electrophoretic separation, they are located by autoradiography. Otherwise, they are stained with methylene blue (Peacock & Dingman, 1967). A few μg of a fragment is needed for the detection by staining as well as for subsequent sequence analysis of fragments of fewer than 150 nucleotides. Fragments are extracted from the gel by homogenization in the presence of phenol (Chia, Randerath, & Randerath, 1973; Randerath, E., et al., 1980) and precipitated with acetonitrile/ethanol (4:1, v/v) (Chia, Randerath, & Randerath, 1973).

If the RNA is 5'- or 3'-terminally labeled with ^{32}P, the terminal fragment can be sequenced by the gel readout procedure. This also applies to fragments that have been end-labeled prior to separation. For sequencing of 5'-terminally ^{32}P-labeled fragments by the thin-layer readout procedure, 5'-terminal $(^{32}P)P_i$ has to be replaced first by nonradioactive phosphate. Fragments for sequencing can also be obtained by a recently introduced method (Donis-Keller, 1979) which allows one to cleave a large RNA at selected sites by using specific deoxyribooligonucleotides for hybridization and RNase H for cleavage of the RNA in the hybridized portions.

mRNAs containing a stretch of A residues at their $3'$ end can be $3'$-end sequenced by the primer extension method, similar to the way in which DNA is sequenced by this method (see Godson, 1980). In general, determination of RNA sequences by primer-directed synthesis and sequencing of their cDNA transcripts offers advantages over the more direct methods for the sequence analysis of large RNAs (see Ghosh et al., 1980).

XI. CONCLUSIONS

Historically, composition and sequence analysis of nucleic acids evolved in three stages, entailing first spectrophotometric techniques, then *in vivo* radioactive labeling, and finally *in vitro* radioactive labeling ("postlabeling") techniques. This development has resulted in a tremendous increase in sensitivity, speed, and scope of analysis which, during the past 5 years, has already generated an overwhelming amount of information on the structure of RNA and DNA in general, leading to an enhanced knowledge of modification patterns of nucleic acids, the effects of various agents on such modifications, the fine structure of genes and their regulatory sequences, and the overall organization of the genetic material in a large number of organisms, from viruses to mammals. In addition, postlabeling methods are now being refined to detect the binding of minute amounts of chemicals, such as mutagens and carcinogens, to RNA and DNA, an application of potentially great significance for the identification of hazardous chemicals in the human environment. It is of interest that methodology in this area has also evolved in the three stages referred to above, with the use of radioactive mutagens and carcinogens currently still being the method of choice in nucleic acid binding studies. We anticipate that this procedure will soon be superseded by postlabeling methods (see Section V above).

The methods developed in the past decade for structural analysis of nucleic acids, while still needing further refinement, will continue to be indispensable tools for the foreseeable future in our quest for a better understanding of evolution, structure, replication, and function (under normal and abnormal conditions) of the genetic material of unicellular and multicellular organisms.

ACKNOWLEDGMENTS

This work was supported by USPHS Grants CA 13591, CA 10893-P8, CA 25590, and CA 32157 and American Cancer Society Grants NP-37 and NP-315. Support by a USPHS Research Career Development Award and an American Cancer Society Faculty Research Award is also gratefully acknowledged. We thank H. P. Agrawal for preparing part of the figures and helping with the preparation of the manuscript, Dr. C. C. Levy for supplying us with a sample of chicken-liver ribonuclease, and Drs. E. J. Murgola and N. Prather for providing some R_F values given in Fig. 5-6.

REFERENCES

Agrawal, H. P., Randerath, K., and Randerath, E. (1981). Tumor mitochondrial transfer ribonucleic acids: The nucleotide sequence of Morris hepatoma 5123D mitochrondial tRNA$^{Asp}_{GUC}$. *Nucl. Acids Res.* 9:2535-2541.

Agrawal, H. P., Gupta, R. C., Randerath, K., and Randerath, E. (1981). The sequence of mitochondrial arginine tRNA (anticodon UCG) from a transplantable rat tumor, Morris hepatoma 5123D. *FEBS Lett.* 130:287-290.

Ames, B. N. (1979). Identifying environmental chemicals causing mutations and cancer. *Science* 204:587-593.

Benton, W. D., and Davis, R. W. (1977). Screening λgt recombinant clones by hybridization to single plaques *in situ. Science* 196:180-182.

Boguski, M. S., Hieter, P. A., and Levy, C. C. (1980). Identification of a cytidine-specific ribonuclease from chicken liver. *J. Biol. Chem.* 255:2160-2163.

Bonner, W. M., and Laskey, R. A. (1974). A film detection method for tritium labeled proteins and nucleic acids in polyacrylamide gels. *Eur. J. Biochem.* 46:83-88.

Braun, R., and Behrens, K. (1969). A ribonuclease from *Physarum.* Biochemical properties and synthesis in the mitotic cycle. *Biochim. Biophys. Acta* 195:87-98.

Brownlee, G. G. (1972). Determination of sequences in RNA. In *Laboratory Techniques in Biochemistry and Molecular Biology*, eds. T. S. Work and E. Work (New York: Elsevier).

Bruce, A. G., and Uhlenbeck, O. C. (1978). Reactions at the termini of tRNA with T_4 RNA ligase. *Nucl. Acids Res.* 5:3665–3677.

Chia, L. S. Y., Randerath, K., and Randerath, E. (1973). Base analysis of ribo-polynucleotides by tritium incorporation following analytical polyacrylamide gel electrophoresis. *Anal. Biochem.* 55:102–113.

Chia, L. S. Y., Morris, H. P., Randerath, K., and Randerath, E. (1976). Base composition studies on mitochondrial 4S RNA from rat liver and Morris hepatomas 5123D and 7777. *Biochim. Biophys. Acta* 425:49–62.

Davies, P. L., van de Sande, J. H., and Dixon, G. H. (1979). Base compositional analysis of nanogram quantities of unlabeled nucleic acids. *Anal. Biochem.* 93:26–30.

De Wachter, R., and Fiers, W. (1972). Preparative two-dimensional polyacrylamide gel electrophoresis of ^{32}P-labeled RNA. *Anal. Biochem.* 49:184–197.

Donis-Keller, H. (1980). Phy M: An RNase activity specific for U and A residues useful in RNA sequence analysis. *Nucl. Acids Res.* 8:3133–3142.

Donis-Keller, H. (1979). Site specific enzymatic cleavage of RNA. *Nucl. Acids Res.* 7:179–192.

Donis-Keller, H., Maxam, A., and Gilbert, W. (1977). Mapping adenines, guanines, and pyrimidines in RNA. *Nucl. Acids Res.* 4:2527–2538.

England, T. E., and Uhlenbeck, O. C. (1978). 3′-Terminal labeling of RNA with T_4 RNA ligase. *Nature* 275:560–561.

England, T. E., Bruce, A. G., and Uhlenbeck, O. C. (1980). Specific labeling of 3′ termini of RNA with T_4 RNA ligase. *Methods Enzymol.* 65:65–74.

Frank, R., Müller, D., and Wolff, C. (1981). Identification and suppression of secondary structures formed from deoxyoligonucleotides during electro-phoresis in denaturing polyacrylamide gels. *Nucl. Acids Res.* 9:4967–4979.

Ghosh, P. K., Reddy, V. B., Piatak, M., Lebowitz, P., and Weissman, S. M. (1980). Determination of RNA sequences by primer directed synthesis and sequencing of their cDNA transcripts. *Methods Enzymol.* 65:580–595.

Gillum, A. M., Urquhart, N., Smith, M., and RajBhandary, U. L. (1975). Nucleo-tide sequence of salmon testes and salmon liver cytoplasmic initiator tRNA. *Cell* 6:395–405.

Godson, G. N. (1980). Primed synthesis methods of sequencing DNA and RNA. *Fed. Proc.* 39:2822–2829.

Gupta, R. C., and Randerath, K. (1979). Rapid print-readout technique for sequencing of RNAs containing modified nucleotides. *Nucl. Acids Res.* 6:3443-3458.

Gupta, R. C., and Randerath, K. (1977a). Use of specific endonuclease cleavage in RNA sequencing. *Nucl. Acids Res.* 4:1957-1978.

Gupta, R. C., and Randerath, K. (1977b). Use of specific endonuclease cleavage in RNA sequencing—An enzymatic method for distinguishing between cytidine and uridine residues. *Nucl. Acids Res.* 4:3441-3454.

Gupta, R. C., Randerath, E., and Randerath, K. (1976a). A double-labeling procedure for sequence analysis of picomole amounts of nonradioactive RNA fragments. *Nucl. Acids Res.* 3:2895-2914.

Gupta, R. C., Randerath, E., and Randerath, K. (1976b). An improved separation procedure for nucleoside monophosphates on polyethyleneimine-(PEI-)cellulose thin layers. *Nucl. Acids Res.* 3:2915-2921.

Gupta, R. C., Randerath, K., and Randerath, E. (1976c). Sequence analysis of small amounts of nonradioactive oligoribonucleotides containing ribose-methylated nucleosides by a combination of [3]H- and ^{32}P-labeling techniques. *Anal. Biochem.* 76:269-280.

Gupta, R. C., Reddy, M. V., and Randerath, K. (1982). ^{32}P-Postlabeling analysis of non-radioactive aromatic carcinogen-DNA adducts. *Carcinogenesis* 3:1081-1092.

Gupta, R. C., Roe, B. A., and Randerath, K. (1980). Sequence of human glycine transfer ribonucleic acid (anticodon CCC). Determination by a newly developed thin-layer readout sequencing technique and comparison with other glycine transfer ribonucleic acids. *Biochem.* 19:1699-1705.

Gupta, R. C., Roe, B. A., and Randerath, K. (1979). The nucleotide sequence of human tRNA[Gly] (anticodon GCC). *Nucl. Acids Res.* 7:959-970.

Hollstein, M., McCann, J., Angelosanto, F. A., and Nichols, W. M. (1979). Short-term tests for carcinogens and mutagens. *Mutat. Res.* 65:133-226.

Khorana, H. G. (1959). Studies on polynucleotides. VII. Approaches to the marking of end groups in polynucleotide chains: The methylation of phosphomonoester groups. *J. Amer. Chem. Soc.* 81:4657-4660.

Krupp, G., and Gross, H. J. (1979). Rapid RNA sequencing: Nucleases from *Staphylococcus aureus* and *Neurospora crassa* discriminate between uridine and cytidine. *Nucl. Acids Res.* 6:3481-3490.

Laskey, R. A. (1980). The use of intensifying screens or organic scintillators for visualizing radioactive molecules resolved by gel electrophoresis. *Methods Enzymol.* 65:363–371.

Laskey, R. A., and Mills, A. D. (1977). Enhanced autoradiographic detection of ^{32}P and ^{125}I using intensifying screens and hypersensitized film. *FEBS Lett.* 82:314–316.

Laskey, R. A., and Mills, A. D. (1975). Quantitative film detection of ^{3}H and ^{14}C in polyacrylamide gels by fluorography. *Eur. J. Biochem.* 56:335–341.

Levy, C. C., and Karpetsky, T. P. (1980). The purification and properties of chicken liver RNase. *J. Biol. Chem.* 255:2153–2159.

Lockard, R. E., Alzner-DeWeerd, B., Heckman, J. E., Tabor, M. W., MacGee, J., and RajBhandary, U. L. (1978). Sequence analysis of 5′(^{32}P)-labeled mRNA and tRNA using polyacrylamide gel electrophoresis. *Nucl. Acids Res.* 5:37–56.

Lu, L.-J. W., and Randerath, K. (1979). Effects of 5-azacytidine on transfer RNA methyltransferases. *Cancer Res.* 39:940–948.

Lu, L.-J. W., Tseng, W.-C., and Randerath, K. (1979). Effects of 5-fluorocytidine on mammalian transfer RNA and transfer RNA methyltransferases. *Biochem. Pharmacol.* 28:489–495.

Lu, L.-J, W., Chiang, G. H., Medina, D., and Randerath, K. (1976a). Drug effects on nucleic acid modification. I. A specific effect of 5-azacytidine on mammalian transfer RNA methylation *in vivo*. *Biochem. Biophsy. Res. Comm.* 68:1094–1101.

Lu, L.-J. W., Chiang, G. H., Tseng, W.-C., and Randerath, K. (1976b). Effects of 5-fluorouridine on modified nucleosides in mouse liver transfer RNA. *Biochem. Biophys. Res. Comm.* 73:1075–1082.

Maxam, A. M., and Gilbert, W. (1977). A new method for sequencing DNA. *Proc. Natl. Acad. Sci. USA* 74:560–564.

Miller, E. C. (1978). Some current perspectives on chemical carcinogenesis in humans and experimental animals: Presidential address. *Cander Res.* 38:1479–1496.

Miller, J. A. (1970). Carcinogenesis by chemicals: An overview—G. H. A. Clowes Memorial Lecture. *Cancer Res.* 30:559–576.

Murao, K., Ishikura, H., Albani, M., and Kersten, H. (1978). On the biosynthesis of 5-methoxyuridine and uridine-5-oxyacetic acid in specific prokaryotic transfer RNAs. *Nucl. Acids Res.* 5:1273–1287.

Nishimura, S. (1972). Minor components in transfer RNA: Their characterization, location, and function. *Prog. Nucl. Acid Res. Mol. Biol.* 12:49–85.

Peacock, A. C., and Dingman, C. W. (1967). Resolution of multiple ribonucleic acid species by polyacrylamide gel electrophoresis. *Biochem.* 6:1818–1827.

Peattie, D. A. (1979). Direct chemical method for sequencing RNA. *Proc. Natl. Acad. Sci. USA* 76:1760–1764.

Pilly, D., Niemeyer, A., Schmidt, M., and Bargetzi, J.-P. (1978). Enzymes for RNA sequence analysis. Preparation and specificity of exoplasmodial ribonucleases I and II from *Physarum polycephalum. J. Biol. Chem.* 253:437–445.

Prensky, W. (1976). The radioiodination of RNA and DNA to high specific activities. *Methods Cell Biol.* 13:121–152.

RajBhandary, U. L. (1980). Recent developments in methods for RNA sequencing using *in vitro* ^{32}P-labeling. *Fed. Proc.* 39:2815–2821.

RajBhandary, U. L., Young, R. J., and Khorana, H. G. (1964). Studies on polynucleotides. XXXII. The labeling of end groups in polynucleotide chains: The selective phosphorylation of phosphomonoester groups in amino acid acceptor ribonucleic acids. *J. Biol. Chem.* 239:3875–3884.

Ralph, R. K., Young, R. J., and Khorana, H. G. (1963). Studies on polynucleotides. XXI. Amino acid acceptor ribonucleic acids (2). The labeling of terminal 5'-phosphomonoester groups and a preliminary investigation of adjoining nucleotide sequences. *J. Amer. Chem. Soc.* 85:2002–2012.

Randerath, E., Agrawal, H. P., and Randerath, K. (1981). Rat liver mitochondrial lysine tRNA (anticodon U*UU) contains a rudimentary D-arm and 2 hypermodified nucleotides in its anticodon loop. *Biochem. Biophys. Res. Comm.* 103:739–744.

Randerath, E., Yu, C.-T., and Randerath, K. (1972). Base analysis of ribopolynucleotides by chemical tritium labeling: A methodological study with model nucleosides and purified tRNA species. *Anal. Biochem.* 48:172–198.

Randerath, E., Chia, L. S. Y., Morris, H. P., and Randerath, K. (1974). Transfer RNA base composition studies in Morris hepatomas and rat liver. *Cancer Res.* 34:643–653.

Randerath, E., Gopalakrishnan, A. S., Gupta, R. C., Agrawal, H. P., and Randerath, K. (1981). Lack of a specific ribose methylation at guanosine 17 in Morris hepatoma 5123D tRNA$_{IGA}^{Ser}$. *Cancer Res.* 41:2863-2867.

Randerath, E., Gupta, R. C., Chia, L. S. Y., Chang, S. H., and Randerath, K. (1979). Yeast tRNA$_{UAG}^{Leu}$. Purification, properties, and determination of the nucleotide sequence by radioactive derivative methods. *Eur. J. Biochem.* 93:79-94.

Randerath, E., Gupta, R. C., Morris, H. P., and Randerath, K. (1980). Isolation and sequence analysis of two major leucine transfer ribonucleic acids (anticodon Mn-A-A) from a rat tumor, Morris hepatoma 5123D. *Biochemistry* 19:3476-3483.

Randerath, K. (1981). 3-([3-^{125}I]Iodo-4-hydroxyphenyl)propionyl carbohydrazide, a new radioiodination reagent for ultrasensitive detection and determination of periodate-oxidized nucleoside derivatives and other carbonyl compounds. *Anal. Biochem.* 115:391-397.

Randerath, K. (1979). Drug effects on tRNA. In *Effects of Drugs on the Cell Nucleus*, eds. H. Busch, S. T. Crooke, and Y. Daskal (New York: Academic Press), pp. 275-299.

Randerath, K. (1973). Continuous directional degradation—A novel method for sequence analysis of polyribonucleotides. *FEBS Lett.* 33:143-146.

Randerath, K. (1971). Application of a tritium derivative method to human brain and brain tumor transfer RNA analysis. *Cancer Res.* 31:658-661.

Randerath, K. (1970). An evaluation of film detection methods for weak β-emitters, particularly tritium. *Anal. Biochem.* 34:188-205.

Randerath, K. (1969). An improved procedure for solid scintillation fluorography of tritium labeled compounds. *Anal. Chem.* 41:991-992.

Randerath, K. (1962a). A simple method for preparing cellulose anion-exchanger powders and papers. *Angew. Chem. Int. Ed.* 1:553-554.

Randerath, K. (1962b). Polyäthylenimin-Cellulose—ein neuer Anionenaustauscher für die Chromatographie von Nucleotiden. *Biochim. Bjophys. Acta* 61:852-854.

Randerath, K., and Randerath, E. (1973). Chemical characterization of unlabeled RNA and RNA derivatives by isotope derivative methods. *Methods Cancer Res.* 9:3-69.

Randerath, K., and Randerath, E. (1971). A tritium derivative method for base analysis of ribonucleotides and RNA. *Proc. Nucl. Acid Res.* 2:796-812.

Randerath, K., and Randerath, E. (1969). Analysis of nucleic acid derivatives at the subnanomole level. III. A tritium labeling procedure for quantitative analysis of ribose derivatives. *Anal. Biochem.* 28:110–118.

Randerath, K., and Randerath, E. (1968). Analysis of nucleic acid derivatives at the subnanomole level. II. Quantitative analysis of ribose derivatives by tritium labeling. *Experimentia* 24:1192–1193.

Randerath, K., and Randerath, E. (1967). Thin-layer separation methods for nucleic acid derivatives. *Methods Enzymol.* 12, Part A:323–347.

Randerath, K., and Randerath, E. (1966). Ion-exchange thin-layer chromatography. XV. Preparation, properties, and applications of paper-like PEI-cellulose sheets. *J. Chromatog.* 22:110–117.

Randerath, K., and Randerath, E. (1964a). Ion-exchange chromatography of nucleotides on poly(ethyleneimine)-cellulose thin layers. *J. Chromatog.* 16:111–125.

Randerath, K., and Randerath, E. (1964b). Enzymatic reactions on ion-exchange thin-layer plates. *Angew. Chem., Int. Ed.* 3:442–443.

Randerath, K., Agrawal, H. P., and Randerath, E. (1981). Tumor mitochondrial transfer RNAs: The nucleotide sequence of mitochondrial $tRNA_{UAG}^{Leu}$ from Morris hepatoma 5123D. *Biochem. Biophys. Res. Comm.* 100:732–737.

Randerath, K., Gupta, R. C., and Randerath, E. (1980). ^{3}H and ^{32}P derivative methods for base composition and sequence analysis of RNA. *Methods Enzymol.* 65:638–680.

Randerath, K., MacKinnon, S. K., and Randerath, E. (1971). An investigation of the minor base composition of transfer RNA in normal human brain and malignant brain tumors. *FEBS Lett.* 15:81–84.

Randerath, K., Reddy, M. V., and Gupta, R. C. 1981). ^{32}P-Labeling test for DNA damage. *Proc. Natl. Acad. Sci. USA* 78:6126–6129.

Randerath, K., Chia, L. S. Y., Gupta, R. C., Randerath, E., Hawkins, E. R., Brum, C. K., and Chang, S. H. (1975). Structural analysis of nonradioactive RNA by postlabeling: The primary structure of baker's yeast $tRNA_{CUA}^{Leu}$. *Biochem. Biophys. Res. Comm.* 63:157–163.

Randerath, K., Chia, L. S. Y., Randerath, E., and Gupta, R. C. (1974). Tritium sequence analysis of polyribonucleotides following periodate-phosphomonoesterase degradation—Analysis of nucleoside methylene dialdehydes derived from tetra-, penta-, and hexanucleotides. *FEBS Lett.* 40:183–186.

Randerath, K., Randerath, E., Chia, L. S. Y., Gupta, R. C., and Sivarajan, M. (1974). Sequence analysis of nonradioactive RNA fragments by periodate-phosphatase digestion and chemical tritium labeling: Characterization of large oligonucleotides and oligonucleotides containing modified nucleosides. *Nucl. Acids Res.* 1:1121-1141.

Reddy, M. V., Gupta, R. C., and Randerath, K. (1981). ^{32}P-Base analysis of DNA. *Anal. Biochem.* 117:271-279.

Richardson, C. C. (1971). Polynucleotide kinase from *Escherichia coli* infected with bacteriophage T_4. *Proc. Nucl. Acid Res.* 2:815-828.

Roe, B. A., Anandaraj, M. P. J. S., Chia, L. S. Y., Randerath, E., Gupta, R. C., and Randerath, K. (1975). Sequence studies on tRNA^{Phe} from human placenta: Comparison with known sequences of tRNA^{Phe} from other normal mammalian tissues. *Biochem. Biophys. Res. Comm.* 66:1097-1105.

Silberklang, M., Gillum, A. M., and RajBhandary, U. L. (1979). Use of *in vitro* ^{32}P labeling in the sequence analysis of nonradioactive tRNAs. *Methods Enzymol.* 59:58-109.

Silberklang, M., Prochiantz, A., Haenni, A. L., and RajBhandary, U. L. (1977). Studies on the sequence of the 3'-terminal region of turnip-yellow-mosaic-virus RNA. *Eur. J. Biochem.* 72:465-478.

Simoncsits, A., Brownlee, G. G., Brown, R. S., Rubin, J. R., and Guilley, H. (1977). New rapid gel sequencing method for RNA. *Nature* 269:833-836.

Simsek, M., Ziegenmeyer, J., Heckman, J., and RajBhandary, U. L. (1973). Absence of the sequence G—T—Ψ—C—G(A)— in several eukaryotic cytoplasmic initiator transfer RNAs. *Proc. Natl. Acad. Sci. USA* 70:1041-1045.

Sivarajan, M., Gupta, R. C., Chia, L. S. Y., Randerath, E., and Randerath, K. (1974). Tritium sequence analysis of oligoribonucleotides: A combination of postlabeling and thin-layer chromatographic techniques for the analysis of partial snake venom phosphodiesterase digests. *Nucl. Acids Res.* 1:1329-1341.

Stanley, J., and Vassilenko, S. (1978). A different approach to RNA sequencing. *Nature* 274:87-89.

Swanstrom, R., and Shank, P. R. (1978). X-Ray intensifying screens greatly enhance the detection by autoradiography of the radioactive isotopes ^{32}P and ^{125}I. *Anal. Biochem.* 86:184-192.

Székely, M. (1971). Fingerprinting nonradioactive nucleic acids with the aid of polynucleotide kinase. *Proc. Nucl. Acid Res.* 2:780-793.

Székely, M., and Sanger, F. (1969). Use of polynucleotide kinase in fingerprinting nonradioactive nucleic acids. *J. Mol. Biol.* 43:607–617.

Tanaka, Y., Dyer, T. A., and Brownlee, G. G. (1980). An improved direct RNA sequence method: Its application to *Vicia faba* 5.8S ribosomal RNA. *Nucl. Acids Res.* 8:1259–1272.

Tseng, W.-C., Medina, D., and Randerath, K. (1978). Specific inhibition of transfer RNA methylation and modification in tissues of mice treated with 5-fluorouracil. *Cancer Res.* 38:1250–1257.

Villarreal, L. P., and Berg, P. (1977). Hybridization *in situ* of SV40 plaques: Detection of recombinant SV40 virus carrying specific sequences of nonviral DNA. *Science* 196:183–185.

Wilson, A. T. (1958). Tritium and paper chromatography. *Nature* 182:524.

6

Characterization of Modified Nucleosides in tRNA

SUSUMU NISHIMURA and YOSHIYUKI KUCHINO

Modified nucleosides are found in various nucleic acid species, such as tRNA, ribosomal RNAs, 5S RNA, small nuclear RNAs, mRNA, and DNA. Among these nucleic acids, tRNA is unique because it contains a variety of modified nucleosides in large quantity. Modified nucleosides in tRNA are classified into two categories: nucleosides with modification of base moieties, and those with ribose methylation, called 2'-O-methylated nucleosides. Some base-modified nucleosides are highly modified and are called hypermodified nucleosides. More than 60 modified nucleosides found in tRNA have been characterized so far (see Fig. 6-1) (Hall & Dunn, 1975; Ishikura, Watanabe, & Ohshima, 1979; Yamada, Murao, & Ishikura, 1981; Nishimura, 1979a; Altwegg & Kubli, 1980; Pang et al., 1982; Kawakami et al., 1979; Kasai et al., 1979; Kuchino et al., 1979a). In addition, the chemical structures of many modified nucleosides have not yet been determined.

Modified nucleosides seem to have important roles in tRNA functions (Nishimura, 1979a). Some hypermodified nucleosides located in the anticodon region function in specific codon recognitions (Nishimura, 1979a). The contents of modified nucleosides in tRNA change specifically during differentiation and transformation of cells, suggesting their relation to the regulatory function of tRNA (Nishimura & Kuchino, 1979). The type of modified nucleoside is related to the source of tRNA. Therefore, it is also important

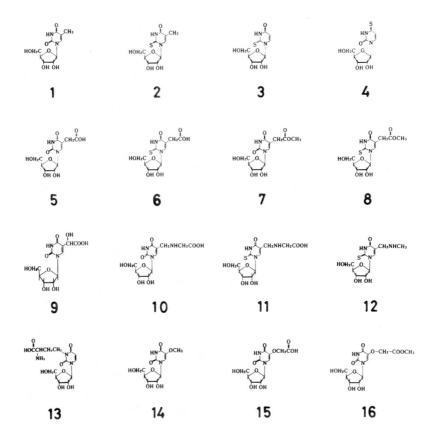

Figure 6-1. Structure of modified nucleosides found in tRNA. 2′-O-Methylated nucleosides are not listed. They are abbreviated as Am, Gm, Um, Cm, etc. for the corresponding nucleosides:

1. Ribothymidine (T, hU)
2. 5-Methyl-2-thiouridine (m^5s^2U)
3. 2-Thiouridine (s^2U)
4. 4-Thiouridine (s^4U)
5. 5-Carboxymethyluridine (cm^5U)
6. 5-Carboxymethyl-2-thiouridine (cm^5s^2U)
7. 5-Methoxycarbonylmethyluridine (mcm^5U)
8. 5-Methoxycarbonylmethyl-2-thiouridine (mcm^5s^2U)
9. 5-Carboxy-hydroxymethyluridine (ch^5U)
10. 5-Carboxymethylaminomethyluridine ($cnmm^5U$)
11. 5-Carboxymethylaminomethyl-2-thiouridine ($cnmm^5s^2U$)
12. 5-Methylaminomethyl-2-thiouridine
13. 3-(3-Amino-3-carboxypropyl)uridine (acp^3U)
14. 5-Methoxyuridine (mo^5U)
15. Uridin-5-oxyacetic acid (V)
16. Uridin-5-oxyacetic acid methyl ester (mV)

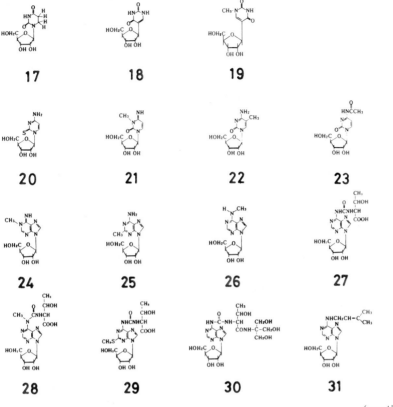

17. Dihydrouridine (D, hU)
18. Pseudouridine (ψ)
19. 1-Methylpseudouridine (m$^1\psi$)
20. 2-Thiocytidine (s^{2}C)
21. 3-Methylcytidine (m^{3}C)
22. 5-Methylcytidine (m^{5}C)
23. N^4-Acethylchtidine (ac^{4}C)
24. 1-Methyladenosine (m^{1}A)
25. 2-Methyladenosine (m^{2}A)
26. N^6-Methyladenosine (m^{6}A)
27. N-[(9-β-D-Ribofuranosylpurin-6-yl)carbamoyl]threonine (t^{6}A)
28. N-[(9-β-D-Ribofuranosylpurin-6-yl)-N-methylcarbamoyl]threonine (mt^{6}A)
29. N-[(9-β-D-Ribofuranosyl-2-methylthiopurin-6-yl)carbamoyl]threonine (s^{2}t^{6}A)
30. N-[N-[9-β-D-Ribofuranosyl-6-yl)carbamoyl]threonyl]2-amido-2-hydroxylmethylpropane-1,3-diol
31. N^6-Δ^2-Isopentenyladenosine (i^{6}A)
32. 2-Methylthio-N^6-Δ^2-isopentenyladenosine (ms^{2}t^{6}A)
33. N^6-cis-4-hydroxyisopentenyladenosine (oi^{6}A)
34. 2-Methylthio-N^6-cis-4-hydroxyisopentenyladenosine (ms^{2}oi^{6}A)
35. N^6-trans-4-hydroxyisopentenyladenosine (Z)

(continued)

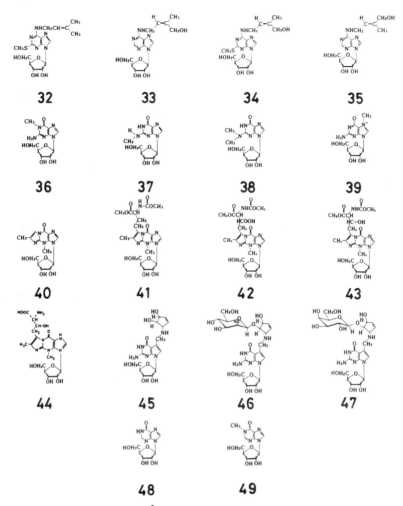

32 **33** **34** **35**

36 **37** **38** **39**

40 **41** **42** **43**

44 **45** **46** **47**

48 **49**

36. 1-Methylguanosine (m^{1}G)
37. N^2-Methylguanosine (m^{2}G)
38. N^2,N^2-Dimethylguanosine (m$_2^2$G)
39. 7-Methylguanosine (m^{7}G)
40. Wyosine (W, Yt)
41. Wybutosine (yW, Y)
42. Peroxywybutosine (O$_2$yW, Yw)
43. Hydroxywybutosine (oyW, Y$_{OH}$)
44. α-Amino-β-hydroxy-4,9-dihydro-4,6-dimethyl-9-oxo-1H-imidazo(1,2-a)purine-7-butyric acid (Y$_{OH}^*$)
45. Queuosine (Q)
46. Mannose-containing queuosine (manQ)
47. Galactose-containing queuosine (galQ)
48. Inosine (I)
49. 1-Methylinosine (m^{1}I)

to analyze modified nucleosides in tRNA from the viewpoint of evolution.

Procedures for characterization of modified nucleosides in tRNA can be classified into two types: (1) Structural characterization of new modified nucleosides, and (2) assignment of modified nucleosides of which the structures are already known. Type (1) is a rather specialized project, and it is difficult to give general experimental procedures in detail because procedures for structural characterization of unknown modified nucleosides differ, depending on the structure of the compound. Therefore, this chapter will be concerned mostly with type (2), although the strategy in type (1) is described briefly in the final section.

IDENTIFICATION OF MODIFIED NUCLEOSIDES BY TWO-DIMENSIONAL THIN-LAYER CELLULOSE CHROMATOGRAPHY AND PAPER CHROMATOGRAPHY

Two-dimensional chromatography is most frequently used for identification of modified nucleosides in tRNA because it is easy and gives high resolution. The method we and some other workers prefer is two-dimensional thin-layer cellulose chromatography with isobutyric acid/0.5 N NH_4OH (5:3, v/v) as solvent in the first dimension and isopropanol/conc. HCl/H_2O (70:15:15, v/v/v) in the second (Nishimura, 1979b). In general, tRNA is completely hydrolyzed to 3′ nucleotides with RNase T_2 before analysis, but 5′ nucleotides obtained by nuclease P1 digestion can also be analyzed. The relative positions on the chromatograms of various 3′ nucleotides of modified nucleosides as well as dinucleotides containing 2′-O-methylated nucleotides ($N_m^1 - N_p^2$, where N^1 and N^2 stand for A, G, U, and C) are shown in Fig. 6-2 (a, b, c). This system is also applicable for analysis of nucleosides. Nucleosides move faster than the corresponding nucleotides in the first dimension, and slower than the latter in the second dimension. When 2–3 A_{260} units of RNase digest of pure tRNA species is applied to a cellulose thin-layer plate (10 × 10 cm), most modified nucleotides present in quantities of 1 mol per tRNA molecule or more can be detected under a UV lamp. Spots of modified nucleotides of ^{32}P-labeled tRNA can be seen more clearly by measurement of radioactivity (see Fig. 6-3). If a large amount of unlabeled tRNA is available, 10–15 A_{260} units of tRNA hydrolyzate can be fractionated by two-dimensional cellulose

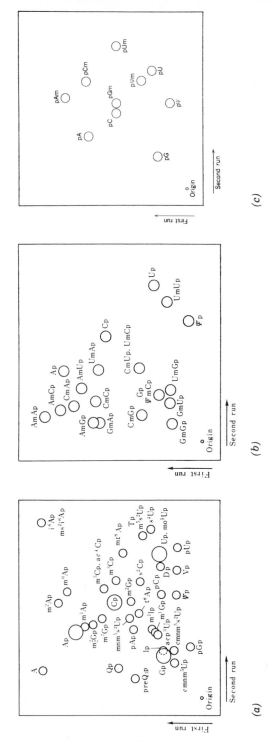

240

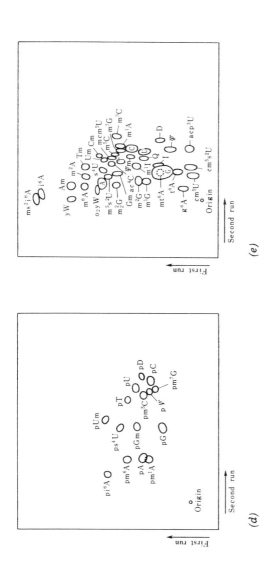

(d)

(e)

Figure 6-2. Chromatographic mobilities of modified nucleosides and nucleotides. (*a*) Positions of modified nucleoside 3'-phosphates. Solvent systems: first dimension, isobutyric acid/0.5 M NH₄OH (5:3, v/v); second dimension, isopropanol/conc. HCl/water (70:15:15, v/v/v). Nucleoside 5'-monophosphates move slightly slower than the corresponding nucleoside 3'-monophosphates in both dimensions. The corresponding nucleosides move faster in the first dimension and slower in the second dimension. PreQ, 7-amino-methyl-7-deazaguanosine. (*b*) Positions of 2'-O-methylated dinucleotides with 3' phosphate (52). The solvent system was as described in (*a*). Data were taken from Hashimoto, Sakai, and Muramatsu (1975). (*c*) Positions of 2'-O-methylated nucleoside 5'-monophosphates (53). The solvent system was as for (*a*). Data were taken from Harada, Sawyer, & Dahlberg (1975). (*d*) Positions of modified nucleoside 5'-phosphates. Solvent systems: first dimension, isopropanol/5% ammonium acetate buffer (pH 3.5) (60:25, v/v) ascending without a cover; second dimension, saturated (NH₄)₂SO₄/1 M sodium acetate buffer (pH 5.5)/isopropanol (40:9:1, v/v/v). (*e*) Positions of modified nucleosides. Solvent systems: first dimension, *n*-butanol/isobutyric acid/conc. NH₃/water (75:37.5:2.5:25, v/v/v/v); second dimension, saturated (NH₄)₂SO₄/0.1 M sodium acetate buffer (pH 6)/isopropanol (79:19:2, v/v/v). g⁶A, *N*-[(9-β-D-ribofuranosylpurin-6-yl)carbamoyl]glycine.

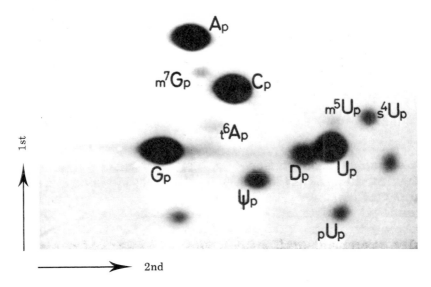

1st

2nd

Figure 6-3.   Chromatogram of ^{32}P-labeled modified nucleotides derived from ^{32}P-labeled *E. coli* tRNA^{Asn} lacking Q.

paper chromatography (Whatman No. 40 or No. 1, 30 × 30 cm) with the same solvent systems for further characterization. Modified nucleotides are eluted from the paper, and their UV spectra at acidic, neutral, and alkaline pH values are measured. Comparison of the spectra with those of authentic samples (Ishikura, Watanabe, & Ohshima 1979; Hall, 1971) can be used to confirm the assignments of modified nucleosides.

In addition to the chromatographic system mentioned above, other systems such as paper electrophoresis are also available (Brownlee, 1972). Figure 6-2(*d*) shows the chromatographic patterns of 5′-modified nucleotides on cellulose paper reported by Feldman and Falter (1971). This system is preferable for identification of modified nucleotides that are labile in strongly acidic conditions and cannot be analyzed in isopropanol/conc. HCl/H$_2$O. For separation of modified nucleosides, Rogg et al. (1976) reported the system shown in Fig. 6-2(*e*). This system gives very good resolution of various nucleosides. However, it cannot be used for separation of modified nucleotides, and the salt used in the second dimension must be removed by some procedure if salt-free nucleoside is required for further study.

IDENTIFICATION OF MODIFIED NUCLEOSIDES BY HIGH-PRESSURE LIQUID CHROMATOGRAPHY (HPLC)

Analysis of modified nucleosides by HPLC has become easier because of recent developments of HPLC in terms of both soft- and hardware. The chief advantage of HPLC is its high sensitivity, and possibly also its high resolution: 0.001 A_{260} unit of a single component can be detected easily if a good machine is used properly. Moreover, a fluorescence detector can be used to achieve higher sensitivity with modified nucleosides that have fluorescence. For example, 0.00005 A_{260} unit (approximately 5 pmol) of the Y base present in tRNA^{Phe} can easily be detected.

A typical example of fractionation of modified nucleosides by HPLC using a column of μBondapak C_{18} was reported by Davis et al. (1979). With use of a volatile buffer such as ammonium formate buffer instead of phosphate buffer for elution, modified nucleosides can be isolated as salt-free material for further analysis. For isolation purposes, elution can even be achieved with a methanol/water mixture without buffer. A disadvantage of HPLC is that the retention time of modified nucleosides varies considerably with various factors such as the batch of column resin, the time after its purchase or preparation of the column, the sample used for analysis, and slight changes in operational conditions. Therefore, values in the literature cannot be used for comparison, and standard samples must be used for final assignments. As an example, Fig. 6-4 shows the characterization of ms^{2}t^{6}A present in *B. subtilis* tRNA^{Lys} by its comparison with authentic ms^{2}t^{6}A. HPLC is also used for identification of modified nucleosides excreted in the urine of cancer patients (Gehrke et al., 1978). Before the assay, the nucleosides must be isolated from the urine by chromatography on a borate column (Uziel, Smith, & Taylor, 1976).

ANALYSIS OF MODIFIED NUCLEOSIDES BY MASS SPECTROMETRY

Mass spectral analysis of modified nucleosides was found to be useful for structural elucidication of unknown modified nucleosides, as will be discussed briefly later. The method can also be used for assignment of known modified nucleosides. Since most modified nucleosides are nonvolatile, they must be derivatized before mass

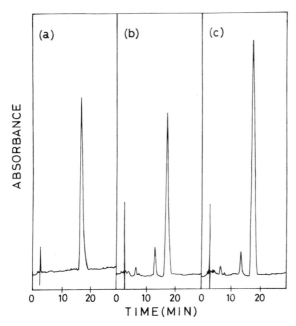

Figure 6-4. Characterization of ms^2t^6A from *B. subtilis* tRNA^{Lys} plus authentic ms^2t^6A; column, μBondapak C_{18} (4 × 600 mm); buffer, 0.01 M $NH_4H_2PO_4$ (pH 5.07) containing 6% methanol; flow rate, 1 ml/min; room temperature.

spectrometry. Trimethylsilylation, permethylation, trifluoroacetylation, and acetylation are commonly used for this purpose (McCloskey, 1974; von Minden & McCloskey, 1973); trimethylsilylation is especially useful, because spectral data are available on many modified nucleosides (Yamaizumi et al., 1979a). If the molecular ion and fragmentation pattern of a given modified nucleoside are identical with reported values, the identification of the component is almost certain. Continuous recording of ions with a high-resolution mass spectrometer certainly enhances the sensitivity and accuracy of the identification.

Usually, 0.05–0.1 A_{260} unit of a pure modified nucleoside is trimethylsilylated in a capillary tube, and an aliquot of the reaction product is analyzed by the direct inlet method in a low-resolution mass spectrometer. A sector-type mass spectrometer is better than a quadruple-type mass spectrometer, because the molecular ions of trimethylsilylated-modified nucleosides are generally above 500, and their intensities are very small compared with those of fragment ions. The following precautions should be taken: (1) since trimethylsilylated derivatives are labile when exposed to moisture, the reaction product must be put directly into the probe, and the reagent liquid must be removed by pumping before the assay; (2) modified nucleosides for mass spectral assay must be salt-free, and their pH must be neutral.

In some cases, a mixture of nucleosides prepared by hydrolysis of tRNA with RNase and phosphomonesterase digestion can be directly analyzed by gas chromatography/mass spectrometry (GC/MS). Many (but not all) trisilylmethylated nucleosides can be passed through the column and fractionated by gas chromatography. However, some modified nucleosides, such as cytidine derivatives, and some hypermodified nucleosides do not readily pass through the column.

One advantage of GC/MS is its high sensitivity: a molecular ion of a single modified nucleoside component present in a pure tRNA sample can be detected after injection of 0.05 A_{260} unit of total tRNA by hydrolyzate into the column. Further improvement in the liquid chromatography/mass spectrometry technique will almost certainly increase the applicability of mass spectrometry for analysis of modified nucleosides.

ANALYSIS OF MODIFIED NUCLEOTIDES BY POSTLABELING PROCEDURES

I. Identification of Modified Nucleotides During tRNA Sequencing

Recent progress in tRNA sequencing by postlabeling procedures has made it possible to obtain the primary structures of tRNAs much faster and more easily than by the conventional Sanger fingerprinting technique (Kuchino, Kasai, et al., 1979; Kuchino et al., 1980; Lockhard et al., 1978; Gupta & Randerath, 1979; Tanaka, Dyer, & Brownlee, 1980). During tRNA sequencing, modified nucleosides in tRNA must be identified. Several postlabeling procedures have been reported, of which our modification (Kuchino, Kasai, et al., 1979; Kuchino et al., 1980) of the original procedure of Stanley and Vassilenko (1978) seems to be the most reliable, at least with respect to identification of modified nucleotides. The modifications we employ are: (1) use of nuclease P1 instead of alkaline hydrolysis to obtain ^{32}P-labeled 5′-nucleotide as terminal nucleotide rather than nucleoside 3′,5′-diphosphate, and (2) use of two-dimensional thin-layer chromatography instead of one-dimensional electrophoresis. These modifications result in clearer assignment of modified nucleosides. Identification of modified nucleotides during sequence analysis has two important advantages: First, the spot of a given modified nucleotide is more clearly seen in the autoradiogram, because the intensities of the major compounds pU, pC, pA, and pG are low. Therefore, modified nucleotides located close to major nucleotides on the thin-layer chromatogram can easily be detected. Second, identification of modified nucleotides in the tRNA sequence is direct proof that they are not contaminants present in the tRNA preparation.

I-1. Experimental

Determination of the Primary Sequence and Location of Modified Nucleotides in tRNA$_{CUR}^{Leu}$ *from* Salmonella Typhimurium. The method consists of (1) partial alkaline hydrolysis of tRNA by formamide to cleave the phosphodiester bond in less than one residue per tRNA molecule; (2) kination of the 5′ end of oligonucleotides by $(\gamma\text{-}^{32}P)ATP$ and T4 polynucleotide kinase; (3) fractionation of the oligonucleotides labeled at the 5′ end by polyacrylamide gel electrophoresis; (4) elution of each labeled oligonucleotide from the gel and its hydrolysis with nuclease P1 to liberate labeled 5′-nucleotide; (5) analysis of the nucleotide by two-dimensional thin-layer chromatography.

(a) Purification of tRNA$_{CUR}^{Leu}$ *from* S. typhimurium. The sample of tRNA used for tRNA sequencing by postlabeling must be pure. Two-dimensional polyacrylamide gel electrophoresis is always performed for final purification of tRNA. Unfractionated tRNA isolated from cells of *S. typhimurium* is fractionated successively by column chromatographies on DEAE-Sephadex A-50 (pH 7.5), BD-cellulose, and RPC-5 as described previously (Nishimura, 1971; Taya & Nishimura, 1973). The resulting tRNA$_{CUR}^{Leu}$, which is more than 85% pure, is then subjected to two-dimensional polyacrylamide gel electrophoresis. One to two A$_{260}$ units of the tRNA$_{CUR}^{Leu}$-rich fraction is applied to 10% gel (thickness, 1 mm; height, 40 cm; width, 20 cm) as a strip 1.5 cm long with xylene cyanol dye. After electrophoresis at 400 V for 6–8 hr (the dye moves to the middle of the gel), the gel is removed from the glass plate, wrapped in a thin polyethylene sheet, and examined with a UV lamp on a fluorescent thin-layer sheet. The major spot of tRNA detected by its strong UV absorption is cut out and loaded on 20% gel (thickness, 1 mm;

height, 40 cm; width, 20 cm). Electrophoresis is performed at 700 V for 30 hr. By this procedure, pure $tRNA_{CUR}^{Leu}$ is obtained. When complete purification is not achieved by this two-dimensional polyacrylamide gel electrophoresis, the tRNA preparation is subjected to a further 15 or 20% polyacrylamide-urea gel.

For elution of tRNA from the gel, the gel containing tRNA is cut out, put in a 1.5-ml Eppendorf centrifuge tube, and crushed with a glass rod. Then 0.5 ml of elution buffer (0.5 M ammonium acetate/0.01 M magnesium acetate/0.1 mM EDTA/0.1% SDS) is added, and the suspension is stood for 8 hr with occasional shaking. The suspension is centrifuged, and the residue is reextracted with 0.5 ml of the same buffer. tRNA is precipitated from the combined eluate by adding 2.5 volumes of ethanol, and kept cold in Dry Ice for 10 min. The tRNA is then collected by centrifugation and dried in a vacuum desiccator.

(b) Sequencing of S. typhimurium $tRNA_{CUR}^{Leu}$ *and assignment of modified nucleotides.* A sample of $tRNA_{CUR}^{Leu}$ (0.01 A_{260} unit, 1 μl in H_2O) is mixed with 10 μl of formamide in a sealed glass capillary tube and heated at 100°C for 5–10 min. The time of heating can be changed, if necessary, to obtain nearly one hit per tRNA molecule. After heating, the reaction mixture is transferred to a 1.5-ml Eppendorf tube, and one-tenth volume of 3 M NaCl and 2.5 volumes of ethanol are added. The tube is cooled in Dry Ice for 10 min, and the precipitate is collected by centrifugation and dried in a vacuum desiccator. Then 5 μl of (γ-^{32}p)ATP (10 mCi/ml; specific activity, 8,000 Ci/mmol), 0.5 μl of 1 M tris-HCl (pH 7.5), 1 μl of 0.15 M $MgCl_2$, 1 μl of 0.15 M 2-mercaptoethanol, and 1 μl of T4 polynucleotide kinase (2.5 units/μl) are added. The reaction mixture is made up to a total volume of 10 μl with water and incubated at 37°C for 30 min. After incubation, the reaction mixture is mixed with an equal volume of 50% glycerol, 0.02% xylene cyanol, and 0.02% bromophenol blue, and half the mixture (10 μl) is applied to 20% polyacrylamide-urea gel (thickness, 0.06 cm; height, 40 cm; width, 20 cm) as a strip 1 cm in length. Electrophoresis is carried out for 10 hr at 700 V. The rest of the reaction mixture (10 μl) is then applied to another slot in the gel, and electrophoresis is continued for 30 hr. After electrophoresis, the glass plate on one side is removed, and the surface of the gel is attached to a used X-ray film. Then, the glass plate on the other side is removed, and the exposed gel surface is wrapped in a thin polyethylene sheet. After putting radioactive markers in the corners of the gel, an autoradiogram is taken (usually with exposure for 20 min to 1 hr) (see Fig. 6.5).

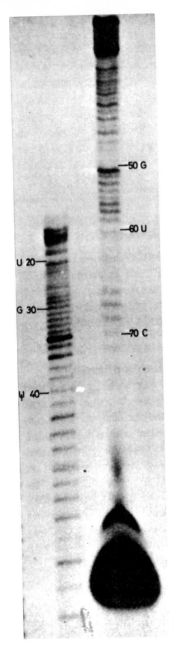

Figure 6-5. Autoradiogram of $(5'\text{-}^{32}P)$-labeled oligonucleotides derived from *S. typhimurium* tRNA$_{CUR}^{Leu}$.

The autoradiogram is placed on the gel, and the portions of the gel corresponding to each dark band are cut out with a razor. Each slice is put in a 1.5-ml Eppendorf tube, and soaked in 0.4 ml of elution buffer (0.5 M ammonium acetate/0.1 M magnesium acetate/ 0.1 mM EDTA/0.1% SDS) with occasional shaking for 5 hr at 37°C. The gel is removed from the tube with forceps. To the eluate, 10 µg of carrier tRNA (unfractionated *E. coli* tRNA) and 2.5 volumes of ethanol are added. The precipitate obtained by centrifugation is dried in a vacuum desiccator, dissolved in 10 µl of 0.02 M ammonium formate buffer (pH 5.3) containing 1 µg of nuclease P1, and incubated at 37°C for 3 hr.

A sample of 1–3 µl of reaction mixture is applied to a thin-layer Avicel cellulose plate (5 × 5 cm) and subjected to two-dimensional chromatography with isobutyric acid/0.5 M ammonia (5:3, v/v) in the first dimension and isopropyl alcohol/conc. HCl/H_2O (70:15:15, v/v/v) in the second. As shown in Fig. 6-6, the nucleotide sequence of residue 2 to 68 of $tRNA_{CUR}^{Leu}$ from *S. typhimurium* was unambiguously determined by reading the main spot in each chromatogram. In the assignment of the modified nucleotides, the following precautions must be taken. Since the intensities of modified nucleotides such as $p\psi$, pm^7G, pD expected as 5'-terminal residues of fragments are low, due to low efficiency of phosphorylation, these spots should be assigned as 5'-terminal nucleotides even though their intensities are much less than those of contaminating pA, pG, pC, and pU.

The nucleotide residue on the 3' side of a 2'-O-methylated nucleotide cannot be determined by the formamide partial-hydrolysis method, since the phosphodiester bond between these two residues is not cleaved by alkaline treatment. However, this problem can be overcome by partial, instead of complete, nuclease P1 digestion of the fragment having 2'-O-methylated nucleotide at the 5' end (with 0.01 g of nuclease P1 instead of 1 µg). In this way, labeled dinucleotide $(^{32}PN_m^1 - N^2)$ is obtained in addition to mononucleotide $(^{32}PN_m^1)$. N^2 can be identified by comparison of the chromatographic mobilities with those of authentic 2'-O-methylated dinucleotides [see Fig. 6-2(*b*)].

The nucleotide sequence near the 3' end cannot be determined by the formamide partial-hydrolysis method. The sequence in this region can easily be determined by the rapid read-off sequencing procedure described by Peattie (1979). The 5' end of tRNA also cannot be determined by the method of formamide partial-hydrolysis because the 5' end of tRNA is naturally phosphorylated. The 5' end of tRNA can be determined by 5'-labeling of tRNA with ^{32}P using polynucleotide kinase after removal of natural phosphate at the 5' end by treatment with alkaline phosphomonoesterase followed by complete digestion with nuclease P1 (Kuchino et al., 1979, 1980).

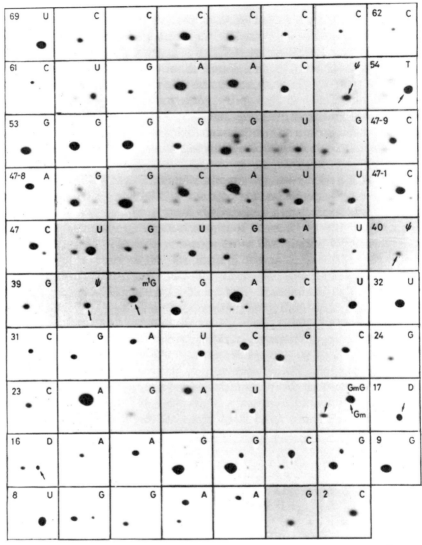

Figure 6-6. Analysis of 5'-terminal nucleotides of fragments of $tRNA^{Leu}_{CUR}$ by two-dimensional thin-layer chromatography. Arrows indicate positions of modified nucleotides.

The data obtained by limit alkaline hydrolysis together with these additional data generally makes it possible to determine unambiguously the total nucleotide sequence of the tRNA. The primary sequence of $tRNA^{Leu}_{CUR}$ from *S. typhimurium* was found to be identical with that of *E. coli* $tRNA^{Leu}_1$, but *S. typhimurium* $tRNA^{Leu}_{CUR}$ contains

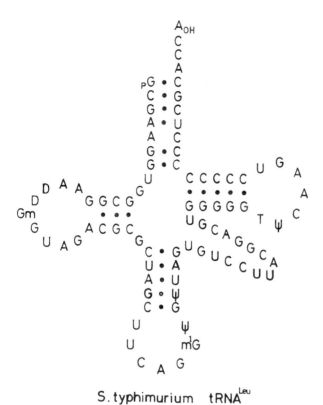

S. typhimurium tRNA^Leu

Figure 6-7. Nucleotide sequence of *S. typhimurium* tRNA$_{CUR}^{Leu}$ arranged in a clover-leaf form.

m^{1}G in position 37, which is not characterized in *E. coli* tRNA$_1^{Leu}$ (see Fig. 6-7).

II. Identification of Modified Nucleotides in Total tRNA by the Postlabeling Procedure

Postlabeling techniques can also be used for identification of modified nucleotides in total tRNA. There are two main methods for this purpose. One is the tritium-labeling method reported by Randerath and Randerath (1971). This method involves (1) formation of nucleoside from tRNA by RNases, nucleases, and phosphomonoesterase digestions, (2) conversion of the ribose moiety of the nucleoside to *cis*-aldehyde by periodate oxidation, (3) introduction of ^{3}H by reduction of *cis*-aldehyde to *cis*-diol by treatment with [^{3}H]borohydride and (4) separation of tritiated derivatives by two-dimensional thin-layer chromatography. Although this procedure is very useful, 2'-O-methylated nucleosides that are resistant to periodate oxidation cannot be detected. The other procedure is kination of 3' nucleotide by $(\gamma\text{-}^{32}P)$ATP as described in the previous section (Silberklang et al., 1977). This method requires considerably less tRNA than the tritium-labeling method.

In this chapter, only the kination method is described in detail. A sample of tRNA (0.01 A_{260} unit) is mixed with 1 μl of 0.03 M sodium acetate buffer (pH 4.6) and 0.05 unit of RNase T_2, and the mixture is made up to 3 μl with H_2O and incubated at $37°C$ for 5 hr. Then 1 μl of buffer (0.5 M tris-acetate (pH 8.0)/0.1 M $MgCl_2$/ 0.1 M 2-mercaptoethanol), 2.5 units of T4 polynucleotide kinase, $(\gamma\text{-}^{32}P)$ATP (8,000 Ci/ml; specific activity, 10 Ci/mmol), and 0.5 μl of 2 mM ATP are added, and the mixture is made up to 10 μl with H_2O. After incubation at $37°C$ for 30 min, remaining ^{32}P-ATP is decomposed by the following procedure. The reaction mixture is incubated with 0.04 unit of apyrase at $37°C$ for 20 min. Then 0.5 μl of 2 mM ATP is added, and incubation is continued at $37°C$ for 10 min. Addition of cold ATP and incubation is repeated once more. Then 2 μl of 0.05 M sodium acetate buffer (pH 4.4) and 2 μl of nuclease P1 solution (50 mg/ml) are added, and the mixture is incubated at $37°C$ for 30 min. A portion of the hydrolyzate (5–10 μl) is spotted on an Avicel-cellulose thin-layer plate (20 × 20 cm), and subjected to two-dimensional thin-layer chromatography with iso-butyric acid/ammonia/H_2O (66:1:33, v/v/v) in the first dimension, and the solvent (100 ml of 0.1 M sodium phosphate buffer, pH 6.8, 60 g of $(NH_4)_2SO_4$, and 2 ml of n-propanol) in the second dimension. The $(\gamma\text{-}^{32}P)$ATP used for postlabeling must be pure, and should not be contaminated with other radioactive materials that interfere with the identification of the modified nucleotides. We prepare $(\gamma\text{-}^{32}P)$ATP for this purpose by direct transfer of ^{32}P-phosphate to ADP by the procedure of Walseth and Johnson (1979). Figure 6-8 shows the radioautogram of the chromatogram of the digest of S. triphimurium tRNA$_{CUR}^{Leu}$.

STRUCTURAL ELUCIDATION OF UNKNOWN MODIFIED NUCLEOSIDES

If the UV spectra or chromatographic behavior of an isolated nucleo-side does not fit with that of any known compound, the nucleoside

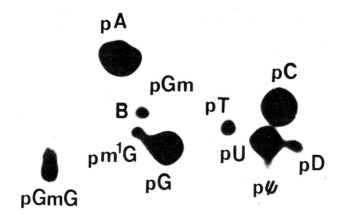

Figure 6-8. Radioautogram of modified nucleotides from *S. typhymurium* tRNA$^{Leu}_{CUR}$ postlabeled with $(\gamma$-^{32}P)ATP.

must be a new, modified nucleoside that has not yet been characterized. In this case, it is obviously interesting to determine its chemical structure. For structural characterization, the material must first be purified. This is not easy, because the amount of a rare modified nucleoside generally is not more than one residue per tRNA species; namely, about 0.02% of the total unfractionated tRNA on a weight basis. The amount of material required for structural studies varies, depending on the modified nucleoside. We isolated 20 mg of queuosine from 100 g of *E. coli* tRNA. In contrast, recently we were able to characterize ms^{2}t^{6}A with only 20 µg of material (Yamaizumi et al., 1979b), and as much as 5 g of rabbit-liver tRNA was required for isolation of even this much material.

For purification of a modified nucleoside, a good method is first to purify a particular tRNA species that contains a wanted modified nucleoside from unfractionated tRNA by successive column chromatographies (Nishimura, 1971). In this way, extensive purification of the modified nucleoside can be achieved. In general, tRNA is hydrolyzed by nuclease P1 or snake venom phosphodiesterase to 5' nucleotides, and the hydrolyzate is subjected to Dowex-1 or DEAE-cellulose column chromatography to purify the modified nucleotide (Cunningham & Gray, 1974; Yokoyama et al., 1979). Alternatively, tRNA is first hydrolyzed by RNase T1 or RNase A. The resulting oligonucleotide containing the modified nucleoside is purified by DEAE-Sephadex A-25 column chromatography (Kasai et al., 1975; Kasai, Murao, et al. 1976) and then completely hydrolyzed by either nuclease P1 or RNase T2 to obtain the modified nucleotide.

This can then be purified by several paper chromatographic procedures, paper electrophoresis, and thin-layer chromatography. The modified nucleotide thus obtained is finally converted to the nucleoside by phosphomonoesterase treatment. The nucleoside can be purified by paper chromatography (solvent A) or preferably by HPLC.

Measurements of UV spectra at different pH values, and of nmr and mass spectra, are routine procedures for obtaining structural information. A sample of 20–40 μg (0.5–1 A_{260} unit) of material is enough for sensitive proton FT-nmr measurement (Yamaizumi et al., 1979b). Mass spectral analysis requires much less material. High-resolution mass spectral data give the elemental composition of the modified nucleoside (Kasai, Murao, et al., 1976; Yamaizumi et al., 1979b). The fragmentation pattern of low-resolution mass spectra is also important for elucidating the structure of the modified nucleoside (Ohashi et al., 1974; Yamaizumi et al., 1979b). From these data, together with information on the physical (pKa and IR) and chemical properties, the structure of a given modified nucleoside can be determined. With some modified nucleosides such as ms^2t^6A, mass spectral data alone without nmr data were sufficient for structural elucidation. Total synthesis of a modified nucleoside provides final proof of its structure. Chemical synthesis of stereo isomers or X-ray analysis of crystals is sometimes necessary for determination of the absolute configuration (Ohgi, Kondo, & Goto, 1979; Yokoyama et al., 1979).

The following literature gives descriptions of the structural elucidations of various modified nucleosides: t^6A (Chheda et al., 1969; Schweizer et al., 1969), ms^2i^6A (Harada et al., 1968), m^6A (Saneyoshi, Harada, & Nishimura, 1969), m^2A (Saneyoshi et al., 1972), V (Murao et al., 1970), ac^4C (Ohashi et al., 1972), mt^6A (Kimura-Harada et al., 1972), acp^3U (Ohashi et al., 1974), Q (Kasai et al., 1975), manQ and galQ (Kasai et al., 1976b), $PreQ_1$ (Okada et al., 1978), $PreQ_0$ (Noguchi et al., 1978), Y (Nakanishi et al., 1970), Y_{OH} (Nakanishi et al., 1971; Kasai et al., 1979; Kuchino, Kato, et al., 1979), ms^2t^6A (Yamaizumi et al., 1979b). Much information is also provided in the review on modified nucleosides by Hall and Dunn (1975).

MATERIALS

Materials used in the experiments were purchased from the following companies.

Uniformly labeled [^{14}C]leucine and carrier-free ^{32}P: New England Nuclear Corp.

RNase T2, RNase T1 and RNase U2: Sankyo Co., Tokyo.

RNase A and Apyrase: Sigma Chemical Co.

Nuclease P1: Yamasa Shoyu Co., Tokyo.

T4 polynucleotide kinase, rabbit-muscle glycerophosphate dehydrogenase, rabbit-muscle triosephosphate isomerase, rabbit-muscle glyceraldehyde-3-P dehydrogenase, Yeast 3-phosphoglycerate kinase and rabbit-muscle lactate dehydrogenase: Boehringer Mannheim Corp.

E. coli alkaline phosphomonoesterase: Worthington Biochemical Corp.

T4 RNA ligase: New England Biolabs. Inc.

Avicel SF cellulose thin-layer plates; Funakoshi Pharmaceutical Co., Tokyo.

Acrylamide and *N,N'*-methylene-bisacrylamide: Eastman Kodak Co.

REFERENCES

Altwegg, M. and Kubli, E. (1980). The nucleotide sequence of glutamate tRNA$_4$ of *Drosophila melanogaster*. *Nucl. Acids. Res.* 8:215–223.

Brownlee, G. G. (1972). *Determination of Sequences in RNA* (Amsterdam/London: North Holland Publishing; New York: American Elsevier).

Chheda, G. B., Hall, R. H., Magrath, D. I., Mozejko, J., Schweizer, M. P., Stasiuk, L., and Taylor, P. R. (1969). Aminoacyl nucleosides. VI. Isolation and preliminary characterization of threonyladenine derivatives from transfer ribonucleic acid. *Biochem.* 8:3278–3282.

Cunningham, R. S. and Gray, M. W. (1974). Derivatives of *N*-[*N*-(9-β-D-ribofuranosylpurin-6-yl)carbamoyl]threonine in phosphodiesterase hydrolyzates of wheat embryo transfer ribonucleic acid. *Biochemistry* 13: 543–553.

Davis, G. E., Gehrke, C. W., Kuo, K. C., and Agris, P. F. (1979). Major and modified nucleosides in tRNA hydrolysates by high-performance liquid chromatography. *J. Chromatogr.* 173:281–298.

Feldman, H., and Falter, H. (1971). Transfer ribonucleic acid from *Mycoplasma laidlawii* A. *Eur. J. Biochem.* 18:573–581.

Gehrke, C. W., Kuo, K. C., Waalkes, T. P., and Borek, E. (1978). Patterns of urinary excretion of modified nucleosides. *Cancer Res.* 39:1150–1153.

Gupta, R. C., and Randerath, K. (1979). Rapid print-readout technique for sequencing of RNA's containing modified nucleotides. *Nucl. Acids Res.* 6:3443-3458.

Hall, R. H. (1971). *The Modified Nucleosides in Nucleic Acid* (New York/ London: Columbia University Press).

Hall, R. H., and Dunn, D. B. (1975). Natural occurrence of the modified nucleosides. In *Handbook of Biochemistry and Molecular Biology*, 3rd Ed., *Nucleic Acids*, Vol. I, ed. G. D. Fasmann (Cleveland: CRC Press), pp. 216-250.

Harada, F., Sawyer, R. C., and Dahlberg, J. E. (1975). A primer ribonucleic acid for initiation of *in vitro* Rous sarcoma virus deoxyribonucleic acid synthesis. *J. Biol. Chem.* 250:3487-3497.

Harada, F., Gross, H. J., Kimura, F., Chang, S. H., Nishimura, S. and RajBhandary, U. L. (1968). 2-Methylthio-N^6-(Δ^2-isopentenyl)adenosine: A component of *E. coli* tyrosine transfer RNA. *Biochem. Biophys. Res. Comm.* 33: 299-306.

Hashimoto, S., Sakai, M., and Muramatsu, M. (1975). 2'-O-methylated oligonucleotides in ribosomal 18S and 28S RNA of a mouse hepatoma, MH134. *Biochem.* 14:1956-1964.

Ishikura, H., Watanabe, K., and Ohshima, T. (1979). Structure elementary composition, molecular weight and UV spectra of bases, nucleosides and nucleotides. In *Biochemical Data Book*, Vol. I, ed. Japanese Biochemical Society (Tokyo: Tokyo Kagaku Dojin), pp. 1032-1062.

Kasai, H., Murao, K., Nishimura, S., Liehr, J. G., Crain, P. F. and McCloskey, J. A. (1976). Structure determination of a modified nucleoside isolated from *Escherichia coli* transfer ribonucleic acid. *N*-[*N*-[(9-β-D-ribofuranosylpurin-6-yl)-carbamoyl]threonyl]-2-amino-2-hydroxymethylpropane-1,3-diol. *Eur. J. Biochem.* 69:435-444.

Kasai, H., Nakanishi, K., Macfarlance, R. D., Torgerson, D. F., Ohashi, Z., McCloskey, J. A., Gross, H. J., and Nishimura, S. (1976). The structure of Q* nucleoside isolated from rabbit liver transfer ribonucleic acid. *J. Amer. Chem. Soc.* 98:5044-5046.

Kasai, H., Ohashi, Z., Harada, F., Nishimura, S., Oppenheimer, N. J., Crain, P. F., Liehr, J. G., von Minden, D. L., and McCloskey, J. A. (1975). Structure of the modified nucleoside Q isolated from *Escherichia coli* transfer ribonucleic acid: 7-(4,5-cis-Dihydroxyl-1-cyclopenten-3-ylaminomethyl)-7-deazaguanosine. *Biochem.* 14:4198-4208.

Kasai, H., Yamaizumi, Z., Kuchino, Y., and Nishimura, S. (1979). Isolation of hydroxy-Y base from rat liver tRNA^{Phe}. *Nucl. Acids Res.* 6:993-1000.

Kawakami, M., Nishio, K., Takemura, S., Kondo, T., and Goto, T. (1979). 5-(Carboxy-hydroxymethyl)uridine, a new modified nucleoside located in the anticodon of tRNA$_2^{Gly}$ from the posterior silk glands of *Bombyx mori*. *Nucl. Acids Res. Symp. Ser.* 6:s53-55.

Kimura-Harada, F., von Minden, D. L., McCloskey, J. A., and Nishimura, S. (1972). *N*-[9-(β-D-Ribofuranosyl)purin-6-yl-*N*-methylcarbamoyl]threonine: A new modified nucleoside isolated from *Escherichia coli* threonine transfer RNA. *Biochem.* 11:3910-3915.

Kuchino, Y., Kasai, H., Yamaizumi, Z., Nishimura, S., and Borek, E. (1979). Under-modified Y base in a tRNA^{Phe} isoacceptor observed in tumor cells. *Biochim. Biophys. Acta* 565:215-218.

Kuchino, Y., Kato, M., Sugisaki, H., and Nishimura, S. (1979). Nucleotide sequence of starfish initiator tRNA. *Nucl. Acids Res.* 6:3459-3469.

Kuchino, Y., Watanabe, S., Harada, F., and Nishimura, S. (1980). Primary structure of AUA specific isoleucine tRNA from *Escherichia coli*. *Biochem.* 19:2085-2089.

Lockhard, R. E.. Alzner-Deweerd, B., Heckman, J. E., MacGee, J., Tabor, M. W., and RajBhandary, U. L. (1978). Sequence analysis of 5'-[^{32}P]-labeled mRNA and tRNA using polyacrylamide gels. *Nucl. Acids Res.* 5:37-56.

McCloskey, J. A. (1974). Mass spectrometry. In *Basic Principles in Nucleic Acid Chemistry*, Vol. I, ed. P. Ts'o (New York/London: Academic Press) pp. 209-309.

Murao, K., Saneyoshi, M., Harada, F., and Nishimura, S. (1970). Uridin-5-oxy acetic acid: A new minor constituent from *E. coli* valine transfer RNA I. *Biochem. Biophys. Res. Comm.* 38:657-662.

Nakanishi, K., Furutachi, N., Funamizu, M., Grunberger, D., and Weinstein, I. B. (1970). Structure of the fluorescent Y base from yeast phenylalaine transfer ribonuleic acid. *J. Amer. Chem. Soc.* 92:7617.

Nakanishi, K., Blobstein, S. H., Fuamizu, M., Furutachi, N., van Lear, G., Grunberger, D., Lanks, K., and Weinstein, I. B. (1971). Structure of the "peroxy-Y base" from liver tRNA^{Phe}. *Nature New Biol.* (London) 234:107.

Nishimura, S. (1979a). Modified nucleosides in tRNA. In *Transfer RNA: Struc-*

ture, Properties, and Recognition, eds. P. R. Schimmel, D. Söll, and J. N. Abelson (Cold Spring Harbor: Cold Spring Harbor Laboratory), pp. 59–79.

Nishimura, S. (1979b). Chromatographic mobilities of modified nucleotides. In Transfer RNA: Structure, Properties, and Recognition, eds. P. R. Schimmel, D. Söll, and J. N. Abelson (Cold Spring Harbor: Cold Spring Harbor Laboratory), pp. 551–552.

Nishimura, S. (1971). Fractionation of transfer RNA by DEAE-Sephadex A-50 column chromatography. In Procedures in Nucleic Acid Research, Vol. 2, eds. G. L. Cantoni and D. R. Davies (New York: Harper & Row), pp. 542–564.

Nishimura, S., and Kuchino, Y. (1979). Transfer RNA and cancer. In Progress in Cancer Biochemistry (Gann Monograph No. 24), eds. T. Sugimura, H. Endo, T. Ono, and H. Sugano (Tokyo: Japanese Scientific Societies Press; Baltimore: University Park Press), pp. 245–262.

Noguchi, S., Yamaizumi, Z., Ohgi, T., Goto, T., Nishimura, Y., Hirota, Y., and Nishimura, S. (1978). Isolation of Q nucleoside precursor present in tRNA of an E. coli mutant and its characterization as 7-(cyano)-7-deazaguanosine. Nucl. Acids Res. 5:4215–4223.

Ohashi, Z., Maeda, M., McCloskey, J. A., and Nishimura, S. (1974). 3-(Amino-3-carboxypropyl)uridine: A novel modified nucleoside isolated from Escherichia coli phenylalanine transfer RNA. Biochem. 13:2620–2625.

Ohashi, Z., Murao, K., Yahagi, T., von Minden, D. L., McCloskey, J. A., and Nishimura, S. (1972). Characterization of C^+ located in the first position of the anticodon of Escherichia coli tRNA^{Met} as N^4-acetylcytidine. Biochim. Biophys. Acta 262:209–213.

Ohgi, T., Kondo, T., and Goto, T. (1979). Total synthesis of optically pure nucleoside Q. Determination of absolute configuration of natural nucleoside Q. J. Amer. Chem. Soc. 101:3629–3633.

Okada, N., Noguchi, S., Nishimura, S., Ohgi, T., Goto, T., Crain, P. F., and McCloskey, J. A. (1978). Structure determination of a nucleoside Q precursor isolated from E. coli tRNA: 7-(aminomethyl)-7-deazaguanosine. Nucl. Acids Res. 5:2289–2296.

Pang, H., Ihara, M., Kuchino, Y., Nishimura, S., Gupta, R., Woese, C. R., and McCloskey, J. A. (1982). Structure of a modified nucleoside in arachaebacterial tRNA which replaces ribothymidine. J. Biol. Chem. 357:1151–1157.

Peattie, D. A. (1979). Direct chemical method for sequencing RNA. *Proc. Natl. Acad. Sci. USA* 76:1760-1764.

Randerath, K., and Randerath, E. (1971). A tritium derivative method for base analysis of ribonucleotides and RNA. In *Procedures in Nucleic Acid Research*, Vol. 2, eds. G. L. Cantoni and D. R. Davies (New York: Harper & Row), pp. 796-812.

Rogg, H., Brambilla, R., Keith, G., and Staehelin, M. (1976). An improved method for the separation and quantitation of the modified nucleosides of transfer RNA. *Nucl. Acids Res.* 3:285-295.

Saneyoshi, M., Harada, F., and Nishimura, S. (1969). Isolation and characterization of N^6-methyladenosine from *E. coli* valine tRNA. *Biochim. Biophys. Acta* 190:264-273.

Saneyoshi, M., Ohashi, Z., Harada, F., and Nishimura, S. (1972). Isolation and characterization of 2-methyladenosine from *Escherichia coli* $tRNA_2^{Glu}$, $tRNA_1^{Asp}$, $tRNA_1^{His}$ and $tRNA^{Arg}$. *Biochim. Biophys. Acta* 262:1-10.

Schweizer, M. P., Chheda, G. B., Baczynskyj, L., and Hall, R. H. Aminoacyl nucleosides (1969). Vii. *N*-(Purin-6-ylcarbamoyl)threonine, a new component of transfer ribonucleic acid. *Biochem.* 8:3283-3289.

Silberklang, M., Prochiantz, A., Haenni, A. L., and RajBhandary, U. L. (1977). Studies on the sequence of the 3′-terminal region of turnip-yellow-mosaic-virus RNA. *Eur. J. Biochem.* 72:465-478.

Stanley, J., and Vassilenko, S. (1978). A different approach to RNA sequencing. *Nature* 274:87-89.

Tanaka, Y., Dyer, T. A., and Brownlee, G. G. (1980). An improved direct RNA sequence method: Its application to *Vicia faba* 5.8S ribosomal RNA. *Nucl. Acids Res* 6:1259-1272.

Taya, Y., and Nishimura, S. (1973). Biosynthesis of 5-methylaminomethyl-2-thiouridine. I. Isolation of a new tRNA-methylase specific for 5-methylaminomethyl-2-thiouridylate. *Biochim. Biophys. Res. Comm.* 51:1062-1068.

Uziel, M., Smith, L. M., and Taylor, S. A. (1976). Modified nucleosides in urine: Selective removal and analysis. *Clin. Chem.* 22:1451-1455.

von Minden, D. L., and McCloskey, J. A. (1973). Mass spectrometry of nucleic acid components, *N,O*-permethyl derivatives of nucleosides. *J. Amer. Chem. Soc.* 95:7480-7490.

Walseth, T. F., and Johnson, R. A. (1979). The enzymatic preparation of $[\alpha\text{-}^{32}P]$ nucleoside triphosphates, cyclic $[^{32}P]AMP$, and cyclic $[^{32}P]GMP$. *Biochim. Biophys. Acta* 526:11-31.

Yamada, Y., Murao, K., and Ishikura, H. (1981). 5-(Carboxymethylaminomethyl)-2-thiouridine, a new modified nucleoside found at the first letter position of the anticodon. *Nucl. Acids Res.* 9:1933-1939.

Yamaizumi, Z., Nishimura, S., Scott, M. F., and McCloskey, J. A. (1979a). Mass spectra of nucleosides and modified nucleosides. In *Biochemical Data Book*, Vol. I, ed. Japanese Biochemical Society (Tokyo: Tokyo Kagaku Dojin) pp. 1062-1082.

Yamaizumi, Z., Nishimura, S., Limburg, K., Raba, M., Gross, H. J., Crain, P. F., and McCloskey, J. A. (1979b). Structure elucidation of high resolution mass spectrometry of a highly modified nucleoside from mammalian transfer RNA. N-[9-β-D-ribofuranosyl-2-methylthiopurin-6-yl)carbamoyl]-threonine. *J. Amer. Chem. Soc.* 101:2224-2225.

Yokoyama, S., Miyazawa, T., Iitaka, Y., Yamaizumi, Z., Kasai, H., and Nishimura, S. (1979). Three-dimensional structure of hyper-modified nucleoside Q located in the wobbling position of tRNA. *Nature* 282:107-109.

Yokomaya, S., Yamaizumi, Z., Nishimura, S., and Miyazawa, T. (1979). ^{1}H-NMR studies on the conformational characteristics of 2-thiopyrimidine nucleotides found in transfer RNAs. *Nucl. Acids Res.* 6:2611-2626.

7

Direct Chemical Method for Sequencing End-labeled Ribonucleic Acids

DEBRA A. PEATTIE

Determining the nucleotide sequence of an RNA molecule is the first step in understanding its fundamental biological properties. The original methods of RNA sequence analysis were based on enzymatic production and chromatographic separation of overlapping oligonucleotide fragments from within an RNA molecule followed by identification of the mononucleotides comprising the oligomer (see Brownlee, 1972, for review). Many of the more recent advances in rapid sequence determination of RNA molecules (Brownlee & Cartwright, 1977; Donis-Keller, Maxam, & Gilbert, 1977; Simoncsits et al., 1977; Lockhard et al., 1978; Ross & Brimacombe, 1978; McGeoch & Turnbull, 1978; Zimmern & Kaesberg, 1978; Kramer & Mills, 1978; Peattie, 1979) have involved one-dimensional electrophoretic separation of ^{32}P-labeled oligoribonucleotides on polyacrylamide gels (Richards & Gratzer, 1964; McPhie, Hounsell, & Gratzer, 1966; Peacock & Dingman, 1967; Maniatis & Efstradiatis, 1980). Only one of these rapid sequencing methods (Peattie, 1979) is a chemical procedure, and it is this direct chemical sequencing method that is described in detail here.

In this sequencing method, RNA molecules end-labeled with ^{32}P are specifically damaged in a limited fashion with base-specific chemical reagents, and the molecules are then chemically fragmented at the sites of damage. The resultant radioactive products constitute a nested set of RNA fragments which can be resolved according to length through a polyacrylamide slab gel. Current gel technology permits at least the first 100–200 bases of the nucleotide sequence to be read directly from an autoradiograph of the gel. Garoff and Ansorge (1981) recently reported consistently reading 350–400 nucleotides from end-labeled DNA molecules, and the procedures they describe should also be applicable to polyacrylamide gels used for reading RNA sequences. Using this method, no primers are required, the RNA substrates are not limited to ones that can be synthesized *in vitro*, and sequence errors resulting from misincorporations do not occur.

This chapter is divided into four sections, which consider:

(i)   principles of the chemical sequencing reactions;
(ii)   radioactively labeling the RNA at one of its termini;
(iii)   procedural details of the chemical sequencing reactions;
(iv)   using the reactions to probe RNA conformation.

There is also an appendix, which contains a concise outline of the sequencing procedure as well as a diagnostic table for aid in correcting occasional aberrations in the chemical RNA sequencing patterns.

I.   PRINCIPLES OF THE CHEMICAL RNA SEQUENCING REACTIONS

The general sequencing strategy is twofold. First, the bases within the RNA molecule are damaged chemically; second, chemical strand scission is induced at the sites of modification. The initial chemical reaction is specific with regard to which base is attacked yet random with regard to the *position* of the base along the polynucleotide strand. In addition, it is a limited reaction such that, on average, only one base is modified per molecule (Maxam & Gilbert, 1977, 1980; Peattie, 1979). This modification weakens or ruptures the glycosyl bond between the base and the ribosyl moiety of the polynucleotide backbone such that chemical strand scission can be induced at the $3'$-adjacent phosphoester linkage under conditions mild enough to preclude general hydrolysis of the polynucleotide. Thus, strand scission at the site of a chemically damaged base generates a nested set of radioactive fragments, and these are resolved according to length by electrophoresis through a polyacrylamide gel. An autoradiograph of such a gel displays a series of bands where each band corresponds to a discrete nucleotide fragment, and the nucleotide sequence is read directly from this banding pattern (see Fig. 7-1).

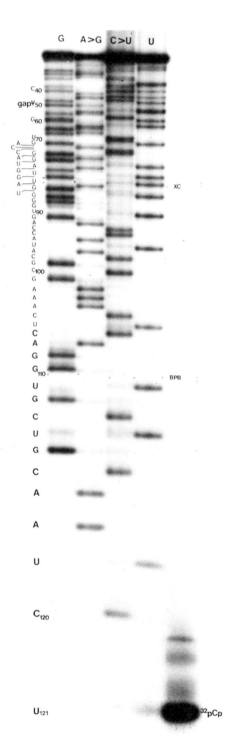

Figure 7-1. Autoradiograph of a sequencing gel displaying the 121 nucleotides of the yeast 5S RNA molecule. The RNA was labeled with (5'-^{32}P)pCp at the 3' terminus and chemically cleaved at guanosines (G), adenosines and guanosines (A > G), cytidines and uridines (C > U), and uridines (U). The samples were electrophoresed on a 0.38-mm-thick polyacrylamide gel (12% acrylamide/0.6% N,N'-methylene-bis(acrylamide)/7 M urea/ 50 mM tris-borate, pH 8.3/ 1 mM EDTA/0.08% ammonium persulfate) at 1.6 kV for 2–2.5 hr, and the gel was exposed to Kodak No-Screen film for 12 hr.

263

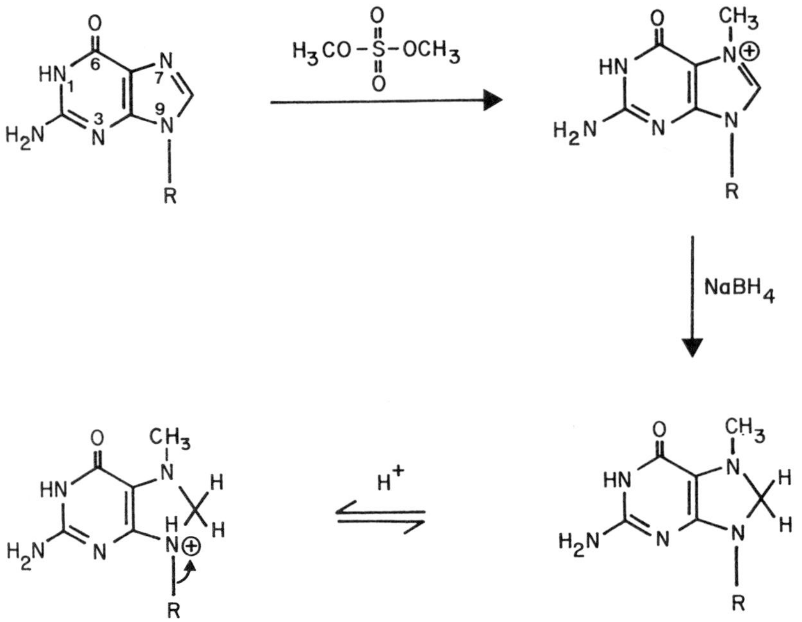

Figure 7-2. Dimethyl sulfate modification at the N-7 atom of guanosine followed by sodium borohydride reduction perturbs the electron resonance of the purine ring and provides a site for aniline-induced strand scission (see Fig. 7-6). $R = \beta$-D-ribofuranosyl.

A. Chemical Base Modification

Guanosine Reaction

The guanosine reaction (Fig. 7-2) uses dimethyl sulfate to alkylate guanosines followed by reduction with sodium borohydride to destabilize the methylated purine rings. Dimethyl sulfate preferentially alkylates the N-7 position of guanosine in both DNA and RNA (Brookes & Lawley, 1961; Lawley & Brookes, 1963); this fixes a positive charge on the N-7 atom and perturbs the electron resonance of the purine ring (Lawley & Brookes, 1963; Kochetov & Budovskii, 1972). The perturbed ring opens in alkali (Kochetov & Budovskii, 1972), making the N-7 methylation a useful reaction for locating guanosines in DNA molecules (Maxam & Gilbert, 1977, 1980), but the random hydrolysis attendant to the alkaline conditions precludes its use for locating guanosines within RNA molecules. In addition, and also unlike the modified DNA base, methyl-7-guanosine in RNA is stable at neutral pH (Haines, Reese, & Todd, 1962) precluding mild depurination. However, the perturbed 7,8-double-bond of this methylated ring system can be reduced easily in a dilute sodium borohydride solution (Pochon et al., 1970; Wintermeyer & Zachau, 1975) and the resultant m^7-dihydroguanosine, which hydrolyzes partially or completely to unknown products, provides a site for aniline-induced strand scission (Wintermeyer & Zachau, 1975).

Adenosine Reaction

Diethyl pyrocarbonate is used in the adenosine reaction (Fig. 7-3) to carbethoxylate base nitrogens. The N-7 atoms of adenosines and guanosines are particularly susceptible to carbethoxylation (Leonard et al., 1971; Vincze et al., 1973; Ehrenberg, Fedorcsak, & Solymosy, 1976) and such modification at these sites destroys the resonance of the heterocyclic ring systems. The imidazole ring opens between atoms N-7 and C-8 (Leonard et al., 1971), thereby creating a site for aniline-induced strand scission (Peattie, 1979).

Uridine Reaction

In the uridine reaction (Fig. 7-4), unprotonated hydrazine opens pyrimidine rings by nucleophilic addition at the 5,6 double-bond (Brown, 1974; Cashmore & Peterson, 1978). It attacks the C-4 and C-6 atoms to produce a new five-membered ring containing C-4, C-5, and C-6 of the hydrazine ring and leaves the hydrazine nitrogens attached to the ribose ring (Brown, 1974; Cashmore & Peterson, 1978). The N-1—C-1' bond then hydrolyzes, and the ring product is released into solution (Levene & Bass, 1927; Baron & Brown, 1955; Cashmore & Peterson, 1978) exposing the ribose ring and leaving it susceptible to aniline attack (Peattie, 1979). Hydrazine attacks uridines much faster than cytidines (Verwoerd & Zillig, 1963); however, 3 M sodium chloride in anhydrous hydrazine allows significant hydrolysis of cytidines (see below).

Cytidine Reactions

The presence of 3 M sodium chloride in anhydrous hydrazine markedly decreases hydrazinolysis of uridines relative to cytidines, and the attack on cytidines becomes the dominant reaction (Fig. 7-5) (Peattie, 1979). This relative suppression, analogous to the decreased hydrazinolysis of thymidines in DNA in the presence of sodium chloride (Maxam & Gilbert, 1977, 1980), is not understood, but

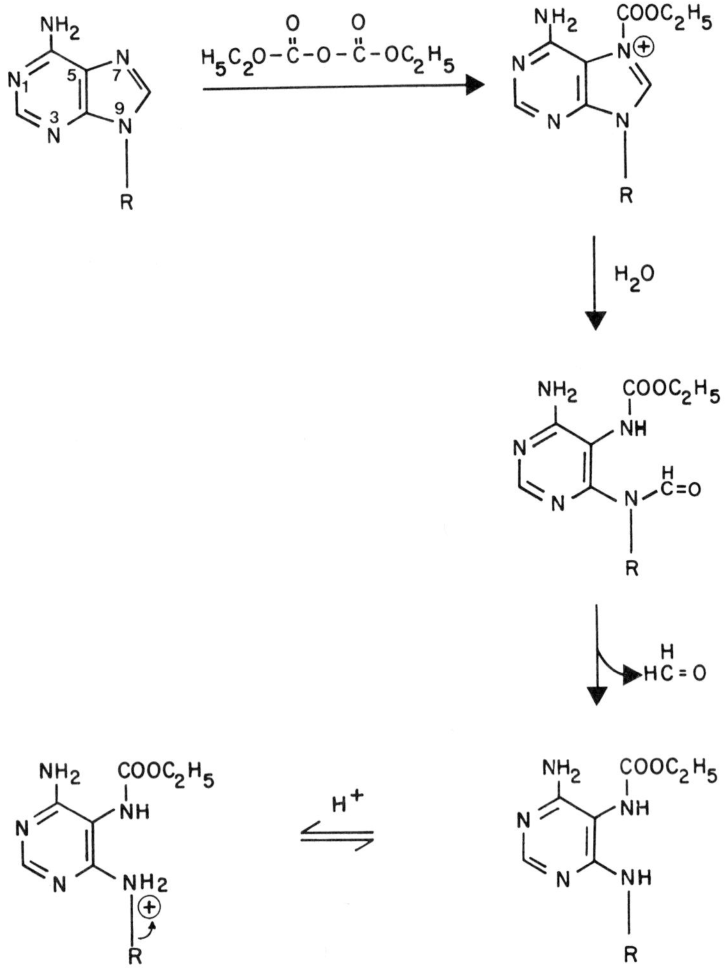

Figure 7-3. Diethyl pyrocarbonate modifies the N-7 atom of adenosine, and the imidiazole ring opens with the release of formaldehyde. This provides a site for aniline-induced strand scission (see Fig. 7-6). $R = \beta$-D-ribofuranosyl.

it results in both cytidines and uridines being susceptible to aniline-induced strand scission (Peattie, 1979).

Cleavage at cytidines and uridines also results when the RNA is methylated with dimethyl sulfate prior to hydrazinolysis (Peattie, 1980; Peattie & Gilbert, 1980). Dimethyl sulfate alkylates the N-3 position of cytidines (Lawley & Brookes, 1963), and the subsequent

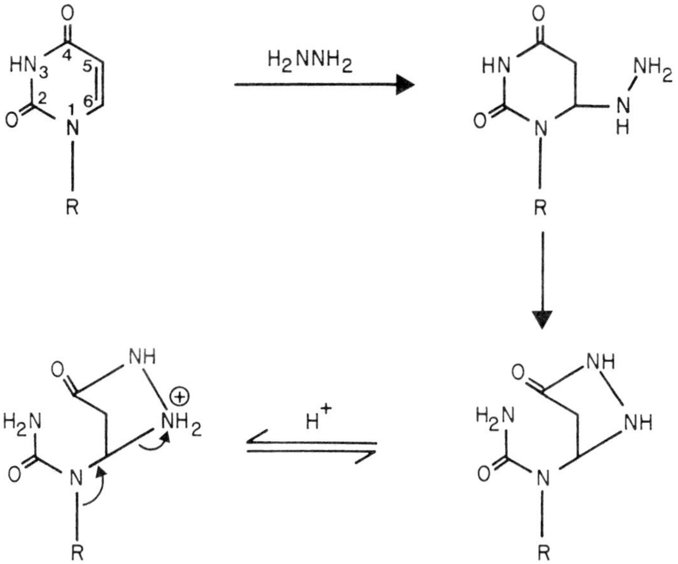

Figure 7-4. Hydrazine provides a site for aniline-induced strand scission (see Fig. 7-6) at uridines by attacking the C-4 and C-6 atoms and releasing the 5-membered pyrazolone ring into solution. $R = \beta$-D-ribofuranosyl.

positive charge on the pyrimidine ring makes m^{3}C susceptible to rapid hydrazinolysis in aqueous hydrazine. The methylated guanosines (formed by N-7 alkylation by the dimethyl sulfate) are not sensitive to the subsequent hydrazine and aniline reactions.

B. Aniline-induced Strand Scission

Aniline (Fig. 7-6) and other primary amines induce strand scission in depurinated or depyrimidinated nucleic acids (Philippsen et al., 1968; Vanyushin & Bur'yanov, 1969; Turchinskii et al., 1970) via aldimine (Schiff base) formation with the open-ring tantomer of the ribose moiety and subsequent β-elimination and cleavage of the phosphodiester bond (Whitfield, 1954; Burton & Peterson, 1960; Yu & Zamecnik, 1960; Khym, 1963; Neu & Heppel, 1964; Whitfield, 1965; Steinschneider & Fraenkel-Conrat, 1966). Aniline hydrochloride catalyzes this reaction with nucleic acids efficiently and optimally at pH 5 (Philippsen et al., 1968; Vanyushin & Bur'yanov, 1969; Turchinskii et al., 1970) without random hydrolysis, but an

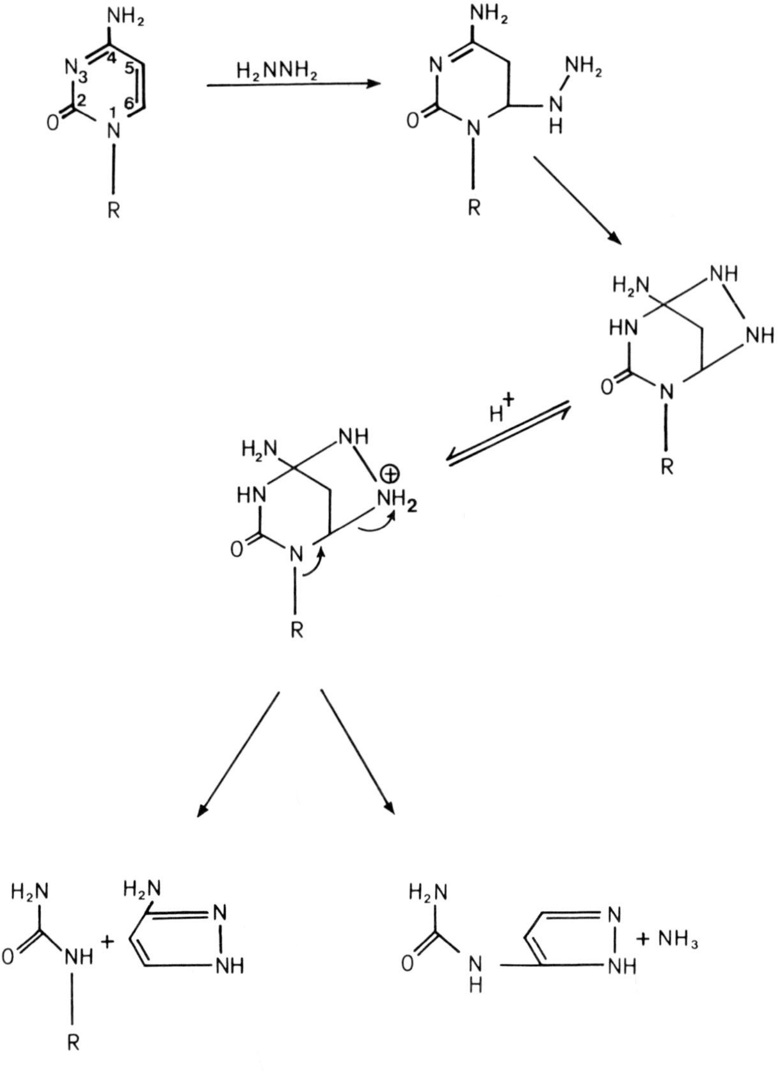

Figure 7-5. Hydrazinolysis of cytidine produces 3-aminopyrazole (lower left) and 3-ureidopyrazole (lower right) and provides a site for aniline-induced strand scission (see Fig. 7-6). $R = \beta$-D-ribofuranosyl.

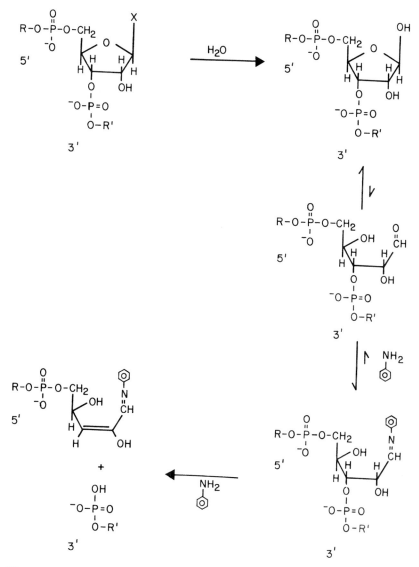

Figure 7-6. Following hydrolysis of the glycosidic bond between the chemically modified base (X) and the ribose ring, the exposed ribose moiety forms its open-ring tautomer. The free aldehyde group of this tautomer reacts with aniline to form an aldimine (Schiff base), and the 3'-phosphoester bond is cleaved via β-elimination. R and R' represent the respective 5' and 3' portions of the RNA strand proximal to the exposed ribose residue.

aniline/acetic acid solution at pH 4.5 effects the same reaction and provides a volatile medium for lyophilization (Peattie, 1979). When used to induce strand scission in DNA, the aniline remains bound to the deoxyribose residue (Vanyushin & Bur'yanov, 1969; Maxam, 1979), and this probably holds for the reaction with RNA as well.

II. END-LABELING THE RNA

The RNA molecule to be sequenced should be labeled with ^{32}P at either its 3' or its 5' terminus.

A. 3'-End-labeling with (5'-^{32}P)pCp and T4 RNA Ligase

T4 RNA ligase catalyzes formation of a phosphodiester bond between an oligonucleotide donor with a 5'-terminal phosphate and an oligo-nucleotide acceptor with a 3'-terminal hydroxyl (England & Uhlenbeck, 1978; Gumport & Uhlenbeck, 1980; Uhlenbeck & Gumport, 1981). Therefore, using (5'-^{32}P)pCp as the donor molecule provides an efficient and easy way to label many RNA molecules at their 3'-hydroxyl termini. The conditions described below are based on those described by Bruce and Uhlenbeck (1978) and are suitable for end-labeling a variety of RNA molecules. These conditions employ an excess of RNA relative to the amount of radioactive label, but although this utilizes the label efficiently it is not necessary. An excess of (5'-^{32}P)pCp can be used if desired. As a rough estimate, 50 pmol of end-labeled RNA is sufficient for many sequencing reactions.

Procedure

Lyophilize x pmol (5'-^{32}P)pCp and $4x$ pmol RNA (use pelleted[1] RNA or RNA stored in water or a volatile buffer) to dryness in a polypropylene snap-cap tube.

---

[1] If the RNA is precipitated immediately before labeling, it can be pelleted quite effectively using 5–10 μg of ribonuclease-free glycogen (oyster glycogen treated with pronase or proteinase K and then phenol-extracted, pelleted, and resuspended in water). Using glycogen is often more convenient than using nucleic acid carriers that can compete for the radioactive substrate, and glycogen does not interfere with the labeling reactions or with the chemical sequencing reactions.

Add

5 µl   2 × concentrated reaction buffer
         100 mM HEPES, pH 7.5
         6.6 mM dithiothreitol
         30 mM magnesium chloride
         40% dimethyl sulfoxide
         0.02 mg/ml serum albumin

x µl   0.01–0.1 M ATP, to obtain a 15-fold excess of ATP relative to the $(5'\text{-}^{32}P)pCp$ concentration

x µl   water, distilled and autoclaved.

Resuspend thoroughly using a Vortexer and spin quickly in an Eppendorf centrifuge. To obtain a final reaction volume of 10 µl, add

x µl   T4 RNA ligase, to obtain a final concentration of approximately 500 units per ml

Centrifuge quickly and incubate at 4°C for 5 hr or at 0°C overnight.

The labeling reaction is terminated effectively by adding an equal volume of 2 × concentrated loading buffer and electrophoresing the sample on a polyacrylamide gel. For a denaturing gel containing 7 M urea, add

10 µl 10 M urea
      40 mM tris-HCl, pH 7.4
      2 mM EDTA
      0.1% xylene cyanol
      0.1% bromophenol blue

Quickly centrifuge the sample, heat it at 60°C for 30 sec, and load it on a polyacrylamide gel slot approximately 1 × 0.075 cm.

After gel electrophoresis, locate the RNA by autoradiography, excise it with a sterile blade, and then elute and precipitate it. To elute the RNA, place the intact piece of polyacrylamide into a 1.5-ml snap-cap tube and add

0.5 M ammonium acetate
0.1 mM EDTA
0.1% sodium dodecyl sulfate

Add sufficient elution solution to cover the gel slice, cap the tube, and place it on a slowly moving rotary apparatus so that the gel slice moves gently in the solution. After elution,[2] transfer the elution solution to another test tube, add 2.5–3 volumes of cold ethanol, chill at $-20°C$ or $-70°C$, and precipitate the RNA. If desired, the efficiency of the end-labeling reaction can be monitored easily before terminating it. This is done using a PEI-cellulose (Macherey-Nagel, Polygram CEL300PEI) thin-layer plate (Maxam & Gilbert, 1980). Prechromatograph the plate in distilled water, air-dry, and cut off a small rectangular piece. Draw a pencil line across the bottom of this small piece, 10 mm from the edge. At two different places on this line, spot small (0.1–0.5 µl) aliquots of the labeling reaction before and after incubation; store the plate at $-20°C$ during the reaction. Spot a third aliquot (taken before *or* after incubation) at the very top of the plate to serve as a reference mark. Dry the plate quickly with a hand-held hair dryer and then chromatograph in 0.3 M sodium dihydrogen phosphate, titrated to pH 3.5 with phosphoric acid. A covered beaker or small screw-cap jar with a small glass rod in the bottom serves well as a chromatography tank. Place the solvent in the beaker or jar (taking care that the final volume will not cover the pencil line on the thin-layer plate) and place the cellulose layer of the plate against the glass rod with the pencil line at the bottom. Chromatograph until the solvent front has run approximately 3/4 of the plate length. Dry the plate, expose it to X-ray-sensitive film, and develop the film.

The RNA remains at the origin while any unincorporated $(5'-^{32}P)pCp$ moves behind the solvent front. If desired, the efficiency of the reaction can be quantitated by (i) scraping the RNA and $(5'-^{32}P)pCp$ spots from the thin-layer plate, suspending the cellulose in Toluene-Omnifluor, and monitoring ^{32}P in a scintillation counter, or (ii) cutting out the regions of the chromatographic plate containing the RNA and $(5'-^{32}P)pCp$ spots and counting the Cerenkov radiation.

If the RNA has been labeled quantitatively, the reaction can be terminated. If there has not been good incorporation, one of three things should be considered:

(i)   the T4 RNA ligase is not working;

---

[2] At room temperature, this procedure effectively elutes small RNA molecules (up to 200 nucleotides) overnight. Higher temperatures (up to 60°C) can be used for more rapid elution, and larger RNA molecules may require that the gel slice be crushed with a ribonuclease-free instrument before adding the elution solution.

(ii) the amount of RNA present is actually less than determined;
(iii) the reaction conditions are not optimal.

If (i) is the problem, more T4 RNA ligase can be added to the reaction. If (ii) is the problem, more RNA can be added to the reaction in a small volume (1-2 μl) of water. In both cases incubation can be continued from this point, and another thin-layer chromatography plate can be run following this second incubation. If (iii) is the problem, the reaction should be terminated while optimal reaction conditions are found (if desired). The labeling conditions can be optimized by performing small-scale reactions using various relative concentrations of $(5'-^{32}P)pCp$, RNA, and ATP. The reactions can each be monitored on the PEI-cellulose plates. However, for practical purposes, it is often *not* necessary to obtain quantitative addition of $(5'-^{32}P)pCp$ to the 3' terminus. Even an inefficient labeling reaction can yield more than enough radioactive material for the chemical sequencing reactions.

B.   5'-End-labeling with ^{32}P

Capped RNA Molecules

Most eukaryotic and viral mRNA molecules possess a $m^7G(5')ppp$-$(5')X$ "cap structure" at their 5' termini (Shatkin, 1976). This nucleotide effectively blocks the 5' terminus and precludes simple dephosphorylation and end-labeling with $(\alpha-^{32}P)ATP$ and T4 polynucleotide kinase; however, other enzymatic and chemical procedures exist which circumvent this problem. Muthukrishnan and Shatkin (1975) have described a chemical means of decapping by oxidizing the 5' terminus with sodium periodate followed by aniline-induced β-elimination, and Efstradiatis et al. (1977) have described an enzymatic method for hydrolyzing the $m^7G(5')ppp$ moiety with tobacco acid pyrophosphatase. In both procedures the initial decapping step can be followed by phosphate removal with phosphatase and radioactive labeling with $(\alpha-^{32}P)ATP$ and T4 polynucleotide kinase (Flavell et al., 1980).

Uncapped RNA Molecules

Cap structures are peculiar to the messenger RNA and heterogeneous nuclear RNA of eukaryotes with the exception of low molecular-weight nuclear RNAs from rat hepatoma nuclei (Shatkin, 1976).

Simsek et al. (1973) and Silberklang et al. (1977) have described a method for removing the terminal phosphate group from uncapped RNA molecules or fragments using phosphatase and replacing it with ^{32}P using (α-^{32}P)ATP and T4 polynucleotide kinase.

C.   Practical Considerations

The 3'-end-labeling reaction using (5'-^{32}P)pCp is particularly useful for labeling RNA molecules that do not label well at their 5' termini. These include molecules such as ribosomal 5S RNAs and tRNAs in which there is extensive intramolecular secondary structure between the two terminal regions. In addition, strand scission at the 3'-penultimate   nucleotide of (5'-^{32}P)pCp-labeled RNA (which is actually the 3' *ultimate* nucleotide of the RNA molecule being sequenced) regenerates the radioactive marker and provides unambiguous identification of the 3'-terminal residue (Fig. 7-7).

When should the RNA be labeled? In theory, and in practice, the RNA fragment or molecule can be end-labeled before *or after* the initial chemical modifications because the phosphodiester back-

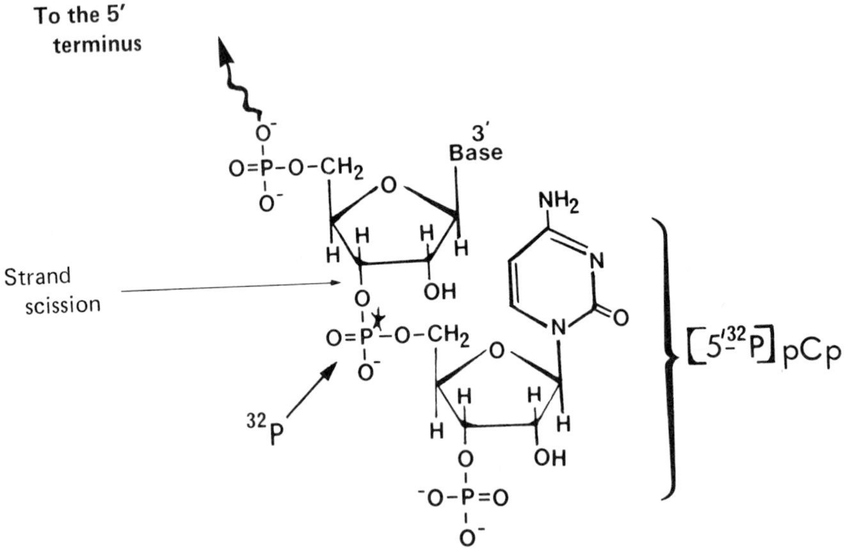

Figure 7-7.   Strand scission at the 3'-ultimate nucleotide of the RNA occurs at the site indicated (also see Fig. 7-6), and free (5'-^{32}P)pCP is regenerated. Using a known (5'-^{32}P)pCp marker thus provides an unambiguous reference for locating the 3'-terminal nucleotide.

bone remains intact until after the aniline-induced strand scission. This property is particularly useful when using the chemical reactions to probe the higher-order conformation of an RNA molecule within a macromolecular complex because it provides a means of probing the complex prior to disrupting it (see Section IV).

The *chemically modified* RNA *can be* manipulated in many various ways before actually inducing strand scission. This property can greatly decrease the amount of effort required to sequence several RNA fragments or molecules. For example, the four initial chemical reactions can be carried out on a *mixture* of end-labeled RNA fragments or molecules (or the RNA could be end-labeled *after* the initial chemical modification reactions) which can then be fractionated on a polyacrylamide gel and eluted and precipitated separately. Each end-labeled sample can then be treated directly with aniline and loaded on a sequencing gel. Thus, four RNA fragments can be sequenced using 4 reaction vessels rather than 16 for the initial chemical modifications (see Fig. 7-8). This strategy works for sequencing mixtures of end-labeled DNA molecules as well (W. Herr, personal communication).

III. BASE-SPECIFIC CHEMICAL CLEAVAGE REACTIONS

Procedures

The chemical procedures described below will specifically cleave any given RNA molecule at one or two of the four major RNA bases and will permit a display of at least 200 nucleotides from a 3′- or 5′-terminal label.

Procedure 1: Limited RNA Cleavage at Guanosines

300 μl   50 mM sodium cacodylate, pH 5.5
         1 mM EDTA
  5 μl   end-labeled RNA, in water

Add the RNA to the buffer in a 1.5-ml polypropylene snap-cap tube and mix. Chill to 0°C on ice. Add

0.5 μl   dimethyl sulfate, reagent grade

Mix briefly on a Vortexer, spin quickly in an Eppendorf 3200/30

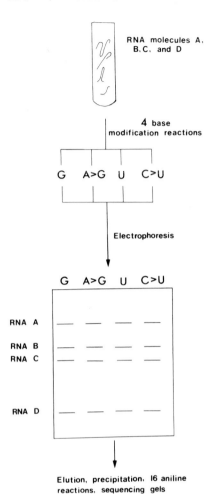

Figure 7-8. The initial chemical modification reactions apply to mixtures of RNA molecules as well as to homogeneous RNA populations, and this property decreases the work required to sequence several different RNA molecules. Here, the end-labeled molecules A, B, C, and D are separated from each other on a polyacrylamide gel *after* the limited base-specific chemical modifications; thus, only 4 such modifications need be done rather than 16. Each molecule is then preprogrammed for cleavage at the sites of base damage, and strand scission with aniline can be induced directly after elution and precipitation from the polyacrylamide gel. The samples can then be run on a standard denaturing polyacrylamide sequencing gel.

microcentrifuge, and heat for 0.5–1 min in a water bath at $90°C$.[3] Chill on ice immediately and add

75 μl   1.0 M tris-acetate, pH 7.5
1.0 M 2-mercaptoethanol
1.5 M sodium acetate
0.1 mM EDTA
0.2 mg/ml carrier tRNA (4°C)

---

[3] If this incubation time is ever increased to 2 min or more (sometimes required for very short RNA fragments or molecules) increase the pH of the sodium cacodylate reaction buffer to pH 7.5. Otherwise the reaction solution will become too acidic as the dimethyl sulfate hydrolyzes into sulfuric acid and methanol, and random depurination will occur.

900 μl   absolute ethanol (4°C)

Mix the solution well by inverting the tube several times, and then chill at −70°C (Dry Ice/ethanol bath) for 5 min. Centrifuge at 12,000 g for 5 min in an Eppendorf microcentrifuge in a cold room.

Remove the supernatant in a fume hood with a drawn-out Pasteur pipette and place it in a waste bottle containing 5 M sodium hydroxide. Resuspend the sample in

200 μl   0.3 M sodium acetate (4°C)

Add

600 μl   absolute ethanol (4°C)

Mix by inversion, chill, centrifuge, and remove the supernatant as above. Add

0.5 ml   absolute ethanol (−20°C)

Chill at −70°C for 2 min, centrifuge for 2 min, and remove the ethanol with a drawn-out Pasteur pipette. Dry the sample under vacuum, and add

10 μl   1.0 M tris-Cl, pH 8.2

Resuspend the sample well and add

10 μl   0.2 M sodium borohydride, freshly made

Centrifuge quickly and incubate at 0°C for 30 min in the dark. Add

200 μl   0.6 M sodium acetate/0.6 M acetic acid, pH 4.5
0.025 μg/ml carrier tRNA (4°C)

Mix on a Vortexer. Add

600 μl   absolute ethanol (4°C)

Mix by inversion, chill, centrifuge, and remove the supernatant. Add

500 μl   absolute ethanol (−20°C)

Chill at −70°C for 2 min, spin for 2 min, and remove the ethanol. Dry the sample under vacuum and add

20 μl   1.0 M aniline, pH 4.5, with acetic acid, freshly made

Vortex the sample briefly, centrifuge it for a few seconds, and incubate at 60°C for 20 min. Quickly freeze the sample at −70°C, and then puncture the tube cap with a hypodermic needle. Briefly freeze the sample again and lyophilize to dryness. Redissolve the RNA in

20 μl   water

Freeze and lyophilize again, and then lyophilize a third time after resuspending the sample in another 20 μl water. For electrophoresis on a denaturing polyacrylamide gel, resuspend the RNA in

2 μl   8 M urea
         20 mM tris-HCl, pH 7.4
         1 mM EDTA
         0.05% xylene cyanol
         0.05% bromophenol blue

Centrifuge for a few seconds, heat at 90°C for 30 sec, and load the sample on the sequencing gel.

Procedure 2: Limited RNA Cleavage at Adenosines and Guanosines

200 μl   50 mM sodium acetate, pH 4.5 with acetic acid
          1 mM EDTA
5 μl   end-labeled RNA, in water

Add the RNA to the buffer in a 1.5-ml snap-cap tube, mix, and chill to 0°C on ice. Add

1 μl   diethyl pyrocarbonate, reagent grade

Mix briefly on a Vortexer, centrifuge quickly, and heat for 10–15 min in a water bath at 90°C. Chill on ice immediately and add

50 μl   1.5 M sodium acetate
          0.2 mg/ml carrier tRNA (4°C)
750 μl   absolute ethanol (4°C)

Mix the solution well by inverting the tube several times and then chill at −70°C for 5 min. Centrifuge in an Eppendorf microcentrifuge (12,000 g) for 5 min in a cold room. Remove the supernatant in a fume hood with a drawn-out Pasteur pipette and place it in a waste bottle containing 5 M sodium hydroxide. Resuspend the sample in

200 μl   0.3 M sodium acetate (4°C)

Add

600 μl   absolute ethanol (4°C)

Mix by inversion, chill, centrifuge, and remove the supernatant as above. Add

500 μl   absolute ethanol (−20°C)

Chill at −70°C for 2 min, centrifuge for 2 min, and remove the ethanol with a drawn-out Pasteur pipette. Dry the sample under vacuum and add

20 μl   1.0 M aniline, pH 4.5, with acetic acid, freshly made

Vortex the sample briefly, centrifuge it for a few seconds, and incubate at 60°C for 20 min. Quickly freeze the sample at −70°C, and then puncture the tube cap with a hypodermic needle. Briefly freeze the sample again and lyophilize to dryness. Redissolve the RNA in

20 μl   water

Freeze and lyophilize again, and then lyophilize a third time after resuspending the sample in another 20 μl water. Resuspend the RNA in

2 μl   8 M urea
        20 mM tris-HCl, pH 7.4
        1 mM EDTA
        0.05% xylene cyanol
        0.05% bromophenol blue

Centrifuge for a few seconds, heat at 90°C for 30 sec, and load the sample on a sequencing gel.

Procedure 3: Limited RNA Cleavage at Uridines

5 μl   end-labeled RNA, in water

Lyophilize to dryness. Add

10 μl   hydrazine/water (1:1, 0°C), freshly made
        use reagent-grade hydrazine (95% minimum by titra-
        tion) directly from the bottle

Mix briefly on a Vortexer, and centrifuge quickly in an Eppendorf microcentrifuge. Incubate on ice for 5–10 min and then add

> 200 μl   0.3 M sodium acetate
> 0.1 mM EDTA
> 0.05 mg/ml carrier tRNA (4°C)
> 600 μl   absolute ethanol (4°C)

Mix the solution well by inverting the tube several times, and then chill at −70°C for 5 min. Centrifuge in an Eppendorf microcentrifuge (12,000 g) for 5 min in a cold room. Remove the supernatant in a fume hood with a drawn-out Pasteur pipette, and place it in a waste bottle containing 2 M ferric chloride. Resuspend the sample in

> 200 μl   0.3 M sodium acetate (4°C)

Add

> 600 μl   absolute ethanol (4°C)

Mix by inversion, chill, centrifuge, and remove the supernatant as above. Add

> 500 μl   absolute ethanol (−20°C)

Chill at −70°C for 2 min, centrifuge for 2 min, and remove the ethanol with a drawn-out Pasteur pipette. Dry the sample under vacuum and add

> 20 μl   1.0 M aniline, pH 4.5, with acetic acid, freshly made

Vortex the sample briefly, centrifuge it for a few seconds, and incubate at 60°C for 20 min. Quickly freeze the sample at −70°C, and then puncture the tube cap with a hypodermic needle. Briefly freeze the sample again and lyophilize to dryness. Redissolve the RNA in

> 20 μl   water

Freeze and lyophilize again, and then lyophilize a third time after resuspending the sample in another 20 μl water. Resuspend the RNA in

> 2 μl   8 M urea
> 20 mM tris-HCl, pH 7.4
> 1 mM EDTA
> 0.05% xylene cyanol
> 0.05% bromophenol blue

Centrifuge for a few seconds, heat at 90°C for 30 sec, and load the sample on a sequencing gel.

Procedure 4: Limited RNA Cleavage at Cytidines and Uridines

5 μl   end-labeled RNA, in water

Lyophilize to dryness. Add

10 μl   3.0 M sodium chloride in anhydrous hydrazine, freshly made (0°C)

use reagent-grade hydrazine (95% minimum by titration) directly from the bottle

Mix briefly on a Vortexer and centrifuge quickly in an Eppendorf microcentrifuge. Incubate on ice for 25–30 min and then add

500 μl   absolute ethanol/water (4:1, −20°C)
0.2 mg/ml carrier tRNA

Mix the solution well by inverting the tube several times and then chill at −70°C for 5 min. Centrifuge at 12,000 g in an Eppendorf microcentrifuge for 5 min in a cold room. Remove the supernatant in a fume hood with a drawn-out Pasteur pipette and place it in a waste bottle containing 2 M ferric chloride. Resuspend the sample in

200 μl   0.3 M sodium acetate (4°C)

Add

600 μl   absolute ethanol (4°C)

Mix by inversion, chill, centrifuge, and remove the supernatant as above. Add

500 μl   absolute ethanol (−20°C)

Chill at −70°C for 2 min, centrifuge for 2 min, and remove the ethanol with a drawn-out Pasteur pipette. Dry the sample under vacuum and add

20 μl   1.0 M aniline, pH 4.5 with acetic acid, freshly made

Vortex the sample briefly, centrifuge it for a few seconds, and incubate at 60°C for 20 min. Quickly freeze the sample at −70°C, and then puncture the tube cap with a hypodermic needle. Briefly freeze the sample again and lyophilize to dryness. Redissolve the RNA in

20 µl   water

Freeze and lyophilize again, and then lyophilize a third time after resuspending the sample in another 20 µl water. Resuspend the RNA in

2 µl   8 M urea
        20 mM tris-HCl, pH 7.4
        1 mM EDTA
        0.05% xylene cyanol
        0.05% bromophenol blue

Centrifuge for a few seconds, heat at 90°C for 30 sec, and load the sample on a sequencing gel.

Procedure 5: Alternative Reaction Scheme for Limited RNA Cleavage at Cytidines and Uridines[4]

300 µl   50 mM sodium cacodylate, pH 5.5
          1 mM EDTA
5 µl   end-labeled RNA, in water

Add the RNA to the buffer in a 1.5-ml polypropylene snap-cap tube, and mix. Chill to 0°C on ice. Add

0.5 µl   dimethyl sulfate, reagent grade

Mix briefly on a Vortexer, spin quickly in an Eppendorf 3200/30 microcentrifuge, and heat for 0.5–1 min in a water bath (see footnote 3) at 90°C. Chill on ice immediately and add

75 µl   1.0 M tris-acetate, pH 7.5
       1.0 M 2-mercaptoethanol
       1.5 M sodium acetate

---

[4] This reaction scheme should be used when there is a m^{7}G residue within the molecule being sequenced (e.g., within a tRNA molecule). The extreme susceptibility of this base to the hydrazine/salt reaction (Procedure 4) can result in quantitative cleavage at this site, making it impossible to localize cytidines beyond this residue. Cytidines and uridines appear as bands of equal intensity when this alternative procedure is used.

0.1 mM EDTA
0.2 mg/ml carrier tRNA (4°C)
900 μl   absolute ethanol (4°C)

Mix the solution well by inverting the tube several times, and then chill at −70°C (Dry Ice/ethanol bath) for 5 min. Centrifuge at 12,000 g for 5 min in an Eppendorf microcentrifuge. Remove the supernatant in a fume hood with a Pasteur pipette and place it in a waste bottle containing 5 M sodium hydroxide. Resuspend the sample in

200 μl   0.3 M sodium acetate (4°C)

Add

600 μl   absolute ethanol (4°C)

Mix by inversion, chill, centrifuge, and remove the supernatant as above. Add

0.5 ml   absolute ethanol (−20°C)

Chill at −70°C for 2 min, centrifuge for 2 min, and remove the ethanol with a drawn-out Pasteur pipette. Dry the sample under vacuum, and add

10 μl   hydrazine/water (1:1, 0°C), freshly made

Mix briefly on a Vortexer and centrifuge quickly in an Eppendorf microcentrifuge. Incubate on ice for 5 min and then add

200 μl   0.3 M sodium acetate
0.1 mM EDTA
0.05 mg/ml carrier tRNA (4°C)
600 μl   absolute ethanol (4°C)

Mix the solution well by inverting the tube several times and then chill at −70°C for 5 min. Centrifuge in an Eppendorf microcentrifuge (12,000 g) for 5 min in a cold room. Remove the supernatant in a fume hood with a Pasteur pipette, and place it in a waste bottle containing 2 M ferric chloride. Resuspend the sample in

200 μl   0.3 M sodium acetate (4°C)

Add

600 μl   absolute ethanol (4°C)

Mix by inversion, chill, centrifuge, and remove the supernatant as above. Add

500 μl   absolute ethanol ($-20°$C)

Chill at $-70°$C for 1 min, centrifuge for 2 min, and remove the ethanol with a drawn-out Pasteur pipette. Dry the sample under vacuum, and add

200 μl   1.0 M aniline, pH 4.5, with acetic acid, freshly made

Vortex the sample briefly, centrifuge it for a few seconds, and incubate at 60°C for 20 min. Quickly freeze the sample at $-70°$C and then puncture the tube cap with a hypodermic needle. Briefly freeze the sample again, and lyophilize to dryness. Redissolve the RNA in

20 μl   water

Freeze and lyophilize again, and then lyophilize a third time after resuspending the sample in another 20 μl water. Resuspend the RNA in

2 μl   8 M urea
20 mM tris-HCl, pH 7.4
1 mM EDTA
0.05% xylene cyanol
0.05% bromophenol blue

Centrifuge for a few seconds, heat at 90°C for 30 sec, and load the sample on a sequencing gel.

Practical Considerations

Occasionally, aberrations occur when running a sequencing gel or when reading the RNA sequence from the autoradiograph of the gel. Several of these are discussed below, and the reader is also referred to Table 7-A-1 in the Appendix.

For a reason that is not understood, 5'-end-labeled RNA produces atypical pyrimidine banding patterns. The uridine patterns are distinct but contain background bands due to strand scission at every base; the cytidine patterns have this same background but lack preferential cleavages at cytidines. This does not preclude chemically sequencing 5'-end-labeled RNA, however, because adenosines, guanosines, and uridines are located and identified unambiguously.

In some cases, the initial modification reactions may need to be varied to increase the overall base modification. This would obtain if (i) the RNA molecule were very small (between about 1–40 nucleotides), (ii) only the first approximate 1–40 nucleotides were desired, or (iii) too much RNA was consistently left unreacted on the sequencing gel. To increase overall base modification in the G, C, and U reactions, lengthen the incubation times (see footnote 3) and keep the reagent concentrations the same. To increase base modification in the A reaction, keep the incubation time the same and increase the diethyl pyrocarbonate concentration. Do not change the temperature for any of the reactions.

When loading the reactions on the sequencing gel, it is helpful to arrange them in the order G, A > G, C > U, U. Centering the two bispecific reactions facilitates reading the autoradiograph of the sequencing gel in a stepwise fashion. If 3′-end-labeled material is being used, the 3′-ultimate nucleotide of the RNA is identified unambiguously by coelectrophoresing an aliquot of (5′-^{32}P)pCp with the samples. The band corresponding to the 3′-ultimate nucleotide is in the same position as the radioactive marker. Identifying the 5′-terminal nucleotide can be difficult when using 3′-end-labeled material because the signal from the intact, unreacted RNA tends to obscure the band corresponding to the ultimate nucleotide. Resolution of the 5′ terminus can be improved by adding a small oligonucleotide extension sequence to the 5′ end of the molecule using T4 RNA ligase (Luehrsen & Fox, 1981).

This chemical RNA sequencing method does not identify unusual bases unambiguously, and standard methods (Hall, 1971; Brownlee, 1972) of identifying these bases should be used. However, the presence of an unusual base is normally signaled by a nucleotide gap on the autoradiograph (e.g., pseudouridine, ribothymidine, methyl-1-adenosine, methyl-5-cytidine) or by a band appearing in the same position in every lane (e.g., methyl-7-guanosine and the Y base of yeast phenylalinine tRNA). The former phenomenon occurs when the unusual base is insensitive to both the chemical modifications and the strand scission reaction. The latter phenomenon occurs when the unusual base is sensitive to the acidic pH of the aniline reaction. In some cases, the ubiquitous band will be much darker in one of the lanes, indicating that the unusual base is sensitive to one of the chemical modification reactions as well as to the aniline-induced strand scission.

Extremely stable regions of secondary structure can induce (i) decreased methylation of guanosines within the structure and (ii) compression of RNA fragments on the sequencing gel (Brown

& Smith, 1977). The first phenomenon is manifested by weak bands in the G lane (Peattie, 1979); however, the A > G reaction unambiguously identifies these bands as corresponding to guanosines. The phenomenon of sequence compression, resulting in a localized "pile up" of fragments on the sequencing gel, can usually be rectified by running an 8.5 M urea polyacrylamide gel at 55–60°C.

IV. USING THE CHEMICAL REACTIONS TO PROBE RNA CONFORMATION

Like proteins, RNA molecules can be defined in terms of different levels of structural order. The primary structure, or nucleotide sequence, is the unique arrangement of covalently linked nucleotides that constitute the molecule, while secondary and tertiary structures result from noncovalent interactions between the nucleotides. The latter interactions, defined as higher-order interactions, organize the RNA molecule into structural units, i.e., they define its overall conformation.

Three of the chemical sequencing reactions have been adapted for probing higher-order interactions within an RNA molecule (Peattie & Gilbert, 1980). These reactions, for guanosines, adenosines, and cytidines, involve essentially the same steps as those required for nucleotide sequencing; however, unlike the sequencing reactions, they reflect base pairing and tertiary interactions within the molecule (Fig. 7-9).

Figure 7-9. The chemical sequencing reagents can also be used for probing the higher-order structure of an RNA molecule. Dimethyl sulfate is used to alkylate the N-7 position of guanosines and the N-3 position of cytidines, and diethyl pyrocarbonate is used to modify the N-7 of adenosines. These reactions result in base-specific susceptibility to an aniline-induced strand scission reaction, but they occur only when the respective atoms are not involved in structural interactions. Hence, the presence of a band indicates a structurally free atom, while the absence of a band indicates a structurally involved atom. This autoradiograph represents different conformers of 3'-end-labeled yeast tRNA[Phe] (Peattie & Gilbert, 1980). The native conformer was probed in 50 mM sodium cacodylate, pH 7.0/10 mM $MgCl_2$ at 37°C and demonstrates structural interactions defined by X-ray crystallographic analysis. The semidenatured conformer was probed in 50 mM sodium cacodylate, pH 7.0/1 mM EDTA at 37°C and displays, in essence, the secondary interactions of the cloverleaf model. The denatured conformer was probed in this same buffer at 90°C and exhibits few higher-order interactions.

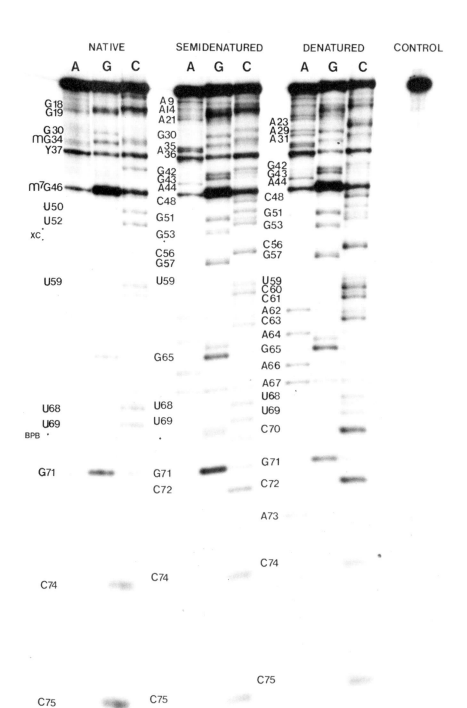

287

The underlying principle of the two procedures is the same: a base-specific reagent weakens or ruptures the glycosyl bond between the base and the ribosyl moiety of the polynucleotide, and this initial chemical modification appears ultimately as a chemically induced strand scission. When using radioactively end-labeled material, such strand scission is manifested as a single band on the autoradiograph of a denaturing polyacrylamide gel. However, the fundamental difference between the nucleotide sequencing reactions and the conformation probing reactions lies within the initial reaction conditions. Only the *initial* chemical attack preprograms the RNA molecule for strand scission, and this initial modification can be sensitive to higher-order interactions within the molecule. Such interactions include hydrogen bonds, ionic interactions, and base stacking interactions. A full-sequence spectrum is generated only if the RNA molecule is initially modified under totally denaturing— i.e., "sequencing"—conditions. If the molecule is probed under native or semidenaturing conditions, only a partial-sequence spectrum is generated because certain sites within the molecule are not available to the chemical reagents.

Guanosine Reaction

Like the G-specific sequencing reaction (Section III, Procedure 1), the guanosine probing reaction reflects N-7 alkylation with dimethyl sulfate (see Fig. 7-2). The N-7 atom is *not* involved in secondary Watson-Crick base pairing (a G—C base pair involves the H-1, H-2, and O-6 guanosines and the O-2, N-3, and H-4 of cytidines), but it is often involved in the general weaker tertiary interactions. Therefore, this reaction monitors tertiary interactions within native or semidenatured RNA molecules. Such interactions, as defined by crystallographic studies of yeast tRNA[Phe] (Ladner et al., 1975; Jack, Ladner, & Klug, 1976, 1977; Quigley & Rich, 1976; Rich & RajBhandary, 1976; Holbrook et al., 1977, 1978; Kim, 1978a, 1978b; Teeter, Quigley, & Rich, 1980), include the hydrogen bonds of non–Watson-Crick base pairs, coordination interactions with various ions (such as Mg^{2+}) or water molecules, and base stacking interactions (Holbrook et al., 1977).

Cytidine Reaction

The cytidine probing reaction, identical in principle to the alternate reaction for cytidine cleavage (Section III, Procedure 5), also depends upon methylation with dimethyl sulfate. This reagent alkylates available N-3 atoms of cytidines as well as the N-7 atoms of guanosines (Lawley & Brookes, 1963), and the resultant methyl-3-cytidines react quite readily with hydrazine. Therefore, methylation followed by hydrazinolysis and aniline-induced strand scission produces cleavages at uridines and at any methyl-3-cytidines. The uridine cleavage occurs as a result of the hydrazine step and does not reflect RNA conformation, because the RNA is denatured at the pH of the hydrazine reaction. This probing reaction monitors tertiary interactions at the cytidine N-3 position, and, because this atom is involved in secondary Watson-Crick interactions, it monitors G—C base pairs as well.

Adenosine Reaction

The adenosine probing reaction, also fundamentally identical to the sequencing reaction (Section III, Procedure 2) involves carbethoxylation of base nitrogens with diethyl pyrocarbonate. This reaction monitors the stacking of adenosines resulting from base pairing or tertiary interactions. It does not similarly monitor guanosines, apparently because guanosine carbethoxylation occurs more readily at the acidic pH used for sequencing rather than at the neutral pH used for probing.

Practical Considerations

These reactions work between 0 and 90°C and from pH 4.5–8.5 in a variety of buffers; however, tris buffers should be avoided since both dimethyl sulfate and diethyl pyrocarbonate modify amino groups. This provides a wide range for optimizing reaction conditions as well as a means of distinguishing subtleties of higher-order conformation. For example, one can follow progressive denaturation or renaturation of an RNA molecule by monitoring it through a gradient of reaction conditions. In addition, this can be displayed on a single autoradiograph by running the reactions in parallel with each other (and with the chemical sequencing reactions) on one polyacrylamide gel (see Fig. 7-9).

The modification reactions work for guanosines (Mirzabekov & Kolchinsky, 1974; Gilbert, Maxam, & Mirzabekov, 1976; Ogata & Gilbert, 1979; Siebenlist, 1980; Siebenlist, Simpson & Gilbert, 1980), cytidines (K. Kirkegaard & J. Sims, personal communications), and adenosines (W. Herr, personal communication) in DNA as well and thereby can also effectively probe DNA-protein interactions. When using DNA, the strand-scission reaction is induced with piperidine as described by Maxam and Gilbert (1977, 1980).

The initial chemical reactions are very limited in extent, ensuring that the integrity of the nucleic acid structure is not significantly affected.

Only the initial chemical reactions determine the subsequent sites of strand scission, and this means that native (or other chosen) conditions need to be maintained *only* for the initial modifications.

The reactions can also monitor RNA conformation in RNA-protein complexes. Diethyl pyrocarbonate tends to dissociate some ribonucleoprotein complexes (Peattie et al., 1981), presumably because it readily modifies histidines (Ehrenberg, Fedoresak, & Solymosy, 1976). This is inconvenient in that it precludes carbethy-oxylation of an intact RNA-protein complex, but the problem can be circumvented by chemically modifying the RNA prior to *in vitro* assembly of the complex (see below). As shown by a study of a tRNA-mRNA-ribosome complex (Peattie & Herr, 1981), however, some RNA-protein complexes are quite stable in the presence of diethyl pyrocarbonate.

The RNA is quite stable after the initial modifications. Thus, the RNA can be modified first and then labeled, purified, and fragmented. This eliminates the need to label the RNA prior to modification and, consequently, provides a means of probing an RNA-protein complex prior to disrupting it. It also means that an RNA molecule can be modified chemically and then tested for biological function. This latter process identifies sites within an RNA molecule that are crucial for functional interaction with a protein and provides a way to study RNA-protein interaction in a complex containing proteins inactivated by diethyl pyrocarbonate (Peattie et al., 1981). Furthermore, very large RNA molecules (such as messengers) can be probed by modifying them first and then cutting them into smaller pieces (Stepanova et al., 1979; Donis-Keller, 1979) for end-labeling, strand scission, and gel analysis.

ACKNOWLEDGMENTS

This work was supported by grant GM09541-17 from the National Institutes of Health, to Walter Gilbert, Department of Biochemistry and Molecular Biology, Harvard University. The expert typing of the manuscript was done by Gina Yiangou while the author was a Visiting Fellow at the Imperial Cancer Research Fund, Lincoln's Inn Fields, London.

REFERENCES

Baron, F., and Brown, D. M. (1955). The structure of cytidylic acids *a* and *b*. *J. Chem. Soc.* 3:2855-2860.

Brookes, P., and Lawley, P. D. (1961). The reaction of mono- and di-functional alkylating agents with nucleic acids. *Biochem. J.* 80:496-503.

Brown, D. M. (1974). Reactions of polynucleotides and nuelcic acids. In *Basic Principles in Nucleic Acid Chemistry*, Vol. 2, ed. P. Ts'o (New York: Academic Press), pp. 1-90.

Brown, N. L., and Smith, M. (1977). The sequence of a region of bacteriophage φX174 DNA coding for parts of genes A and B. *J. Mol. Biol.* 116:1-30.

Brownlee, G. G. (1972). *Determination of Sequences in RNA* (New York: American Elsevier).

Brownlee, G. G., and Cartwright, E. M. (1977). Rapid gel sequencing of RNA by primed synthesis with reverse transcriptase. *J. Mol. Biol.* 114:93-117.

Bruce, A. G., and Uhlenbeck, O. C. (1978). Reactions at the termini of tRNA with T4 RNA ligase. *Nucl. Acids Res.* 5:3665-3677.

Burton, K., and Peterson, G. B. (1960). The frequencies of certain sequences of nucleotides in deoxyribonucleic acid. *Biochem. J.* 75:17-27.

Cashmore, A. R., and Peterson, G. B. (1978). The degradation of DNA by hydrazine: Identification of 3-ureidopyrazole as a product of the hydrazinolysis of deoxycytidine residues. *Nucl. Acids Res.* 5:2485-2491.

Donis-Keller, H. (1979). Site specific enzymatic cleavage of RNA. *Nucl. Acids Res.* 7:179-192.

Donis-Keller, H., Maxam, A. M., and Gilbert, W. (1977). Mapping adenines, guanines, and pyrimidines in RNA. *Nucl. Acids Res.* 4:2527-2538.

Efstradiatis, A., Vournakis, J. N., Donis-Keller, H., Chaconas, G., Dougall, D. K., and Kafatos, F. C. (1977). End labeling of enzymatically decapped mRNA. *Nucl. Acids Res.* 4:4165-4174.

Ehrenberg, L., Fedorcsak, I., and Solymosy, F. (1976). Diethyl pyrocarbonate in nucleic acid research. In *Prog. Nucleic Acid Res. Mol. Biol.*, Vol. 16, ed. W. E. Cohn (New York: Academic Press), pp. 189-262.

England, T. E., and Uhlenbeck, O. C. (1978). 3'-Terminal labeling of RNA with T4 RNA ligase. *Nature* 275:560-561.

Flavell, A. J., Cowie, A., Arrand, J. R., and Kamen, R. (1980). Localization of three major capped 5′ ends of polyoma virus late mRNA's within a single tetranucleotide sequence in the viral genome. *J. Virol.* 32:902-908.

Garoff, H., and Ansorge, W. (1981). Improvements of DNA sequencing gels. *Analyt. Biochem.*

Gilbert, W., Maxam, A. M., and Mirzabekov, A. D. (1976). Contacts between the LAC repressor and DNA revealed by methylation. In *Control of Ribosome Synthesis*, eds. N. C. Kjelgaard and O. Maaloe (New York: Academic Press), pp. 139-148.

Gumport, R. I., and Uhlenbeck, O. C. (1980). T4 RNA ligase as a nucleic acid synthesis and modification reagent. In *Gene Amplification and Analysis*, Vol. II: *Analysis of Nucleic Acid Structure by Enzymatic Methods*, eds. J. G. Chirikjian and T. S. Papas (New York: Elsevier/North Holland).

Haines, J. A., Reese, C. B., and Todd, A. R. (1962). The methylation of guanosine and related compounds with diazomethane. *J. Chem. Soc.* Part IV: 5281-5288.

Hall, R. H. (1971). *The Modified Nucleosides in Nucleic Acids* (New York: Columbia University Press).

Holbrook, S. R., Sussman, J. L., Warrant, R. W., Church, G. M., and Kim, S.-H. (1977). RNA-ligand interactions. I: Magnesium binding sites in yeast tRNA^{Phe}. *Nucl. Acids Res.* 4:2811-2820.

Holbrook, S. R., Sussman, J. L., Warrand, R. W., and Kim, S.-H. (1978). Crystal structure of yeast phenylalanine transfer RNA. II: Structural features and functional implications. *J. Mol. Biol.* 123:631-660.

Jack, A., Ladner, J. E., and Klug, A. (1976). Crystallographic refinement of yeast phenylalanine transfer RNA at 2.5 A resolution. *J. Mol. Biol.* 108: 619-649.

Jack, A., Ladner, J. E., Rhodes, D., Brown, R. S., and Klug, A. (1977). A crystallographic study of metal-binding to yeast phenylalanine transfer RNA. *J. Mol. Biol.* 111:315-328.

Khym, J. X. (1963). The reaction of methylamine with periodate-oxidized adenosine 5′-phosphate. *Biochem.* 2:344-350.

Kim, S.-H. (1978a). Crystal structure of yeast tRNA^{Phe}: Its correlation to the solution structure and functional implications. In *Transfer RNA*, ed. S. Altman (Cambridge, Mass.: MIT Press), pp. 248-293.

Kim, S.-H. (1978b). Three-dimensional structure of transfer RNA and its functional implications. *Adv. Enzymol.* 46:279-315.

Kochetov, N. K. and Budovskii, E. I. (1972). *Organic Chemistry of Nucleic Acids* (New York: Plenum).

Kramer, F. R., and Mills, D. R. (1978). RNA sequencing with radioactive chainterminating ribonucleotides. *Proc. Natl. Acad. Sci. USA* 75:5334-5338.

Ladner, J. E., Jack, A., Robertus, J. D., Brown, R. S., Rhodes, D., Clark, B. F. C., and Klug, A. (1975). Structure of yeast phenylalanine transfer RNA at 2.5 A resolution. *Proc. Natl. Acad. Sci. USA* 72:4414-4418.

Lawley, P. D., and Brookes, P. (1963). Further studies on the alkylation of nucleic acids and their constituent nuelcotides. *Biochem. J.* 89:127-138.

Leonard, N. J., McDonald, J. J., Henderson, R. E. L., and Reichmann, M. E. (1971). Reaction of diethyl pyrocarbonate with nucleic acid components: Adenosine. *Biochem.* 10:3335-3342.

Levene, P. A., and Bass, L. W. (1927). The action of hydrazine hydrate on uridine. *J. Biol. Chem.* 71:167-172.

Lockhard, R. E., Alzner-Deweerd, B., Heckman, J. E., MacGee, J., Tabor, M. W., and RajBhandary, U. L. (1978). Sequence analysis of $5'[^{32}P]$-labeled MRNA and tRNA using polyacrylamide gels. *Nucl. Acids Res.* 5:37-56.

Luehrsen, K. R., and Fox, G. E. (1981). Secondary structure of eukaryotic cytoplasmic 5S ribosomal RNA. *Proc. Natl. Acad. Sci. USA* 78:2150-2154.

McGeoch, D. J., and Turnbull, N. T. (1978). Analysis of the $3'$-terminal nucleotide sequence of vesicular stomatitis virus N protein mRNA. *Nucl. Acids Res.* 5:4007-4024.

McPhie, P., Hounsell, J., and Gratzer, W. B. (1966). The specific cleavage of yeast ribosomal ribonucleic acid with nucleases. *Biochem.* 5:988-993.

Maniatis, T., and Efstradiatis, A. (1980). Fractionation of low molecular weight DNA or RNA in polyacrylamide gels containing 98% formamide or 7 M urea. *Methods Enzymol.* 65:299-305.

Maxam, A. M. (1979). Nucleotide sequence of DNA. Ph.D. dissertation, Harvard University.

Maxam, A. M., and Gilbert, W. (1980). Sequencing end-labeled DNA with base-specific chemical cleavages. In *Methods in Enzymology*, Vol. 65, eds. L. Grossman and K. Moldave (New York: Academic Press), pp. 499–560.

Maxam, A. M., and Gilbert, W. (1977). A new method for sequencing DNA. *Proc. Natl. Acad. Sci. USA* 74:560–564.

Mirzabekov, A. D., and Kolchinsky, A. M. (1974). Localization of some molecules within the grooves of DNA by modification of their complexes with dimethyl sulphate. *Molec. Biol. Res.* 1:385–390.

Muthukrishnan, S., and Shatkin, A. J. (1975). Reovirus genome RNA segments: resistance to S_1 nuclease. *Virology* 64:96–105.

Neu, H. C., and Heppel, L. A. (1964). Nucleotide sequence analysis of polyribonucleotides by means of periodate oxidation followed by cleavage with an amine. *J. Biol. Chem.* 239:2927–2934.

Ogata, R. T., and Gilbert, W. (1979). DNA-binding site of *lac* repressor probed by dimethyl sulfate methylation of *lac* operator. *J. Mol. Biol.* 132:709–728.

Peacock, A. C., and Dingman, C. W. (1967). Resolution of multiple ribonucleic acid species by polyacrylamide gel electrophoresis. *Biochem.* 6:1818–1827.

Peattie, D. A. (1980). Chemically probing RNA sequence and conformation. Ph.D. dissertation, Harvard University.

Peattie, D. A. (1979). Direct chemical method for sequencing RNA. *Proc. Natl. Acad. Sci. USA* 76:1760–1764.

Peattie, D. A., and Gilbert, W. (1980). Chemical probes for higher-order structure in RNA. *Proc. Natl. Acad. Sci. USA* 77:4679–4682.

Peattie, D. A., and Herr, W. (1981). Chemical probing of the tRNA-ribosome complex. *Proc. Natl. Acad. Sci. USA* 78:2273–2277.

Peattie, D. A., Douthwaite, S., Garrett, R. A., and Noller, H. F. (1981). *Proc. Natl. Acad. Sci. USA.*

Philippsen, P., Thiebe, R., Wintermeyer, W., and Zachau, H. G. (1968). Splitting of phenylalanine specific tRNA into half molecules by chemical means. *Biochem. Biophys. Res. Comm.* 33:922–928.

Pochon, F., Pascal, Y., Pitha, P., and Michelson, A. M. (1970). Photochimie des polynucleotides. IV: Photochimie de quelques nucleosides puriques methyles. *Biochim. Biophys. Acta.* 213:273–281.

Quigley, G. J., and Rich, A. (1976). Structural domains of transfer RNA molecules. *Science* 194:796-806.

Rich, A., and RajBhandary, U. L. (1976). Transfer RNA: Molecular structure, sequence, and properties. *Ann. Rev. Biochem.* 45:805-860.

Richards, E. G., and Gratzer, W. B. (1964). An investigation of soluble ribonucleic acid by zone electrophoresis. *Nature* 204:878-879.

Ross, A., and Brimacombe, R. (1978). Application of a rapid gel method to the sequencing of fragments of 16S ribosomal RNA from *Escherichia coli.* *Nucl. Acids Res.* 5:241-256.

Shatkin, A. J. (1976). Capping of eucaryotic mRNAs. *Cell* 9:645-653.

Siebenlist, U. (1979). RNA polymerase unwinds an 11-base pair segment of a phage T7 promoter. *Nature* 279:651-652.

Siebenlist, U., Simpson, R. B., and Gilbert, W. (1980). *E. coli* RNA polymerase interacts homologously with two different promoters. *Cell* 20:269-281.

Silberklang, M., Prochiantz, A., Haenni, A.-L., and RajBhandary, U. L. (1977). Studies on the sequence of the 3'-terminal region of Turnip-Yellow-Mosaic-Virus RNA. *Eur. J. Biochem.* 72:465-478.

Simoncsits, A., Brownlee, G. G., Brown, R. S., Rubin, J. R., and Guilley, H. (1977). New rapid gel sequencing method for RNA. *Nature* 269:833-836.

Simsek, M., Ziegenmeyer, J., Heckman, J., and RajBhandary, U. L. (1973). Absence of the sequence G—T—Ψ—C—G (A)— in several eukaryotic cytoplasmic initiator transfer RNAs. *Proc. Natl. Acad. Sci. USA* 70: 1041-1045.

Steinschneider, A., and Fraenkel-Conrat, H. (1966). Studies of nucleotide sequences in tobacco mosaic virus ribonucleic acid. IV: Use of aniline in stepwise degradation. *Biochem.* 5:2735-2743.

Stepanova, O. B., Metelev, V. G., Chichkova, N. V., Smirnov, V. D., Rodionova, N. P., Atabekov, J. G., Bogdanova, A. A., and Shabarov, Z. A. (1979). Addressed fragmentation of RNA molecules. *FEBS Lett.* 103:197-199.

Teeter, M. M., Quigley, G. T., and Rich, A. (1980). Metal ions and transfer RNA. In *Nucleic Acid-Metal Ion Interactions*, ed. T. G. Spiro (New York: Wiley Interscience), pp. 146-177.

Turchinskii, M. F., Gus'kova, L. I., Khazai, I., Budovskii, I., and Kochetov,

N. K. (1970). Chemical method of specific degradation of ribonucleic acids with selectively removed bases. III: Cleavage of phosphoester bonds in ribose 2'- and 3'-phosphates catalyzed by amines. *Mol. Biol.* (USSR) 4:343-348.

Uhlenbeck, O. C., and Gumport, R. I. (1981). In *The Enzymes* (in preparation).

Vanyushin, B. F., and Bur'yanov, Y. I. (1969). Degradation of DNA to purine sequences by hydrolysis of apyrimidic acid with aniline. *Biochemistry USSR* 34:574-582.

Verwoerd, D. W., and Zillig, W. (1963). A specific partial hydrolysis procedure for soluble RNA. *Biochim. Biophys. Acta.* 68:484-486.

Vincze, A., Henderson, R. E. L., McDonald, J. J., and Leonard, N. J. (1973). Reaction of diethyl pyrocarbonate with nucleic acid components: Bases and nucleosides derived from guanine, cytosine, and uracil. *J. Amer. Chem. Soc.* 95:2677-2682.

Whitfeld, P. R. (1965). Application of the periodate method for the analysis of nucleotide sequence to tobacco mosaic virus RNA. *Biochim. Biophys. Acta.* 108:202-210.

Whitfeld, P. R. (1954). A method for the determination of nucleotide sequence in polyribonucleotides. *Biochem. J.* 58:390-396.

Wintermeyer, W., and Zachau, H. G. (1975). Tertiary structure interactions of 7-methylguanosine in yeast tRNA[Phe] as studied by borohydride reduction. *FEBS Lett.* 58:306-309.

Yu, C.-T., and Zamecnik, P. C. (1960). A hydrolytic procedure for ribonucleosides and its possible application to the sequential degradation of RNA. *Biochim. Biophys. Acta.* 45:148-154.

Zimmern, D., and Kaesberg, P. (1978). 3'-Terminal nucleotide sequence of encepholomycarditis virus RNA determined by reverse transcriptase and chain-terminating inhibitors. *Proc. Natl. Acad. Sci. USA* 75:4257-4261.

APPENDIX

Outline for Four Base-specific Reactions for Chemical Sequencing of End-labeled RNA Molecules

G	A > G	U	C > U[a]
300 μl G Buffer	200 μl A,G Buffer		
		5 μl RNA-^{32}P	5 μl RNA-^{32}P
5 μl RNA-^{32}P	5 μl RNA-^{32}P	Lyophilize	Lyophilize
Chill to 0°C	Chill to 0°C	10 μl Hz/H$_2$O	10 μl Hz/NaCl
0.5 μl DMS	1 μl DEP	Mix, spin	Mix, spin
90°C, 0.5-1 min	90°C, 5-10 min	0°C, 5-10 min	0°C, 20-30 min
0°C	0°C	—	—
75 μl G Pptn	50 μl A,G Pptn	200 μl U Pptn	—
900 μl EtOH	750 μl EtOH	600 μl EtOH	500 μl C,U Pptn
Pellet	Pellet	Pellet	Pellet
200 μl 0.3 M NaAc	200 μl 0.3 M NaAc	200 μl 0.3 M NaAc	200 μl 0.3 M NaAc
600 μl EtOH	600 μl EtOH	600 μl EtOH	600 μl EtOH
Pellet	Pellet	Pellet	Pellet
EtOH wash, dry	EtOH wash, dry	EtOH wash, dry	EtOH wash, dry
10 μl 1.0 M tris-Cl pH 8.2	—	—	—
10 μl 0.2 M NaBH$_4$	—	—	—
0°C, 30 min/dark	—	—	—
200 μl NaBH$_4$ Stop Mix	—	—	—
600 μl EtOH	—	—	—
Pellet	—	—	—
EtOH wash, dry	—	—	—
20 μl 1 M aniline pH 4.5	20 μl 1 M aniline pH 4.5	20 μl 1 M aniline pH 4.5	20 μl 1 M aniline pH 4.5
60°C, 20 min/dark	60°C, 20 min/dark	60°C, 20 min/dark	60°C, 20 min/dark
Lyophilize	Lyophilize	Lyophilize	Lyophilize
20 μl water	20 μl water	20 μl water	20 μl water
Lyophilize	Lyophilize	Lyophilize	Lyophilize
20 μl water	20 μl water	20 μl water	20 μl water
Lyophilize	Lyophilize	Lyophilize	Lyophilize
2-3 μl loading buffer	2-3 μl loading buffer	2-3 μl loading buffer	2-3 μl loading buffer
90°C, 30 sec	90°C, 30 sec	90°C, 30 sec	90°C, 30 sec
Load on gel	Load on gel	Load on gel	Load on gel

[a]An alternative C + U reaction is described in Section III, Procedure 5.

G Reaction

G Buffer: 50 mM sodium cacodylate-HCl, pH 5.5
1 mM EDTA

Do not autoclave this buffer. The pH drops drastically, and the volatile and poisonous compound cacodyl forms upon heating.

G Pptn solution: 1.0 M tris-acetic acid, pH 7.5
(for precipitation) 1.0 M 2-mercaptoethanol
1.5 M sodium acetate
0.1 mM EDTA
0.2 mg/ml carrier tRNA

NaBH₄ stop: 0.6 M sodium acetate, pH 4.5
0.6 M acetic acid, pH 4.5
0.025 mg/ml tRNA

Comments: The 0.2 M NaBH₄ solution must be fresh.

A > G Reaction

A,G Buffer: 50 mM sodium acetate, pH 4.5, with acetic acid
1 mM EDTA

A,G Pptn solution: 1.5 M sodium acetate
(for precipitation) 0.2 mg/ml carrier tRNA

Comments: The diethyl pyrocarbonate should be stored at 4°C or −20°C (it remains a liquid at −20°C), preferably in aliquots of 1 ml or so. This both slows down its hydrolytic decomposition (to ethanol and carbon dioxide) and facilitates pipetting the small volumes required. A newly purchased bottle of diethyl pyrocarbonate does not necessarily guarantee that it has not decomposed. If the diethyl pyrocarbonate does not boil about a Pasteur pipette inserted in the bottle (making it quite difficult to pipette the liquid), it is probably no good; in this case, a new bottle should be opened.

U Reaction

Hz/H₂O: Hydrazine/water (1:1), cool to 0°C; use Eastman hydrazine (95% minimum by titration) directly from the bottle

U precipitation: 0.3 M sodium acetate
(Pptn) solution 0.1 mM EDTA
0.05 mg/ml carrier tRNA

Comments: The Hz/H_2O solution should be fresh.

Use a fresh pipette to remove each ethanol supernatant; otherwise, hydrazine is carried over from each precipitation and significant blurriness can occur on the sequencing gel.

C > U Reaction

Hz/NaCl: NaCl, oven-dried, to 3.0 M in hydrazine. *Keep anhydrous.* Cool to 0°C.

C,U Pptn solution: ethanol/water (4:1); store at $-20°C$
(for precipitation)   0.02 mg/ml carrier tRNA

Comments: The Hz/NaCl solution should be fresh and cooled to 0°C. The salt can crystallize out if the solution is left on ice too long.

If water is present during this reaction, the C cleavage decreases significantly and the U cleavage predominates.

Use a new pipette to remove each supernatant from the ethanol precipitations.

Aniline Reaction

To obtain small volumes ($\leqslant$0.5 ml) of a 1 M aniline solution, use x ml aniline (11 M), $3x$ ml glacial acetic acid, and $7x$ ml water. Add the aniline to the water/acetic acid solution, and rinse the dispensing capillary several times. Redistilled aniline (stored at $-20°C$) is optimal for this reaction, but reagent-grade aniline (which varies in color from yellow to brown) will work.

Loading Buffer

8 M urea
20 mM tris-Cl, pH 7.4
1 mM EDTA
0.05% xylene cyanol
0.05% bromophenol blue

Carrier tRNA

The tRNA should be treated with pronase or proteinase K before extracting with neutralized phenol and precipitating. The tRNA can be included in the precipitation solution as indicated here, or it can be added to the labeled RNA as carrier before the initial chemical modification reactions. In the latter case, the tRNA should be omitted from the precipitation solutions.

Table 7-A-1. Diagnostic Table for Aid in Correcting Aberrant RNA Sequencing Patterns

Problem	Potential causes	Suggestions
(1) Random hydrolysis in G lane coupled with lack of guanosine specificity.	(a) The sodium cacodylate reaction buffer has been autoclaved, and the pH has fallen too low.	(a) Use autoclaved water and clean reagents to make up the reagent buffer. Do not autoclave buffer.
	(b) The methylation reaction was done for 2 min or longer, and the pH has fallen too low.	(b) For longer methylation reactions, make the reaction buffer pH 7.5.
(2) Shadow bands are present		
(a) in every reaction lane.	(a) the RNA is heterogenous at its labeled end.	(a) Resolve the different RNA species from each other more completely. A long (40 cm) polyacrylamide gel (7–8.5 M urea) is usually sufficient for this.
(b) in selected reaction lanes and only noticeable for rather short oligonucleotides in the lower half of the polyacrylamide gel.	(b) Unavoidable internal sequence heterogeneity due to gene heterogeneity close to the site of label.	(b) The predominant sequence is usually clear and can be read without error. If the different RNA species must be separated from each other, the nondenaturing polyacrylamide gel system described by Korn & Gurdon[1] is likely to work.

Observation	Cause	Remedy
(c) in one reaction lane and at a fixed position, usually when the fragment is running in the lower half of the gel.	(c) This seems to occur occasionally in tRNA nucleotide sequences. This phenomenon has not been explained satisfactorily, and the best solution is to ignore it.	
(3) Displacement and distortion of bands running with one or both of the dye markers.	The concentration of one or both of the dyes is too high, and the fragments are being physically displaced.	Decrease the concentration of one or both of the dyes; 0.05% is the recommended concentration.
(4) The dye markers enter the gel slowly and do not enter cleanly and sharply.	The samples were not lyophilized well enough, and there is anilinium acetate in the sample.	Lyophilize the samples *to dryness* 3 times after the aniline reaction.
(5) All the bands within a string of G's (with the exception of the G nearest the site of label) are faint.	The bases lie within a G—C rich region of the RNA molecule, and the 30-60 sec methylation reaction is not long enough to denature this region well.	(a) Preheat the sample at 90°C before adding the dimethyl sulfate in order to open up this region of the molecule before chemically modifying it. (b) Convert all cytidines within the molecule to uridines via the bisulfite reaction[2,3] then do the U reaction.[4] A string of G—U base pairs melts more easily than its G—C analog.

Table 7-A-1. *(Continued)*

Problem	Potential causes	Suggestions
(6) The pyrimidine lanes are smeared such that the sequence cannot be read.	(a) The same Pasteur pipette is being used to remove the supernatants from each ethanol precipitation. This carries hydrazine over from each precipitation and leads to smearing.	(a) Use a new Pasteur pipette to remove every supernatant.
	(b) The hydrazine is bad.	(b) Open a new bottle of hydrazine, preferably one from another lot. A new bottle should be opened every 6–8 weeks as a preventative measure; however, different lots of hydrazine are variable, and even fresh hydrazine does not ensure quality.
(7) Lack of preferential cleavage at C's relative to U's in the hydrazine/ NaCl C reaction.	(a) The NaCl concentration is less than 3 M in the hydrazine due to inaccurate weighing of the NaCl or due to water in the NaCl.	(a) Make sure that the salt is being weighed correctly and make sure that the salt is well dried before use. 110°C over-night dries the salt satisfactorily.
	(b) Problems with the hydrazine.	(b) Open another bottle of hydrazine, preferably one from a different lot. Keep the reaction vessel on ice at all times, including the intervals involved in the ethanol precipitation procedure.
	(c) Too much Mg^{2+} in the stock RNA solution.[5]	(c) Reprecipitate the RNA without

Mg^{2+}, and wash with cold ethanol.

(8) The C's of 5'-end-labeled RNA are not cleaved preferentially, and a random hydrolysis ladder is generated.

Most likely results from a side-reaction of the hydrazine with the exposed ribose moiety at the 3' end of the 5' fragment.

No solution as yet. Read the A's, G's, U's (see Problem 10) as normal and use the hydrolysis ladder as a template for locating the cytidines.

(9) The C reaction cleaves the RNA at an internal residue such that pyrimidines beyond this site are difficult or impossible to read.

A m^7G, extremely susceptible to the hydrazine/NaCl reaction, occurs within the nucleotide sequence.

Use alternative C reaction (Section III, Procedure 5 in text).

(10) U pattern of 5'-end-labeled RNA is contaminated with a background of random hydrolysis.

Appears to result as a side-reaction with the hydrazine (see Problem 8)

No real solution. Read the prominent bands as uridines and ignore the readily identifiable background bands.

(11) There is a gap in the nucleotide sequence pattern

(a) There is an unusual nucleotide in the RNA sequence which is insensitive to the chemical reactions. The possibilities include ψ, ribo T, m^1A, and m^5C.

(a) The possibilities can be narrowed by sequencing the corresponding DNA fragment since all RNA base modifications occur after transcription. Follow normal procedures[6,7] for identifying unusual ribonucleotides

(b) Seen sometimes in tRNA sequences such a gap can be due to aberrant mobilities of tRNA fragments differing only by one nucleotide. Unlike the gaps induced by unusual bases (see 11a) these are often not integral multiples of normal band spacing.

(b) Apparently an inherent property of the RNA molecule. No solution known.

303

Table 7-A-1. *(Continued)*

Problem	Potential causes	Suggestions
(12) A band appears in all four sequencing lanes which is (a) of equal intensity in all lanes (b) of equal tensity in 3 lanes but especially predominant in the fourth	There is probably an unusual nucleotide in the RNA sequence (a) which is sensitive to the aniline strand scission reaction (b) which is sensitive to the aniline strand scission reaction *and* to one of the chemicals used for the initial modification. These bases include m^7G, ms^2i^6A, and Y.	The possibilities can be narrowed by sequencing the corresponding DNA fragment since RNA base modifications occur after transcription. Follow normal procedures[6,7] for identifying unusual ribonucleotides.
(13) Weak bands in the A > G pattern	The diethyl pyrocarbonate is old and has decomposed	Open a new bottle of diethyl pyrocarbonate, and store it at $-20°$C.

[1] Korn, L. J., and Gurdon, J. B. (1981). *Nature* 289:461–465.

[2] Hayatsu, H., Wataya, Y., Kai, K., and Iida, S. (1970). *Biochem.* 9:2858–2865.

[3] Shapiro, R., Servis, R. E., and Welcher, M. (1970). *J. Amer. Chem. Soc.* 92:724–726.

[4] C. Lee and J. Fresco, personal communication.

[5] R. Baer, personal communication.

[6] Brownlee, G. G. (1972). *Determination of Sequences in RNA* (New York: American Elsevier).

[7] Hall, R. (1971). *The Modified Nucleoside in Nucleic Acids* (New York: Columbia University Press).

8

The Study of Protein-DNA Binding
Specificity: DNase Footprinting

ALBERT SCHMITZ and DAVID J. GALAS

INTRODUCTION

Protein-DNA interactions are of fundamental biological importance in gene expression and control, replication and recombination. An important step in elucidating the functions of a DNA binding protein lies in determining the exact position and extent of its target sequence (or sequences) on a DNA molecule. Several methods have been described previously for identifying a protein binding site *in vitro*, all of which have various drawbacks. These methods involve the protection of a region of the DNA molecule from attack by a protein molecule bound to the DNA *in vitro*, and include protection against methylation of purines by dimethylsulfate (Gilbert, Maxam, & Mirzabekov, 1976; Johnsrud, 1978; Ogata & Gilbert, 1978), extensive digestion by endonucleases (Matsukage, Murakami, & Kameyama, 1969; Le Talaer & Jeanteur, 1971; Heyden, Nüsslein, & Schaller, 1972; Gilbert & Maxam, 1973; Pribnow, 1975; Walz & Pirotta, 1975; Pirotta, 1975; Schaller, Gray, & Hermann, 1975; Liu & Wang, 1978) or Exonuclease III (Simpson, 1980: Tsurimoto & Matsubara, 1981), and protection from UV-induced cleavage (Ogata & Gilbert, 1977; Simpson, 1979, 1980). The disposition of a protein molecule on the DNA has also been studied by determining the particular phosphates on the DNA backbone that can inhibit binding when they have been modified by ethylnitrourea (Simpson, 1980; Maxam & Gilbert, 1980).

We describe in this chapter a simple method that allows the determination of the exact position and extent of a protein binding site with a minimum amount of work and as few artifacts as possible. Since this technique, called footprinting (Galas & Schmitz, 1978), was suggested it has been used extensively in a variety of protein-DNA systems, many of which would have been very difficult to analyze without it. A compilation of many of the DNA binding sites determined by the footprinting method is presented in Table 8-1. In a simple conjoining of the Maxam-Gilbert sequencing method and the technique of DNase-protected fragment isolation, footprinting takes advantage of the relatively low DNA cleavage specificity of DNase I in a partial digestion, and the ability of a DNA-bound protein to prevent cleavage of the backbone between the base pairs it covers. In principle, any endonuclease with sufficiently low cleavage specificity could be used.

We will describe in this chapter the footprinting procedures commonly used and illustrate several features of protein-DNA interactions, studied by footprinting, that may be useful to researchers in using this method.

BOVINE PANCREATIC DEOXYRIBONUCLEASE I (DNase I)

DNase I is a monomeric glycoprotein with a polypeptide chain consisting of 257 amino acids, corresponding to a molecular weight of about 30,000 daltons (Lindberg, 1967; Liao et al., 1973). DNase I acts as an endonuclease cleaving single- and double-stranded DNAs, producing nucleotides with a 5'-phosphate and 3'-hydroxyl group (Laskowski, 1961; Bernardi et al., 1975). The specificity as well as the mechanism and pH optimum of the cleavage reactions depend on the nature of the activating metal ions (Lehman, Roussos, & Pratt, 1962; Shack & Bynum, 1964; Melgar & Goldtwait, 1968; Junowicz & Spencer, 1973a, 1973b; Clark & Eichhorn, 1974). The chelating agent EDTA strongly inhibits DNase I cutting. Chymotrypsin is the most frequent contaminant in DNase I preparations. The divalent metal ion Ca^{2+} binds to DNase I and thereby stabilizes the protein against proteolytic degradation by chymotrypsin, probably by inducing a conformational change (Lizarraga et al., 1978; Tullis & Rubin, 1980).

DNase I will break only one strand of the duplex DNA per encounter in the presence of the metal ion Mg^{2+} only, while the substitution of Mg^{2+} by Ca^{2+} or Mn^{2+} can result in the cleavage of both strands in a single event (Bollum, 1965; Melgar & Goldwait, 1968). The pH optimum for DNase action is near 7 in the presence of these ions (Lehman, Roussos, & Pratt, 1962; Shack & Bynum, 1964).

The sequence specificity of DNase I cleavage is rather low, but it clearly prefers some sites over others; the nature of this specificity will be discussed in a later section. For the moment, it is enough to note that almost every phosphodiester linkage is cleaved to a detectable extent. The notable exceptions are some stretches of A's or T'S (Ross et al., 1979).

DESCRIPTION OF THE METHOD

A protein binding site on a DNA molecule can be detected if the protein is able to prevent cleavage between the bases in the binding site during DNase I digestion. This principle has been applied, for example, to the isolation of promoter-containing DNA fragments protected by RNA polymerase against extensive DNase I digestion (Matsukage, Murakami, & Kameyama, 1969; Le Talaer & Jeanteur, 1971; Siebenlist & Gilbert, 1980; Schaller et al., 1975; Pirotta, 1975; Walz & Pirotta, 1975; Pribnow, 1975). The protected fragment can be isolated and then sequenced (Maxam & Gilbert, 1977, 1980).

The footprinting method modifies this procedure in two respects (Galas & Schmitz, 1978). First, the DNA is digested only partially by DNase I, so as to cut each DNA molecule only once, on average. Second, the DNA fragment used is radioactively labeled on only one end of the double-stranded DNA molecule. A DNA-bound protein can protect the phosphodiester bonds against DNase I attack by sterically excluding contact. The cleavage products of reactions carried out in the presence and absence of the DNA-binding protein are then fractionated according to size by electrophoresis through a denaturing polyacrylamide gel (Maxam & Gilbert, 1977), and the gel is autoradiographed. The film will reveal a ladder of bands corresponding to the end-labeled, single-stranded DNA fragments differing, in most positions, by one nucleotide in length. The protein binding site can be visualized as a blank section in the ladder (a "footprint") due to the absence of end-labeled fragments terminating at nucleotides that were protected by the DNA-binding protein against DNase I attack.

The bases of the binding site can be identified by running Maxam-Gilbert sequencing reactions of the same DNA fragment as

Table 8-1. Proteins and Their Binding Sites on DNA as Determined by Footprinting

Protein	Binding site	Reference
lac repressor	*lac* operator	Galas & Schmitz (1978)
		Schmitz & Galas (1979)
RNA polymerase (*E. coli*)	*lac* promoter	Schmitz & Galas (1979)
Int protein (phage λ)	phage λ *att* site	Ross et al., 1979
	bacterial *att* site	
cAMP receptor protein (*E. coli*)	*gal* control region	Taniguchi et al. (1979)
phage λ repressor	right operator of λ DNA	Johnson et al. (1979)
phage P22 repressor	phage P22 left and right operators	Poteete et al. (1980)
lac repressor I12-X86	nonoperator binding sites	Schmitz & Galas (1980)
5S RNA transcription factor (*Xenopus leavis*)	internal control region of 5S RNA genes (*Xenopus borealis*)	Engelke et al. (1980)
cAMP receptor protein, *ara* C protein	*ara* control region	Ogden et al. (1980)
		Lee et al. (1981)

RNA polymerase (*E. coli*)	*ara* control region	Lee et al. (1981)
phage λ O protein	phage λ origin of replication	Tsurimoto & Matsubara (1981)
DNA gyrase (*E. coli*)	plasmid ColEI DNA	Morrison & Cozzarelli (1981)
DNA gyrase (*M. luteus*)	plasmid pBR322 DNA	Kirkegaard & Wang (1981)
DNA gyrase (*E. coli*)	plasmid ColEI and pBR322 DNA	Fisher et al. (1981)
cAMP receptor protein (*E. coli*)	*lac* control region	Schmitz (1981)
Alanine aminoacyll tRNA synthetase (*E. coli*)	protein's own promoter	Putney & Schimmel (1981)
RNA polymerase (*E. coli*)	β-lactamase promoter of Tn3	Russell & Bennett (1981)
5S RNA transcription factor (*Xenopus laevis*)	deletion mutants of 5S RNA	Sakonju et al. (1981)
cAMP receptor protein and RNA polymerase (*E. coli*)	pBR322 promoter	Queen & Rosenberg (1981)
lexA protein (*E. coli*)	*lexA* & *recA* control region	Little et al. (1981)
		Brent & Ptashne (1981)
Tn1000 (λ δ) repressor-resolvase	internal resolution site of Tn3	Lauth, Reed, & Grindley (unpublished)

size markers alongside the footprinting reaction products (Maxam & Gilbert, 1977, 1980). The precise definition of the binding site, of course, cannot be made without knowing exactly how the steric exclusion of DNase I action occurs—the relative shapes, sizes, and stand-off distances of DNase I and the binding protein along the DNA.

THE REACTIONS

Double-stranded DNA can be end-labeled by adding a radioactively labeled phosphate to its 5' end using T_4-kinase, for example (Maxam & Gilbert, 1977), or by filling-in a suitable staggered end using radioactive deoxynucleoside triphosphate and DNA polymerase "large fragment" ("Klenow" fragment) (Maxam & Gilbert, 1980).

The volume of the footprinting reaction is kept sufficiently small (100 µl) to permit high DNA and protein concentration, if necessary, and easy handling of the sample in subsequent steps. The buffer used has to be adjusted to the specific requirements of the DNA-binding protein with respect to pH, salt concentrations, temperature, etc., which ultimately have to be determined experimentally, without neglecting the requirements for DNase I activity: a pH near 7, activating divalent metal ions (Ca^{2+}, Mg^{2+}, or Mn^{2+}), and the absence of inhibitors such as high concentrations of the chelating agents EDTA and citrate (see Laskowski, 1966). The incubation time necessary to achieve the desired partial digest of the DNA sample should be sufficiently short to avoid dissociation of prebound protein molecules during the cleavage reaction, and eventually to permit kinetic experiments to be performed. One could measure, for example, the dissociation half-life of a protein from one or more target sites on the DNA (Schmitz & Galas, 1980). Extensive purification of the protein whose target site on a DNA molecule has to be determined can be partially neglected if there exist independent criteria that define the binding characteristics of the protein, such as a specific reduction in the protein-DNA interactions on addition of a ligand (IPTG for the *lac* repressor-operator complex) or dependence of DNA binding on a ligand (cyclic AMP for specific CRP-promoter DNA interactions). The footprint of a protein should be obtained on both strands of the DNA before one can consider that the binding site has been fully defined. Of the protein sites determined to date, almost all are significantly asymmetric on the DNA strands.

TYPICAL FOOTPRINTING REACTION

The protocol used to visualize the principal interaction site of the *lac* repressor on a *lac* operator-containing DNA fragment, for a final reaction volume of 100 μl, is as follows:

Add 5 μl (about 10^4 counts Cherenkov/100 base pairs) of the double-stranded DNA fragment (radioactively labeled on one 5' end, and dissolved in 10 mM tris-HCl, pH 7.9 (22°C), 0.1 mM EDTA) to 90 μl of binding buffer at 22°C in a siliconized, 1.5-ml plastic centrifuge tube.

Binding buffer: 10   mM tris-HCl, pH 7.9 (22°C)
                         10   mM $MgCl_2$
                         10   mM KCl
                          5   mM $CaCl_2$
                       0.1 mM DTT

Mix briefly and add 5 μl of a suitable dilution in binding buffer of a pure repressor preparation. The dilution factor must be determined empirically, since the specific operator-binding activity of repressor may vary in different preparations. Mix gently, and incubate at 22°C for 10 min.

Add 5 μl of a DNase I solution (Worthington, DPFF) that has been freshly diluted from 1 mg/ml to 5 μg/ml in binding buffer. Mix gently, and incubate at 22°C.

After 30 sec of DNase I digestion, add 25 μl of stop solution.

Stop solution: 3   M ammonium acetate
                   0.25 M EDTA
                   0.12 mg/ml sonicated calf-thymus DNA

Mix well and add 0.4 ml of cold ethanol to precipitate the DNA. Mix and chill at −70°C (Dry Ice/ethanol bath) for 5 min. Centrifuge in microcentrifuge for 5 min (12,000 g). Remove the supernatant with a plastic pipette tip. Dissolve the pellet in 100 μl of a solution containing

    0.3 M Na-acetate
    0.1 mM EDTA
    10   mM $MgCl_2$

Add 0.3 ml of ethanol, mix, and centrifuge again after chilling for 5 min at −70°C. Discard the supernatant, and repeat this precipitation *twice* more. Then carefully wash the pellet by gently adding

0.4 ml of cold 80% ethanol. Do not mix, but centrifuge directly for 1 min. Remove the supernatant, and dry the sample under vacuum for 5 min. Dissolve the DNA in:

50 mM NaOH
1 mM EDTA
5 M urea
0.05 % xylene cyanol
0.05 % bromophenol blue

Load the sample on a denaturing polyacrylamide gel (Maxam & Gilbert, 1980).

The same reactions are carried out in the absence of repressor by adding 5 μl of binding buffer to the reaction mixture instead of a repressor dilution. Maxam-Gilbert sequencing reactions are done in parallel with the same DNA fragment to allow the identification of the bands in the footprinting ladder (Maxam & Gilbert, 1977, 1980).

An untreated DNA sample should always be electrophoresed through the denaturing gel to verify the integrity of the DNA, since single-strand nicks, for instance, might create artificial protein binding sites.

High concentrations of protein in the footprinting reaction mixtures might result in the distortion of fragment migration during electrophoresis. This can be eliminated by a phenol extraction of the proteins after stopping the footprinting reaction.

Add 150 μl of water-saturated, distilled phenol to the sample. Mix well for 30 sec, then centrifuge for 2 min in the microcentrifuge. Remove the upper (aqueous) layer carefully with a plastic pipette tip and transfer to a new, siliconized plastic centrifuge tube. Precipitate the DNA by adding 0.45 ml of ethanol and proceed as described above.

If the binding buffer for the protein being studied contains high concentrations of gycerol, the phenol extraction has to be done after the first DNA precipitation in the footprinting protocol to obtain distinct phases.

Assignment of the DNase I-produced Bands

For the identification of the DNA fragments generated by DNase I, it is necessary to know the mobility of the DNase I-produced fragment relative to the chemically produced fragment with the same number of base pairs (from the Maxam-Gilbert reactions). Consider that the chemical cleavages leave both product fragments with a terminating phosphate group on both the 3' and 5' end after eliminating the modified nucleotide, whereas the DNase I reaction leaves a phosphate group on the 5' end and a hydroxyl group on the 3' end after hydrolysis of the connecting phosphodiester bond without eliminating a base. The relative mobilities of the fragments produced by different means depend on which end of the fragment is labeled in the experiment. In no case is the change as great as adding or subtracting a base. However, the labeled products of the sequencing reaction for a 5'-labeled fragment will have a higher mobility than the corresponding products (same number of base pairs) of the DNase I reaction, and migrate faster on the gel because of the extra charge provided by the 3' phosphate of the chemically produced fragment (McConnell, Searcy, & Sutcliffe, 1978; Sutcliffe & Church, 1978; and contrary to the statement in Galas & Schmitz, 1978). On the other hand, if the fragment is labeled on the 3' end (by filling-in a staggered end with radioactive-labeled deoxynucleotide triphosphates, for example) we obtain a set of products from the DNase I reaction that match the mobilities of the sequencing fragments exactly. These relationships are illustrated in Fig. 8-1.

OTHER ENZYMES SUITABLE FOR FOOTPRINTING REACTIONS

In principle, one should be able to obtain footprints of DNA-binding proteins using any endonucleolytic, DNA-hydrolyzing agent that cleaves under physiological conditions, is large enough to be sterically restricted by a DNA-bound protein, and is sufficiently nonspecific to permit good definition of the binding site by suppression of the cleavage reactions. Other enzymes that have been used include micrococcal nuclease [molecular weight of 17,000 daltons and activities of an endo 5'-phosphodiesterase and exonuclease (Alexander, Hepper, & Hurwitz, 1961; Ross et al., 1979)], and neocarzinostatin [a polypeptide of 10,700 daltons, which is highly specific for A and T bases (D'Andrea & Haseltine, 1978; Ross et al., 1979]. Reaction of DNA with these enzymes creates fragments carrying a 3' monophosphate which co-migrate with the sequencing reaction products from a 5'-labeled fragment. DNase II is much less useful for defining protein sites because its pH optimum is near 4.5 (Bernardi & Griffe, 1964). Other agents that may be useful, but not yet tried to our knowledge, include bleomycin (D'Andrea & Haseltine, 1978) and any of a number of nonspecific endonucleases found in a variety of tissues and organisms.

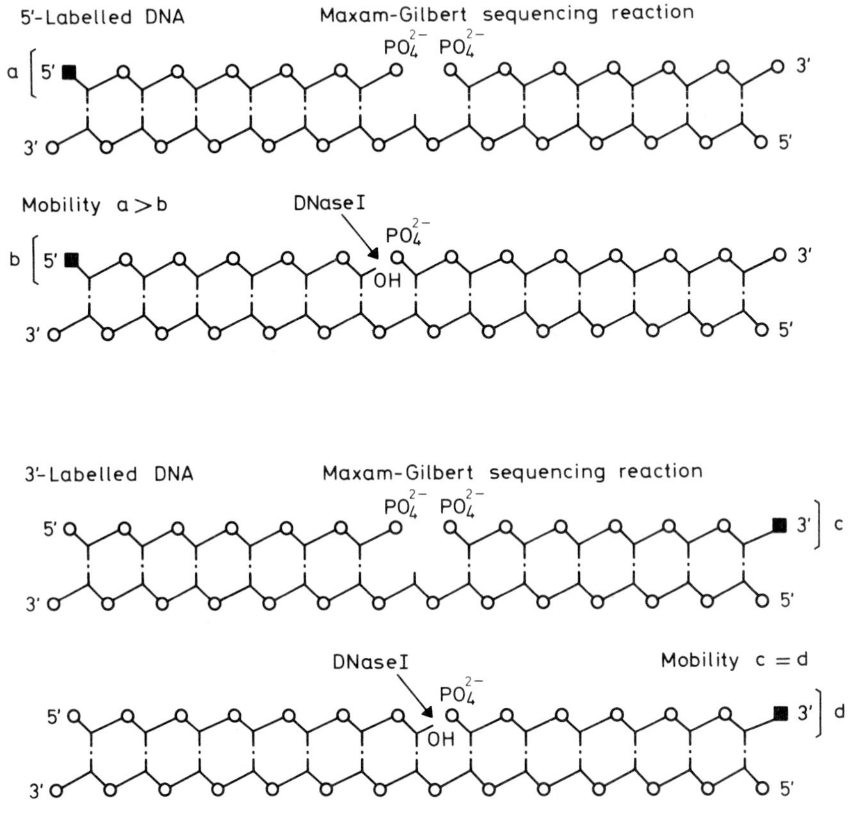

Figure 8-1. Schematic comparison of the nature of the cleavage induced by DNase I and the Macam-Gilbert sequencing reactions. The open circles represent phosphate groups of the DNA backbone, and the filled-in squares represent the ^{32}P-labeled phosphate groups. The upper two drawings depict the cleavages that yield 5'-end-labeled fragments of the same length. The mobilities of these fragments (*a* and *b*) are such that *a* moves faster during electrophoresis than *b*. The lower two drawings depict the same situation for 3'-labeled fragments because of the asymmetry of the DNase I cleavage; however, these fragments have the same mobility. Note that DNase I must cleave at different positions to give the same size fragment (3' and 5' label) as the Maxam-Gilbert sequencing reactions.

INTERACTIONS ANALYZED BY FOOTPRINTING

1. Footprint of lac Repressor on Operator DNA

The footprinting method was initially applied to the well-characterized *lac* repressor-operator system to evaluate the potential of the method (Galas & Schmitz, 1978). A double-stranded DNA fragment spanning the *lac* promoter-operator region of *E. coli* labeled on one 5′ end was incubated with pure repressor protein to saturate the operator. This complex was partially digested with DNase I, and the products were electrophoresed through a denaturing polyacrylamide gel alongside Maxam-Gilbert base-specific sequencing reactions. The left-hand panel in Fig. 8-2 shows the autoradiograph of this gel. The two lanes on the left show the sequencing reactions of the fragment. Lane 4 (counted from the left) represents the partial DNase I digest in the absence of repressor and shows that the specificity of DNase I cleavage is low enough to allow cutting between essentially all the bases of this DNA fragment. The footprint of repressor is evident in lane 3 by the suppression of bands in the ladder corresponding to the position of the operator sequence, as indicated by the bracket on the right-hand side.

The right-hand panel in Fig. 8-2 shows the repressor footprint on the same DNA fragment, but labeled on the 5′ end of the opposite strand (upper strand in Fig. 8-3). The protection of DNA-bound repressor against DNase I cleavage is again clearly visible in lane 3 when compared to lane 4 (counted from the left). The extent is indicated by the brackets on the right-hand side. But there is one striking feature in this protection pattern when compared to that on the other strand: a repressor-induced *enhancement* for DNase I cleavage between two bases of the operator sequence, as indicated by the arrow on the right (Figs. 8-2 and 8-3). The molecular mechanism that results in this enhancement is not clear, but it could be caused by a repressor-induced change in the DNA conformation (Wang et al., 1974; Pfahl, 1978) and/or by protein-protein interactions between operator-bound repressor and DNase I leading to a locally increased DNase I concentration.

An important question in using the footprinting method is: How accurately does a footprint define the extent of a protein interaction site on a DNA molecule? The definition will be strongly influenced by steric inhibition of DNase I attack at the very edges of the protein interaction site, resulting in an overestimation of the size of the target site. A comparison between the extent of the

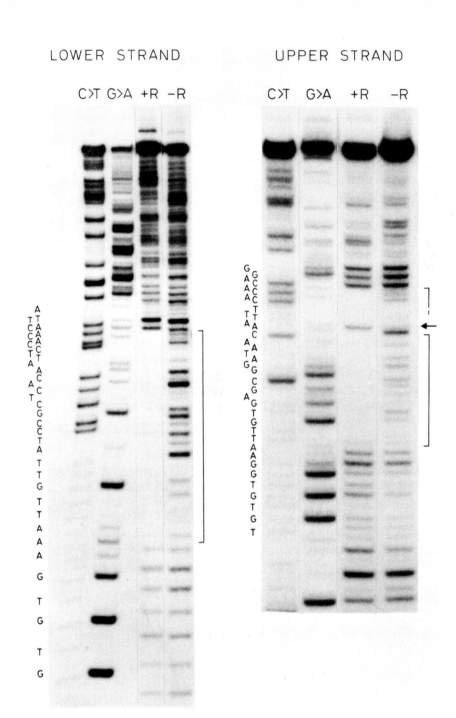

operator, as defined by the footprinting method and by other independent techniques, can help define the relationship between the physical contact of protein and DNA and the footprint and can give us an idea of how close to a bound protein DNase I can cleave. Figure 8-3 shows the DNA sequence of an operator-containing DNA fragment. The footprinting protection data are indicated by brackets, and the arrow pointing to the repressor-induced site for enhanced DNAase I indicates cutting. Boldface T's represent sites where crosslinking between repressor and $5'$-bromodeoxyuridine-substituted DNA occurs after UV radiation (Ogata & Gilbert, 1977). The very top shows single base-pair exchanges which reduce the affinity of operator for repressor (o^c) and result in an increased basal level of β-galactosidase *in vivo* (Smith & Sadler, 1971; Gilbert et al., 1975). The plot on the bottom shows the purine bases (A and G) whose changes in methylation by dimethylsulfate have been attributed to the close contacts between repressor and these sites (Gilbert, Maxam, & Mirzabekov, 1976). The sites protected against partial DNase I cleavage, indicated by brackets in Fig. 8-2, reveal the staggered ends of the protected region, which might be explained as a direct consequence of the binding mode of the protein. The repressor could interact with the operator by touching just one face of the DNA (Fig. 8-4) (Schmitz & Galas, 1979; Siebenlist, Simpson, & Gilbert, 1980). The staggered ends would then result from DNase I cutting between the most exposed bases at the extremities of the protected DNA region because of the double-helical structure of the operator as shown in Fig. 8-4. We cannot rule out, however, that the inherent tendency of DNase I to make staggered cuts on naked DNA contributes to the asymmetry of some footprints (Lutter, 1977; Hefron, So, & McCarthy, 1978).

On the upper strand, the repressor protects a stretch of DNA against DNase I attack that covers all the bases thought to be in close contact with the repressor by *three* independent criteria. The site defined by the footprint exceeds this site on the upper strand by only two bases at the $5'$ end and by three bases at the $3'$ end (Fig.

Figure 8-2. The footprint of *lac* repressor on both strands of operator DNA. Partial DNase I digestion products of $5'$-end-labeled, operator-containing DNA fragments bound to I12-X86 repressor (Schmitz et al., 1978) were run on a 20% denaturing polyacrylamide gel next to the indicated Maxam-Gilbert sequencing reaction products and were autoradiographed. The brackets indicate the DNA sequence that is protected by repressor against DNase I attack, and the arrow the enhanced DNase I cleavage site.

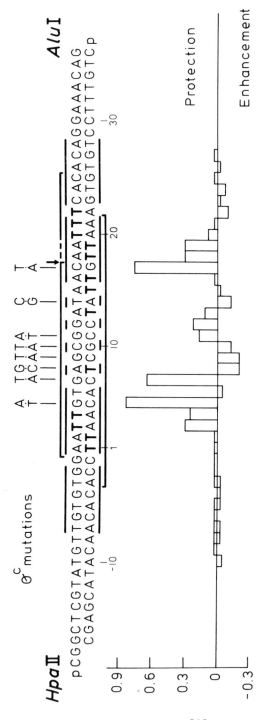

Figure 8-3. Characterization of the *lac* operator by four different methods: *1.* The repressor footprinting data are indicated by the brackets alongside the DNA sequence. The arrow above the upper strand points to the repressor-induced site for enhanced DNase I cleavage. *2.* Base changes which weaken the repressor-operator interactions (*o*^{c} mutations) are shown above the sequence (Smith & Sadler, 1971; Gilbert et al., 1975). *3.* Boldface T's represent sites where crosslinking between repressor and 5'-bromo-deoxyuridine-substituted DNA occurs after UV radiation (Ogata & Gilbert, 1977). *4.* Repressor-induced changes in the methylation pattern of purines (A and G) are shown in the diagram below the sequence (Gilbert, Maxam, & Mirzabekov, 1976). The ordinate represents the logarithm of the methylation rate (methylation in the presence of repressor/methylation without repressor). Redrawn from Ogata and Gilbert (1979), with the addition of the *o*^{c} mutations, the footprinting data, and the UV-induced cross-linking sites. The lines along the sequence mark the twofold symmetry present in this DNA sequence (Gilbert et al., 1975).

318

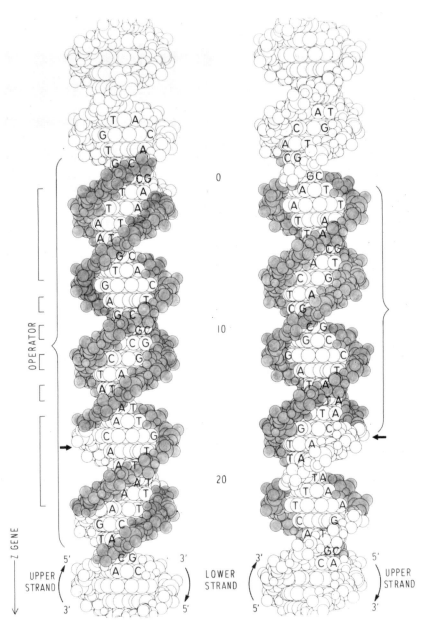

Figure 8-4. Drawing of the protection pattern of repressor on operator DNA in B-form. The two views are front and back (180° rotation) of the same DNA segment. The darkened backbone represents the region protected from DNase I attack. The square brackets indicate the operator symmetry, and the curved brackets the protected region on the facing site of the DNA as seen in each view. The arrow indicates the site of enhanced DNase I cleavage.

8-3). A similar picture emerges from the DNase I protection pattern on the lower DNA strand (Fig. 8-3). Again, all important bases ("close contact" sites) are covered, with the possible exception of A at position 22, which shows only a weak enhancement of methylation by dimethylsulfate and lies just outside the DNase I-protected region. It should be recalled, however, that we are using the convention here that a base is protected against DNase I attack when the phosphodiester bond on its 5′ side is protected. The adenine at position 22 is protected on its 3′ side from DNase I attack.

One might conclude from this comparison that the protection of a site on a DNA molecule against DNase I attack by a bound protein can define this interaction site to within a few bases on each end of the DNA strands (in the case of the repressor-operator interactions, within two to three bases). In this sense, DNase I is a tool of unexpected precision. The precision will, of course, depend on the mode of the protein-DNA interactions and will therefore be subject to some ambiguity. Specific binding of a protein to a site on DNA that involves wrapping of the DNA around the protein (as with gyrase, or in nucleosomes), or other special cases, might obscure the accurate determination of the primary binding site. The repressor-operator interaction provides examples of several phenomena encountered in footprinting studies and illustrates the potential precision of the method in defining unknown sites.

2. Footprints of RNA Polymerase on the lac Promoter

A more complex system, analyzed by the footprinting method, is represented by the interaction of RNA polymerase with a promoter. The *lac* system was used in this study because, while much information had accumulated about the interaction of RNA polymerase with the *lac* promoter, there were still several unanswered questions about the interaction itself and how polymerase action is inhibited by the repressor.

The restriction fragment used for these studies contains the complete promoter-operator region of a promoter mutant (called UV5) which renders *lac* operon transcription independent of the action of the cyclic AMP receptor protein (CRP), presumably by increasing the affinity of RNA polymerase for the promoter. The CRP protein, when complexed with cyclic 5′-3′-adenosine monophosphate (cAMP), stimulates *lac* operon transcription from the wild-type promoter by binding to the −60 base-pair region (60 base pairs upstream of the major transcription start; Simpson, 1980).

The partial DNase I digestion of the promoter DNA, complexed with RNA polymerase, was initially carried out in binding buffer containing 25% (v/v) glycerol and 125 mM KCl, conditions that had been shown to affect favorable RNA polymerase binding to promoter DNA (Nakanishi et al., 1974). We found during the course of the experiments, however, that the footprint of RNA polymerase on the UV5 *lac* promoter-operator DNA is essentially unaffected by the presence of glycerol and of salt (10 mM to 125 mM KCl) under the conditions used (Schmitz & Galas, 1979). Figure 8-5 (left-hand panel, lane 2) shows the footprint of RNA polymerase, saturated with sigma factor, on the lower strand of the UV5 promoter (see also Fig. 8-6). The protection pattern is complete (100% inhibition of cutting at some sites) and contains interspersed bands resulting from RNA polymerase-induced sites for enhanced DNase I cutting. These enhancements are even more pronounced than the one example in the repressor footprint on the operator. The RNA polymerase footprint covers the *lac* operator and extends over 68 base pairs.

The right-hand panel (lane 2) of Fig. 8-5 shows the footprint of RNA polymerase on the upper strand of the promoter-operator region DNA (comparing lanes 1 and 2). This protection pattern is similar in extent, and in exhibiting interspersed bands, to the footprint on the lower strand. It differs, however, in the weak protection of the operator region shown in the diagram of Fig. 8-6 (subsequent footprints on other promoters have shown similar weak protection in the region near transcription initiation; Russell & Bennett, 1981). The footprint of RNA polymerase shows staggered ends, as for the repressor-operator, but is more asymmetric. This is not unexpected, because of the unidirectional nature of transcription. The large extent of the footprint on the DNA (more than 74 bases) can easily be accounted for by the high molecular weight, and consequent size, of the subunits constituting RNA polymerase (total molecular weight about 550,000 daltons). All known point mutations affecting CRP-independent transcription of the *lac* operon map within the protected region (Dickson et al., 1975). Figure 8-6 shows a diagram comparing the RNA polymerase footprint with the close-contact points between the enzyme and the bases of the promoter-operator region as determined by methylation protection and crosslinking experiments (Johnsrud, 1978; Simpson, 1979). As in the case of the repressor-operator interactions, all close contact points between RNA polymerase and promoter DNA, as determined by these independent methods, are contained within the footprint. The most striking difference between the interaction site defined by footprinting and that defined by other methods is the significantly

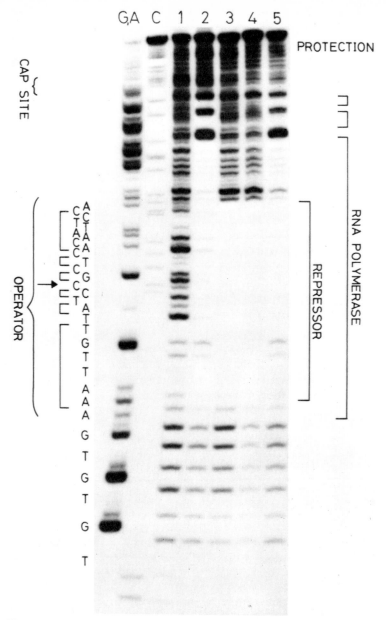

Figure 8-5. Footprint of RNA polymerase and repressor, individually and in competition, on the *lac* control region DNA (Schmitz & Galas, 1979) 5′-end-labeled on the lower DNA strand (left-hand panel) and upper strand (right-hand panel; Fig. 8-6). CRP (cyclic AMP receptor protein, Emmer et al., 1970) and CAP (catabolite gene activating protein, Zubay, Schwartz, & Beckwith, 1970) are the same.

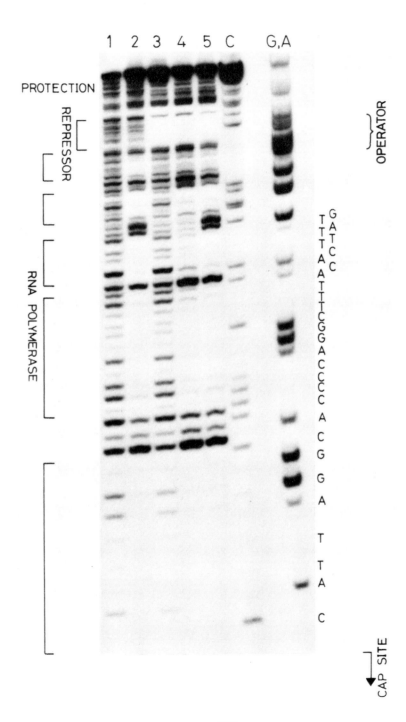

PROTECTION

REPRESSOR

RNA POLYMERASE

OPERATOR

CAP SITE

323

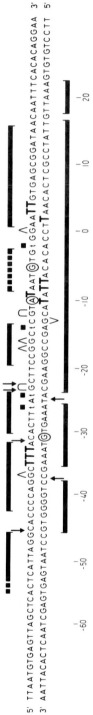

Figure 8-6. Characterization of the *lac* promoter by three different criteria: *1.* The RNA polymerase footprinting data are indicated by the heavy brackets alongside the DNA sequence. The arrows point to RNA polymerase-induced sites for enhanced DNase I cleavage (Schmitz & Galas, 1979). *2.* RNA polymerase-induced changes in DNA cleavage at 5-bromo-uracil-substituted thymine sites after UV irradiation (Simpson, 1979). Bold-face T's represent enhanced cleavage in the presence of RNA polymerase; small T's with a black square, reduced cleavage. *3.* RNA polymerase-induced changes in the methylation pattern of purines (A and G) are indicated by circled bases (blocked methylation), half circles (partial blockage), and carets above the bases (enhanced methylation) (Johnsrud, 1978).

324

greater extent of the footprint at both ends. Since we know from the repressor study that it is possible to define the site rather precisely by footprinting, the large protected regions here could be caused by a fundamental difference between these proteins and their modes of interaction with DNA. As we will illustrate with further examples, it is possible for a protein to bind specifically to a DNA site and not affect any changes in the methylation pattern (Schmitz & Galas, 1980; Kirkegaard & Wang, 1981). It may be, then, that in this case the polymerase is indeed physically covering the entire region defined by the footprint.

The existence of DNase I-enhanced-cutting sites within the RNA polymerase footprint on opposite strands represents a major obstacle to the isolation of the entire RNA polymerase-bound DNA fragment after extensive DNase I digestion (Le Talaer & Jeanteur, 1971; Schaller et al., 1972; Pirotta, 1975). The failure to demonstrate that RNA polymerase can rebind to fragments isolated in this manner (Schaller, Gray, & Hermann, 1975; Walz & Pirotta, 1975) is probably caused by the fact that the RNA polymerase-induced sites for enhanced DNase I cutting within the promoter lead to the isolation of only part of the DNA sequence covered by RNA polymerase.

Once the footprints of RNA polymerase and repressor on the *lac* promoter-operator region have been determined separately, one can address the question of how *lac* operon transcription control is exerted by the repressor. Can it be accounted for by the structural overlap of the binding sites? We recognized that competition experiments with footprinting could determine whether binding of repressor and RNA polymerase to the promoter-operator region are mutually exclusive events (Eron & Block, 1971; Chen et al., 1971).

The left-hand panel of Fig. 8-5 shows a set of footprinting experiments in which the lower strand of promoter-operator DNA was end-labeled (see also Fig. 8-6). Repressor was allowed to bind to the operator, and the RNA polymerase was added for a time interval sufficient to allow binding to the promoter in the absence of repressor (lane 4). Comparison of lane 4 with lane 2 (only RNA polymerase added) and lane 3 (only repressor added) shows that repressor, when bound to the operator, prevents RNA polymerase binding. This is clear from the absence of RNA polymerase-specific protection in lane 4 in the region toward the CRP site on this lower strand. The inverse experiment, in which RNA polymerase is prebound to the promoter and then repressor is added, is shown in lane 5. It cannot be determined directly from the footprint whether repressor has replaced some RNA polymerase molecules from the operator due to the overlap of the binding sites in this region. Repressor does, however, impair RNA polymerase binding to the promoter DNA as shown by the repressor-induced appearance of two strong, and some weak, interspersed bands just 3' of the operator toward the CRP (CAP) site, which are suppressed in the footprint of RNA polymerase (lane 2). The same set of experiments carried out with a DNA fragment labeled at the 5' end of the upper strand (Fig. 8-6) is shown in the right-hand panel of Fig. 8-5. The reaction in lane 4 allows the comparison of the protection pattern of the competition experiment with repressor added before RNA polymerase, with the footprint or RNA polymerase only (lane 2), and with repressor only (lane 3). We found, to our surprise, that prebound repressor is unable to prevent RNA polymerase interactions with the promoter. This is clearly shown by the RNA polymerase-specific footprint toward the CRP site which is different, however, from the one obtained in the presence of RNA polymerase only (judged by the intensity changes of the interspersed bands). Lane 5 in Fig. 8-5 shows the outcome of the experiment in which repressor is added after the formation of the RNA polymerase-promoter complex. In this case, there is no change in the footprint of RNA polymerase extending from the operator toward the CRP site as compared to that of polymerase only. It is difficult, however, to determine whether repressor has replaced RNA polymerase in the operator, because of the overlap of the footprints in this region. It is not clear whether the weak protection by RNA polymerase in the operator is caused by lower affinity for this part of the site or to a mixed population of RNA polymerase molecules, some of which have an altered operator-binding activity. In spite of these uncertainties we can conclude that repressor can interact with the operator when RNA polymerase is bound to the remaining promoter sequence (Fig. 8-5, lane 4). The competition experiments were also performed with a mutant repressor (I12-X86), which binds 10^4 times more strongly to operator than wild-type repressor (Schmitz, Coulondre, & Miller, 1978), to exclude the possibility of repressor dissociation during the DNase I digestion. The results were identical (data not shown).

Figure 8-7 shows a diagram compiling the data of the competition experiments. The collected data lead to the following interpretation:

1. A repressor molecule, prebound to the operator, blocks extensive RNA polymerase-promoter interactions,
2. RNA polymerase is still able to interact with parts of the promoter sequence, but
3. these interactions differ from those of RNA polymerase alone (Fig. 8-5, right-hand panel, lane 4), and
4. RNA polymerase prebound to the promoter-operator region

still allows some interactions of repressor with the operator, which do not greatly influence the RNA polymerase footprint between the operator and the CRP site, as shown by the presence of the repressor-induced site for enhanced DNase I cleavage in lane 5 of Fig. 8-5 (right-hand panel).

With regard to point 4, we cannot rule out, however, that the observed concomitant binding of RNA polymerase and repressor is due to the presence of a mixed population of RNA polymerase molecules, differing in affinity for the operator. Although it is now clear that the footprint of RNA polymerase on some other promoters is very similar to its footprint on the *lac* UV5 promoter (Russell & Bennett, 1981), the competitive interactions with other proteins, like the repressor, may be strongly dependent on the particular promoter sequence.

3. Footprints of CRP on the lac Control Region DNA

The sensitivity of the footprinting method is convincingly demonstrated by the analysis of the interactions of the cyclic AMP receptor protein (CRP) with the wild-type *lac* promoter-operator region.

A CRP binding site in the −60 base-pair region has been defined by sequencing mutational changes in the DNA, which affect CRP-specific *lac* transcription stimulation, and its interactions with its target site on the DNA (Dickson et al., 1975; Majors, 1975; Reznikoff & Abelson, 1978; Beckwith, 1978). The approximate extent of the CRP interaction site has been determined by physicochemical methods, including protection of bases against chemical modification and protection of DNA against digestion by Exonuclease III (Simpson, 1980). The footprint of CRP obtained on the *lac* promoter-operator region is schematically presented in Fig. 8-8 (Schmitz, 1981). The DNA fragment protected against partial DNase I digestion by purified CRP, complexed with cAMP, is indicated by the brackets; the protein-induced sites for enhanced DNase I cutting are indicated by arrows inside the brackets.

The footprint covers all the bases within the region that had been shown by other methods to be important for CRP-DNA interactions (Majors, 1975a; Dickson et al., 1975; Dickson et al., 1977; Reznikoff & Abelson, 1978). The extent of the footprint coincides, at the 3′ ends, with the sites protected against Exonuclease III digestion (arrows outside the brackets; Simpson, 1980). The sensitivity of the method also allowed us to detect a second CRP interaction site (CRP site 2) on this DNA fragment which had not

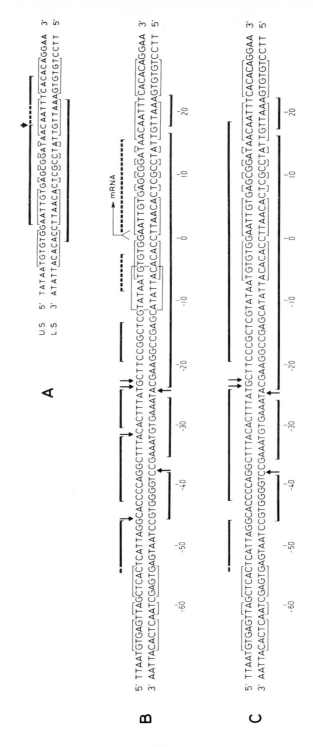

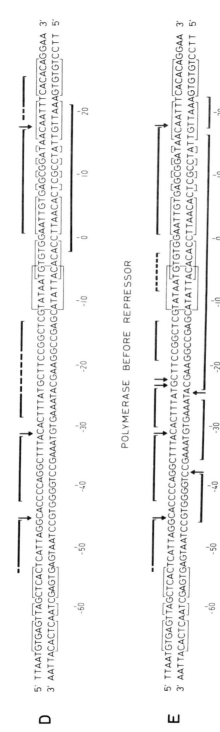

REPRESSOR BEFORE POLYMERASE

D
5' TTAAT GTGAGT TAGCTCACTCATT AGGCACCCCAGGCTTTACACTTTATGCTTCCGGCTCG TATAAT GTGTGGAATTGT GAGCG GAT AACAATTT CACACAGGAA 3'
3' AATTA CACTCAAT CGAGTGAGTAATCCGTGGGGTCCGAAATGTGAAATGACGAAGGCCGAGC ATATT ACACCTTAACACT CGCC T ATTGTTAAAGTGTGTCCTT 5'

POLYMERASE BEFORE REPRESSOR

E
5' TTAAT GTGAGT TAGCTCACTCATT AGGCACCCCAGGCTTTACACTTTATGCTTCCGGCTCG TATAAT GTGTGGAATTGT GAGCG GAT AACAATTT CACACAGGAA 3'
3' AATTA CACTCAAT CGAGTGAGTAATCCGTGGGGTCCGAAATGTGAAATGACGAAGGCCGAGC ATATT ACACCTTAACACT CGCC T ATTGTTAAAGTGTGTCCTT 5'

Figure 8-7. Diagrams of sequences protected against DNase I digestion in the *lac* control region DNA (Schmitz & Galas, 1979). The solid brackets represent protected regions, the broken-line brackets represent less well-protected regions, and the arrows represent sites of enhanced DNase I cleavage. The fine-line brackets around base-pair +10 indicate the symmetry of the operator, and those around base −60 indicate the symmetry of the CRP site. *A*: Repressor footprint. *B* and *C*: Footprints of different RNA polymerase preparations. *D*: Protection pattern of repressor followed by RNA polymerase. *E*: Protection pattern of RNA polymerase followed by repressor.

previously been described. CRP site 2 covers the operator itself (Fig. 8-8). The binding to site 2 is also cAMP-dependent. CRP sites 1 and 2 are similar in extent and orientation to the characteristic pattern of DNase I-enhanced cutting sites. CRP interacts less well, however, with site 2 than site 1 as determined by the salt-dependence of the binding reactions. The footprint of repressor on the operator is also shown in Fig. 8-8 to indicate the differences among the three sites. The footprints of the two proteins are very similar, differing principally in the characteristic pattern of induced DNase I cutting sites.

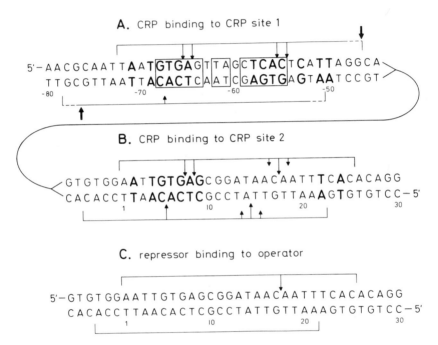

Figure 8-8.   Schematic representation of the CRP and repressor footprints on wild-type *lac* promotor-operator DNA as indicated by the brackets alongside the DNA sequence (Schmitz, 1981). Uncertainties in the extent of the protection at the very ends of the footprints are indicated by interrupted lines. Long arrows within the brackets point to protein-induced enhanced DNase I cutting sites; short arrows point to sites with an unchanged cleavage rate. The two arrows outside the brackets in *A* indicate the blocking site against Exonuclease III attack (Simpson, 1980).   *A.* Footprint of CRP on site 1 (−60 base-pair region). Boldface letters indicate matches with the consensus sequence (Taniguchi, O'Neill, & de Crombrugghe, 1979). Boxed sequences indicate dyad symmetries. *B.* Footprint of CRP on site 2 (*lac* operator). Boldface letters indicate matches between the sequences of CRP site 1 and 2.   *C.* Footprint of repressor on the *lac* operator (Schmitz & Galas, 1979). The brackets include sites where the extent of protection due to the weakness of bands is difficult to judge.

Taniguchi, O'Neill, and de Crombugghe (1979) have derived a consensus sequence for CRP binding sites by comparing the CRP interaction sequences of the *lac* CRP site 1, *ara* and *gal* operons. The consensus sequence is AAA<u>GTGTG</u>ACA (5' to 3'). The underlined region is apparently most important for specific CRP-DNA interactions. The matches with the consensus sequence, which is present as an inverted repeat in the CRP site 1, are indicated by boldface letters in the sequence of CRP site 1 in Fig. 8-8. The weaker binding of CRP to site 2 (operator) might be explained by the absence of the inverted repeat of the consensus sequence in this site. It should be noted, however, that both CRP footprints are very similar in extent, probably reflecting interactions of one dimeric CRP molecule in both sites. It should also be noted in this context that the interactions of CRP with site 1 and 2 are increased in strength by the presence of glycerol, a DNA destabilizing agent (Nakanishi et al., 1974). Site 2 is essentially glycerol-dependent. The use of the footprinting method to discover new binding sites for a protein, even in the region of known sites, is demonstrated by the discovery of the CRP site 2 in the operator. Comparing footprints of the same protein in several different sites should be very useful in attempts to understand the nature of specific recognition of DNA sequences by proteins. Footprints of proteins with multiple binding sites have now been reported: notably *lac* repressor (Schmitz & Galas, 1980), *int* protein of phage λ (Ross et al., 1979), the *ara*C protein (Ogden et al., 1980; Lee, Gielow, & Wallace, 1981), and DNA gyrase (Kirkegaard & Wang, 1981; Morrison & Cozzarelli, 1981; Fisher et al., 1981).

4. Footprint of DNA Gyrase

The footprint of DNA gyrase presents some particularities not encountered in the interactions previously discussed. We therefore turn now to a discussion of this protein-DNA interaction.

Enzymes capable of breaking double-stranded DNA in a reversible way have been classified as type-II DNA topoisomerases (Gellert et al., 1976; Cozzarelli, 1980b). DNA gyrase of *E. coli* is a multimeric type-II topoisomerase thought to act by a mechanism that involves the concerted breaking and rejoining of two opposite DNA strands (see Cozzarelli, 1980b). Passing DNA through the break can result in topoisomerase-specific reactions, such as relaxation and supercoiling, knotting and unknotting, and catenation and decatenation of DNA circles (Ross et al., 1979; Morrison & Cozzarelli, 1981). Gyrase has been found to interact strongly with discrete sites on the DNA. Binding of ATP to the DNA-gyrase complex can induce

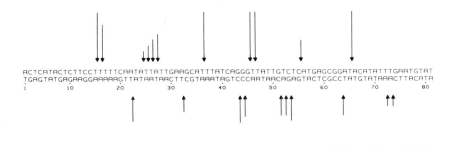

Figure 8-9. Topography of gyrase-DNA interaction (redrawn from Kirkegaard & Wang, 1981). The underlined sequence indicates a site on pBR322 DNA which is protected by gyrase against staphylococcal nuclease digestion. The thin arrows

a supercoiling reaction without hydrolyzing the coenzyme; further supercoiling reactions, on the other hand, require the energy provided by ATP hydrolysis (Sugino et al., 1978). Gyrase introduces cuts which are staggered by 4 bases on the complementary strands with 5'-phosphate-protruding and 3'-hydroxyl-recessive ends (Morrison & Cozzarelli, 1981; Kirkegaard & Wang, 1981; Fisher et al., 1981).

E. coli gyrase protects a DNA fragment of about 145 base pairs against extensive digestion by micrococcal nuclease (Liu & Wang, 1978). Protection of DNA, which has been labeled by nick translation, against DNase I attack results in the protection of DNA fragments differing in length by multiples of about 10 bases (Liu & Wang, 1978). This phenomenon was investigated further by two groups using the footprinting method (Kirkegaard & Wang, 1981; Morrison & Cozzarelli, 1981). Figure 8-9 shows an example of the footprint of gyrase on plasmid pBR322 DNA, as determined by Kirkegaard and Wang (1981, site 3 in their paper; Fig. 8-2 here). Gyrase protects a DNA region of about 150 base pairs against DNase I attack, as indicated by the bar at the bottom of Fig. 8-9. The central region of about 40 base pairs is flanked on either side by a protected region containing regularly spaced, gyrase-induced sites for enhanced DNase I cutting. These sites are separated by 10-11 bases on each strand and staggered by 2-4 bases on opposite strands. This kind of footprint is reminiscent of the DNase I protection pattern of nucleosomes and is consistent with a model in which the flanking regions, of about 50 base pairs, are wrapped around the gyrase molecule (Liu & Wang, 1978; Kierkegaard & Wang, 1981). The spacing of 10-11 bases of the DNase I-enhanced cutting sites on each strand would reflect the periodicity of exposed sites of the DNA helix, and the

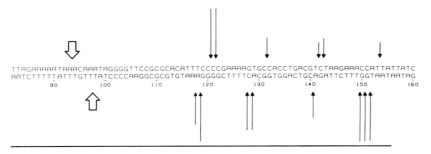

point to gyrase-induced sites for enhanced DNase I cleavage (the arrow length is proportional to the degree of enhancement). The two hollow arrows indicate gyrase-effected cleavage sites upon alkali-treatment.

spacing of 2-4 base pairs between cuts on complementary strands would reflect the register of sites on opposite strands protruding from the enzyme surface.

NUCLEASE SPECIFICITIES

Although it is clear that the specificity of DNase I is sufficiently low to permit precise definition of protein footprints in most sequences, the issue of the sequence specificity of this enzyme (and other endonucleases) may be important in future studies. If a certain sequence is recognized and cleaved preferentially, it is of great interest to the application of the present method to know whether it is the bases themselves that are recognized or whether it is a conformation of the DNA, preferentially assumed by that sequence, that constitutes the feature recognized by the enzyme. If conformation is involved to any great extent, the binding of a protein molecule to the DNA has the potential to change the spectrum of specificities in regions near the binding site by altering, perhaps in a subtle way, the conformation of the DNA.

It has recently become clear that the general A—T-rich region preference of micrococcal nuclease and its preference for single-stranded DNA are manifestations of some more subtle recognition processes (Von Hippel & Felsenfeld, 1964; Dingwall, Lomonosoff, & Laskey, 1981; Hörz & Altenburger, 1981). The fact that micrococcal nuclease has been used extensively in the analysis of nucleosome phasing, under the assumption that there were no important sequence specificities of the nuclease, should be taken as fair warning to investigators planning to use DNA-cleaving agents to define protein binding sites (Kornberg, 1981). There are, in fact, marked sequence preferences for this enzyme—much stronger than for DNase I—that could lead to significant artifacts in certain types of experiments defining nucleosome positions, for example. It is not yet clear, however, whether the specificity is related to a conformational peculiarity of the sequences preferred by micrococcal nuclease. Dingwall, Lomonosoff, and Laskey (1981) reported that DNase I exhibits a similar kind of specificity (cleaves the same sequences quite well) but that it is much less pronounced. Our examination of the preferred cutting site for DNase I in several distinct DNA sequences has not led to the identification of any kind of consensus sequence for nuclease recognition. On the other hand, there are a few preferred sites with striking similarities. A site within the sequence GATA, for example, is cut preferentially when this sequence occurs in several different places at the distal end of the *lac*I gene (Schmitz & Galas, 1980). It is cleaved with similar preference when it occurs in the *lac* operator (Galas & Schmitz, 1978). We have noted other preferences in sequence, but because of the apparent complexity and low specificity it is not possible to say yet what features contribute to the overall cleavage rate at a given site. It has been suggested that poor base-overlap in stacking is a factor in determining vulnerability to micrococcal nuclease (Dingwall, Lomonosoff, & Laskey, 1981), but there is no direct evidence of this. In general, we know very little about the subject. It is equally clear, however, that there is much to be gained by increasing our understanding of the interaction of nucleases with specific DNA sequences. They may be useful as probes for classes of DNA sequences (not exact sequences as for restriction enzymes), or for conformations of the DNA (Rhodes & Klug, 1980), and ultimately for the subtle conformation changes that may accompany DNA-protein interactions.

EXTENSIONS OF THE METHOD

It is now well established that the conformation of the DNA can have profound effects on genetic control in some cases (see Smith, 1981; Kano et al., 1981). The prime example of this sort of effect is provided by studies of the effect of supercoiling on gene expression of plasmid-carried genes. The fact that some promoters are sensitive to supercoiling, for example, makes it clear that, to understand the nature of DNA-protein interactions in gene control, it will be essential to be able to study the interactions when the DNA site is in a super-coiled molecule, or in a state other than linear, in any case. The footprinting method as described here is limited to linear DNA molecules because of the necessity for single-end labeling. Unlabeled DNA could be used, however, to determine the sensitivity of sites to DNA-cleaving agents if one could subsequently identify the proper set of DNA fragments in the large mixed population produced in the partial cleavage reaction. Wu (1980) has used the method of trans-ferring the entire population of fragments from the gel in which they have been fractionated to nitrocellulose, and then hybridizing to a specific, radioactively labeled restriction fragment that will pick out the proper nested set of fragments. The fragments in the nuclease protection reaction must be produced by the partial nuclease diges-tion, together with a complete restriction enzyme digestion at a specific site to define a unique direction for the nested set of frag-ments picked out by the radioactive probe. The general method should be applicable in many other cases where the specific inter-action of a given protein is being investigated by footprinting. The study of gene-control-related interactions in supercoiled DNA, as mentioned above, would be of great interest.

CONCLUDING REMARKS

We have illustrated in the preceding sections the wide potential applicability of the footprinting method. The major advantage of this technique is its simplicity in determining the extent of a protein binding site on both strands of a DNA molecule. It defines a binding site through the steric inhibition of nuclease cleavage by the DNA-binding protein. The site thus defined can therefore reflect the mutually exclusive binding events of proteins involved in regulation of transcription due to overlapping interaction sites. It is superior to the isolation of binding sequences after extensive DNase I digestion, since the artifacts inherent to this latter technique, caused by protein-induced sites of enhanced DNase I cutting within the protected site, are avoided.

The use of Exonuclease III digestion in probing a protein inter-action site is restricted to the possible determination of the 3' end of the protected fragment only, and by the reduced sensitivity in defining two nearby binding sites on the same fragment with differing affinities for a protein.

Methylation protection, crosslinking studies, and the determination of phosphates whose ethylation interferes with protein binding are superior to the footprinting method in determining single specific bases affected by protein binding. The footprinting method can, however, be much more sensitive than methylation protection, as shown by the fact that some protein interaction sites defined by DNase I protection do not exhibit any methylation protection at all, although they contain potential sites for methylation (the *ara* CRP site; Lee, Gielow, & Wallace, 1981), the pseudo-operator sites for *lac* repressor (Schmitz & Galas, 1980), and a DNA gyrase binding site (Kirkegaard & Wang, 1981), for example. All protein binding sites characterized on both strands have been found to exhibit protein-induced, DNase I-enhanced cutting sites (or unaffected sites) within the footprint. While these enhancements could be caused either by a DNA conformation change or by protein-protein interactions, the presence of the enhancements for many different proteins in many different DNA sites suggests that it is a change in the DNA conformation that is responsible. Surprisingly, it is these enhancements that often provide the most useful preliminary indicators of the existence of a weak footprint. Thus, they are of great practical value in searching for binding sites, or in searching for proper binding conditions. The enhanced cutting sites can also help to discriminate between a footprint and artifacts that can be caused by high protein concentrations, which may result in nonspecific retention of DNA fragments in the gel slots or uneven cleavage of the DNA among the reaction samples.

ACKNOWLEDGMENTS

This work was supported by a grant to J. H. Miller from the Swiss National Fund (F.N. 3.179.77). We are grateful to Otto Jenni for his skillful rendering of the figures.

REFERENCES

Alexander, M., Hepper, L. A., and Hurwitz, J. (1961). The purification and properties of micrococcal nuclease. *J. Biol. Chem.* 236:3014–3019.

Beckwith, J. R. (1978). The genetic system. In *The Operon*, eds. J. H. Miller and W. S. Reznikoff (Cold Spring Harbor, New York: Cold Spring Harbor Laboratory), pp. 11-30.

Bernardi, G., and Griffe, M. (1964). Studies on acid desoxyribonuclease. II. Isolation and characterization of spleen-acid deoxyribonuclease. *Biochem.* 3:1419.

Bernardi, A., Gaillard, C., and Bernardi, G. (1975). The specificity of five DNases as studied by the analysis of 5'-terminal doublets. *Eur. J. Biochem.* 52:451-456.

Bollum, F. J. (1965). Degradation of the homopolymer complex polydeoxy-adenylate-polydeoxythimidylate, polydeoxyinosinate-polydeoxycytidylate, and polydeoxyguanylate-polydeoxycytidylate by deoxyribonuclease I. *J. Biol. Chem.* 240:2599-2601.

Brent, R., and Ptashne, M. (1981). Mechanism of action of the lexA gene product. *Proc. Natl. Acad. Sci. USA* 78:4204-4208.

Chen, B., Crombrugghe, B., Anderson, W. B., Gottesman, M. E., Pastan, I., and Perlman, R. L. (1971). On the mechanism of action of *lac* repressor. *Nature New Biol.* 233:67-70.

Clark, P., and Eichhorn, G. L. (1974). A predictable modification of enzyme specificity. Selective alteration of DNA bases by metal ions to promote cleavage specificity by desoxyribonuclease. *Biochem.* 13:5098-5102.

Cozzarelli, N. R. (1980a). DNA gyrase and the supercoiling of DNA. *Science* 207:953-960.

Cozzarelli, N. R. (1980b). DNA topoisomerases. *Cell* 22:327-328.

Cozarrelli, N. R. (1980c). DNA gyrase and the supercoiling of DNA. *Science* 207:953-960.

D'Andrea, A. D., and Haseltine, W. (1978). Sequence specific cleavage of DNA by the anti-tumor antibiotics neocarcinostatin and bleomycin. *Proc. Natl. Acad. Sci. USA* 75:3608-3612.

Davidson, J. N. (1972). In *The Biochemistry of the Nucleic Acids*, 7th Ed. (New York: Academic Press), p. 192.

Dickson, R. C., Abelson, J. N., Barnes, W. M., and Reznikoff, W. S. (1975). Genetic regulation: The *lac* control region. *Science* 187:27-35.

Dickson, R. C., Abelson, J., Johnson, P., Reznikoff, W. S., and Barnes, W. M.

(1977). Nucleotide sequence changes produced by mutations in the *lac* promoter of *Escherichia coli. J. Mol. Biol.* 111:65–75.

Dingwall, C. D., Lomonosoff, G. P., and Laskey, R. A. (1981). High sequence specificity of micrococcal nuclease. *Nucl. Acids Res.* 9:2659–2673.

Emmer, M., Crombrugghe, B., Pastan, I., and Perlman, R. (1970). Cyclic AMP receptor protein of *E. coli*: Its role in the synthesis of inducible enzymes. *Proc. Natl. Acad. Sci. USA* 66:480–487.

Engelke, D. R., Ng, S., Shastry, B. S., and Roeder, R. G. (1980). Specific interaction of a purified transcription factor with an internal control region of 5S RNA genes. *Cell* 9:717–728.

Eron, L., and Block, R. (1971). Mechanism of initiation and repression of in vitro transcription of the lac operon of Escherichia coli. *Proc. Natl. Acad. Sci. USA* 68:1828–1832.

Fisher, M., Mizuuchi, K., O'Dea, M. H., Ohmori, H., and Gellert, M. (1981). Site-specific interaction of DNA gyrase with DNA. *Proc. Natl. Acad. Sci. USA* 78:4165–5169.

Galas, D. J., and Schmitz, A. (1978). DNAase footprinting: A single method for the detection of protein DNA binding specificity. *Nucl. Acid Res.* 5:3157–3170.

Gellert, M., Mizuuchi, K., O'Dea, M. H., and Nash, H. A. (1976). DNA gyrase: An enzyme that introduces superhelical turns into DNA. *Proc. Natl. Acad. Sci. USA* 73:3872–3876.

Gilbert, W. (1976). Starting and stopping sequences for the RNA polymerase. In *RNA Polymerase*, eds. R. Losik and M. Chamberlin (Cold Spring Harbor, New York: Cold Spring Harbor Laboratory), pp. 193–205.

Gilbert, W., and Maxam, A. (1973). The nucleotide sequence of the *lac* operator. *Proc. Natl. Acad. Sci. USA* 70:3581–3584.

Gilbert, W., Maxam, A., and Mirzabekov, A. (1976). Contacts between *lac* repressor and DNA revealed by methylation. In *Control of Ribosome Synthesis*, eds. N. O. Kjeldegaard and O. Maaløe (Copenhagen: Munksgaard), pp. 139–148.

Gilbert, W., Gralla, J., Majors, J., and Maxam, A. (1975). Lactose operator sequences and the action of *lac* repressor. In *Protein-Ligand Interactions*, eds. H. Sund and G. Blauer (Berlin: Walter-de Gruyter), pp. 193–210.

Hatayama, T., Goldberg, I. Takeshita, M., and Grollman (1978). Nucleotide specificity in DNA scission by neocarzinostatin. *Proc. Natl. Acad. Sci. USA* 75:3603-3607.

Hefron, F., So, M., and McCarthy, B. J. (1978). In vitro mutagenesis of a circular DNA molecule by using synthetic restriction sites. *Proc. Natl. Acad. Sci. USA* 75:6012-6016.

Heyden, B., Nüsslein, C., and Schaller, H. (1972). Single RNA polymerase binding site isolated. *Nature New Biol.* 240:9-12.

Hörz, W., and Altenburger, W. (1981). Sequence specific cleavage of DNA by micrococcal nuclease. *Nucl. Acids Res.* 9:2643-2658.

Johnson, A. D., Meyer, B. J., and Ptashne, M. (1979). Interactions between DNA-bound repressors govern regulation by λ phage repressor. *Proc. Natl. Acad. Sci. USA* 76:5061-5065.

Johnsrud, L. (1978). Contacts between *Escherichia coli* RNA polymerase and a *lac* operon promoter. *Proc. Natl. Acad. Sci. USA* 75:5314-5318.

Junowicz, E., and Spencer, J. H. (1973a). Studies on bovine pancreatic deoxyribonuclease A. I. General properties and activation with different bivalent metals. *Biochim. Biophys. Acta* 312:72-84.

Junowicz, E., and Spencer, J. H. (1973b). Studies on bovine pancreatic deoxyribonuclease A. II. The effect of different bivalent metals on the specificity of degradation of DNA. *Biochim. Biophys. Acta.* 312:85-102.

Kano, Y., Miyashita, I., Nakamura, H., Kuroki, K., Nagata, A., and Imamoto, F. (1981). *In vivo* correlations between DNA supercoiling and transcription. *Gene* 13:173-184.

Kirkegaard, K., and Wang, J. C. (1981). Mapping the topography of DNA wrapped around gyrase by nucleolytic and chemical probing of complexes of unique DNA sequences. *Cell* 23:721-279.

Kornberg, R. (1981). The location of nucleosomes in chromatin: Specific or statistical? *Nature* 292:579-580.

Laskowski, M. (1966). Pancreatic Deoxyribonuclease I. In *Procedures in Nucleic Acids Research*, eds. G. L. Cantoni and D. R. Davies (New York: Harper & Row), pp. 85-101.

Laskowski, M. (1961). Deoxyribonucleases. In *The Enzymes*, eds. P. D. Boyer, H. Landy, and K. Myrbäck (New York: Academic Press), pp. 123-147.

Lee, N. L., Gielow, W. O., and Wallace, R. G. (1981). Mechanism of *araC* autoregulation and the domains of two overlapping promoters, p_c and p_{BAD} in the L-arabinose regulatory region of *Escherichia coli. Proc. Natl. Acad. Sci. USA* 78:752-756.

Lehman, R., Roussos, G. G., and Pratt, E. A. (1962). The deoxyribonucleases of *Escherichia coli. J. Biol. Mol.* 237:819-828.

Le Talaer, J., and Jeanteur, Ph. (1971). Purification and base composition analysis of phage lambda early promoters. *Proc. Natl. Acad. Sci. USA* 68:3211-3215.

Liao, Ta-Hsiu, Salnikow, J., Moore, St., and Stein, W. H. (1973). Bovine pancreatic desoxyribonuclease A. *J. Biol. Chem.* 248:1489-1493.

Lindberg, U. (1967). Molecular weight and amino acid composition of desoxyribonuclease I. *Biochem.* 6:335-342.

Little, J. W., Mount, D. W., and Yanisch-Perron, C. R. (1981). Purified lex A protein is a repressor of the rec A and lex A genes. *Proc. Natl. Acad. Sci. USA* 78:4199-4203.

Liu, L. F., and Wang, J. C. (1978). DNA-DNA gyrase complex: The wrapping of the DNA duplex outside the enzyme. *Cell* 15:979-984.

Liu, L. F., Liu, C. C., and Alberts, B. M. (1980). Type II DNA topoisomerases: Enzymes that can unknot a topologically knotted DNA molecule via a reversible double-strand break. *Cell* 19:697-707.

Lizarraga, B., Sanchez-Romero, D., Gils, A., and Melgar, E. (1978). The role of Ca^{2+} on pH-induced hydrodynamic changes of bovine pancreatic deoxyribonuclease A. *J. Biol. Chem.* 253:3191-3195.

Lutter, L. C. (1977). Deoxyribonuclease I produces staggered cuts in the DNA of chromatin. *J. Mol. Biol.* 117:53-69.

Majors, J. (1975a). Initiation of in vitro mRNA synthesis from the wild type *lac* promoter. *Proc. Natl. Acad. Sci. USA* 72:4394-4398.

Majors, J. (1975b). Specific binding of CAP factor to *lac* promoter DNA. *Nature* 256:672-674.

Maxam, A., and Gilbert, W. (1980). Sequencing end-labelled DNA with base-specific chemical cleavages. *Methods Enzymol.* 65:499-560.

Maxam, A., and Gilbert, W. (1977). A new method for sequencing DNA. *Proc. Natl. Acad. Sci. USA* 74:560-564.

McConnell, D. J., Searcy, D. G., and Sutcliffe, J. G. (1978). A restriction enzyme ThaI from the thermophilic mycoplasma thermoplasma acidophilium. *Nucl. Acids Res.* 5:1729-1739.

Melgar, E., and Goldtwait, D. A. (1968). Deoxyribonucleic acid nucleases. *J. Biol. Mol.* 243:4409-4416.

Morrison, A., and Cozzarelli, N. R. (1981). Contacts between DNA gyrase and its binding site on DNA: Features of symmetry and asymmetry revealed by protection from nucleases. *Proc. Natl. Acad. Sci. USA* 78:1416-1420.

Mutsukage, A., Murakami, S., and Kameyama, T. (1969). The isolation and characterization of DNA region bound to *Escherichia coli* RNA polymerase. *Biochim. Biophys. Acta* 179:145-157.

Nakanishi, S., Adhya, S., Gottesman, M., and Pastan, I. (1974). Activation of transcription at specific promoters by glycerol. *J. Biol. Chem.* 249:4050-4056.

Ogata, R. T., and Gilbert, W. (1979). DNA-binding site of *lac* repressor probed by dimethylsulfate methylation of lac operator. *J. Mol. Biol.* 132:709-728.

Ogata, R. T., and Gilbert, W. (1978). An amino-terminal fragment of *lac* repressor binds specifically to *lac* operator. *Proc. Natl. Acad. Sci. USA* 75:5851-5854.

Ogata, R. T., and Gilbert, W. (1977). Contacts between the *lac* repressor and thymines in the *lac* operator. *Proc. Natl. Acad. Sci. USA* 74:4973-4976.

Ogden, S., Haggerty, D., Stoner, C. M., Kolodrubetz, D., and Schleif, R. (1980). The *Escherichia coli* L-arabinose operon: Binding sites of the regulatory proteins and a mechanism of positive and negative regulation. *Proc. Natl. Acad. Sci. USA* 77:3346-3350.

O'Neill, M. C., Amass, B., and Crombrugghe, B. (1981). Molecular model of the DNA interaction site for the cyclic AMP receptor protein. *Proc. Natl. Acad. Sci. USA* 78:2213-2217.

Pfahl, M. (1978). Effects of DNA denaturants on the *lac* repressor-operator interactions. *Biochim. Biophys. Acta* 520:285-290.

Pirotta, V. (1975). Sequence of the O_R operator of phage λ. *Nature* 254:114-121.

Poteete, A. R., Ptashne, M., Ballivet, M., and Eisen, H. (1980). Operator-sequences of bacteriophage P22 and 21. *J. Mol. Biol.* 137:81-91.

Pribnow, D. (1975). Nucleotide sequence of an RNA polymerase binding site at an early T_7 promoter. *Proc. Natl. Acad. Sci. USA* 72:784-788.

Price, P. A., Liu, T. Y., Stein, W. H., and Moore, S. (1969). Properties of chromatographically purified bovine pancreatic deoxyribonuclease. *J. Biol. Chem.* 244:917-923.

Putney, S. D., and Schimmel, P. (1981). An aminoacyl tRNA synthetase binds to a specific DNA sequence and regulates its gene transcription. *Nature* 291:632-635.

Queen, C., and Rosenberg, M. (1981). A promoter of pBR322 activated by cAMP receptor protein. *Nucl. Acids Res.* 9:3365-3377.

Reznikoff, W. S., and Abelson, J. N. (1978). The *lac* promoter. In *The Operon*, eds. J. H. Miller and W. S. Resnikoff (Cold Spring Harbor, New York: Cold Spring Harbor Laboratories), pp. 221-243.

Rhodes, D., and Klug, A. (1980). Helical periodicity of DNA determined by enzyme digestion. *Nature* 286:573-578.

Ross, W., Landy, A., Kikuchi, Y., and Nash, H. (1979). Interaction of *Int* protein with specific site on λ *att* DNA. *Cell* 18:297-307.

Russel, D. R., and Bennett, G. N. (1981). Characterization of the β-lactamase promoter of pBR322. *Nucl. Acids Res.* 9:2517-2533.

Sakonju, S., Brown, D. D., Engelke, D., Sun-Yu, Ng, Shastry, B. S., and Roeder, R. G. (1981). The binding of a transcription factor to deletion mutants of a 5s ribosomal RNA gene. *Cell* 23:665-669.

Schaller, H., Gray, C., and Hermann, K. (1975). Nucleotide sequence of an RNA polymerase binding site from the DNA of bacteriophage fd. *Proc. Natl. Acad. Sci. USA* 72:737-741.

Schmitz, A. (1979). Cyclic AMP receptor protein interacts with lactose operator DNA. *Nucl. Acids Res.* 9:277-292.

Schmitz, A., and Galas, D. J. (1980). Sequence-specific interactions of the tight-binding I12-X86 *lac* repressor with non-operator DNA. *Nucl. Acids Res.* 8:487-506.

Schmitz, A., and Galas, D. J. (1979). The interaction of RNA polymerase and *lac* repressor with the *lac* control region. *Nucl. Acids Res.* 6:111-136.

Schmitz, A., Coulondre, C., and Miller, J. H. (1978). Genetic studies of the *lac* repressor: V. Repressor which binds operator more tightly generated by suppression and reversion of nonsense mutations. *J. Mol. Biol.* 123: 431–456.

Shack, J., and Bynum, B. (1964). Interdependence of variables in the activation of deoxyribonuclease I. *J. Biol. Chem.* 239:3843–3848.

Siebenlist, U., and Gilbert, W. (1980). Contacts between *Escherichia coli* RNA polymerase and an early promoter of phage T7. *Proc. Natl. Acad. Sci. USA* 77:122–126.

Siebenlist, U., Simpson, R. B., and Gilbert, W. (1980). *E. coli* RNA polymerase interacts homologously with two different promoters. *Cell* 20:269–281.

Simpson, R. B. (1980). Interaction of the cAMP receptor protein with the *lac* promoter. *Nucl. Acids Res.* 8:759–766.

Simpson, R. B. (1979). Contacts between *Escherichia coli* RNA polymerase and thymines in the *lac* UV5 promoter. *Proc. Natl. Acad. Sci. USA* 76:3233–3237.

Smith, G. R. (1981). DNA supercoiling: Another level for regulating gene expression. *Cell* 24:599–600.

Smith, T. F., and Sadler, J. R. (1971). The nature of lactose operator constitutive mutations. *J. Mol. Biol.* 59:273–305.

Sugino, N. P., Higgins, P. O., Brown, C. L., Peebles, N. R., and Cozzarelli, N. R. (1978). Energy coupling in DNA gyrase and the mechanism of action of novobiocin. *Proc. Natl. Acad. Sci. USA* 75:4838–4842.

Sutcliffe, J. G. and Church, G. M. (1978). The cleavage site of the restriction endonuclease Ava II. *Nucl. Acids Res.* 5:2313–2319.

Taniguchi, T., O'Neill, M., and de Crombrugghe, B. (1979). Interaction site of *Escherichia coli* cyclic AMP receptor protein on DNA of galactose operon promoter. *Proc. Natl. Acad. Sci. USA* 76:5090–5094.

Tsurimoto, T., and Matsubara, K. (1981). Purified bacteriophage λ protein binds to four repeating sequences of λ replication origin. *Nucl. Acids Res.* 9:1789–1799.

Tullis, R. H., and Rubin, H. (1980). Calcium protects DNaseI from proteinase K: A new method for the removal of contaminating RNase from DNaseI. *Anal. Biochem.* 107:260–264.

Von Hippel, P. H., and Felsenfeld, G. (1964). Micrococcal nuclease as a probe of DNA conformation. *Biochem.* 3:27–32.

Walz, A., and Pirotta, V. (1975). Sequence of the P_R promoter of phage λ. *Nature* 254:118-121.

Wang, J. C., and Liu, L. F. (1979). DNA topoisomerases: Enzymes that catalyze the concerted breaking and rejoining of DNA backbone bonds. In *Molecular Genetics*, Part III, ed. J. H. Taylor (New York: Academic Press), pp. 65-88.

Wang, J. C., Barkley, M. D., and Bourgeois, S. (1974). Measurements of unwinding of *lac* operator by repressor. *Nature* 251:247-248.

Wu, C. (1980). The 5' ends of Drosophila heat shock genes in chromatin are hypersensitive to DNaseI. *Nature* 286:854-860.

Zubay, G., Schwartz, D., and Beckwith, J. (1970). Mechanism of activation of catabolite-sensitive genes: A positive control system. *Proc. Natl. Acad. Sci. USA* 66:104-110.

ADDENDUM

The following list of articles, not referred to in the text, is intended to aid researchers who wish to use these methods for their own purposes by providing a representative set of references to recent work in the literature on a wide variety of DNA-protein interactions.

Aiba, H. (1983). Auto-regulation of the *Escherichia coli* CRP gene: CRP is a transcriptional repressor for its own gene. *Cell* 32:141-149.

Berlin, V., and Yanofsky, C. (1983). Release of transcript and template during transcription termination at the TRP operon attenuator. *J. Biol. Chem.* 258:1714-1719.

Fischer, P. A., and Korn, D. (1981). Ordered sequential mechanism of substrate recognition and binding by KB cell DNA polymerase alpha. *Biochem.* 20:4560-4569.

Friden, P., Newman, T., and Freundlich, M. (1982). Nucleotide sequence of the ilvB promoter-regulatory region: A biosynthetic operon controlled by attenuation and cyclic AMP. *Proc. Natl. Acad. Sci. USA* 79:6156-6160.

Grindley, N. D., Lauth, M. R., Wells, R. G., Wityk, R. J., Salvo, J. J., and Reed, R. R. (1982). Transposon-mediated site-specific recombination: Identification of three binding sites for resolvase at the res sites of gamma delta and Tn3. *Cell* 30:19-27.

Hansen, U., Tenen, D. G., Livingston, D. M., and Sharp, P. A. (1981). T antigen repression of SV40 early transcriptin from two promoters. *Cell* 27:603-612.

Hansen, J., Pschorn, W., and Ristow, H. (1982). Functions of the peptide antibodies tyrocidine and gramicidin: Induction of conformational and structural changes of superhelical DNA. *Eur. J. Biochem.* 126:279-284.

Ho, Y., and Rosenberg, M. (1982). Characterization of the phage lambda regulatory protein cII. *Ann. Microbiol.* A133:215-218.

Kanazawa, H., Mabuchi, K., and Futai, M. (1982). Nucleotide sequence of the promoter region of the gene cluster for promoter-translocating ATPase from *Escherichia coli* and identification of the active promoter. *Biochem. Biophys. Res. Comm.* 107:568-575.

Klemenz, R., Stillman, D. J., and Geiduschek, E. P. (1982). Specific interactions of Saccharomyces cerevisiae proteins with a promoter region of eukaryotic tRNA genes. *Proc. Natl. Acad. Sci. USA* 79:6191-6195.

Kosiba, B. E., and Schleif, R. (1982). Arabinose-inducible promoter from *Escherichia coli*: Its cloning from chromosomal DNA, identification as the araFG promoter and sequence. *J. Mol. Biol.* 156:53-66.

Lee, N. L., Gielow, W. O., and Wallace, R. G. (1981). Mechanism of araC autoregulation and the domains of two overlapping promoters, Pc and PBAD, in the L-arabinose regulatory region of *Escherichia coli. Proc. Natl. Acad. Sci. USA* 78:752-756.

Mikryukov, N. A., Karginov, V. A., and Vasilenko, S. K. (1982). Precise localization of the *Escherichia coli* RNA polymerase binding sites in the region of ampicillin and tetracycline resistance promoter of the pBR322 plasmid. *Bioorg. Khim. 8:269-271.*

Misra, T. K., Grandgenett, D. P., and Parsons, J. J. (1982). Avian retrovirus pp32 DNA-binding protein. I. Recognition of specific sequences on retrovirus DNA terminal repeats. *J. Virol.* 44:330-343.

Moran, C. P., Lang, N., and Losick, R. (1981). Nucleotide sequence of a Bacillus subtilis promoter recognized by Bacillus subtilis RNA polymerase containing sigma-37. *Nucl. Acids. Res.* 9:5979-5990.

Ovchinnikov, Y. A., Efimov, V. A., Chakhmakhcheva, O. G., Shiba, N. P., and Lipkin, V. M. (1981). Interaction of DNA-dependent RNA polymerase *Escherichia coli* with analogs of the promoter region of bacteriophage fd-DNA containing 5-bromouracil residues in the template strand. *Bioorg. Khim.* 7:485.

Piette, J., Cunin, R., Boyen, A., Charlier, D., Crabeel, M., Van Vliet, F., Glansdorff, N., Squires, C., and Squires, C. L. (1982). The regulatory operon of the divergent argECBH operon in *Escherichia coli* K-12. *Nucl. Acids Res.* 10:8031-8048.

Queen, C., and Rosenberg, M. (1981). A promoter of pBR322 activated by a cAMP receptor protein. *Nucl. Acids Res.* 9:3365-3377.

Sakonju, S., Brown, D. D., Engelke, D., Ng, S. Y., Shastry, B. S., and Roeder, R. G. (1981). The binding of a transcription factor to deletion mutants of a 5S ribosomal RNA gene. *Cell* 23:665-669.

Sakonju, S., and Brown, D. D. (1982). Contact points between a positive transcription factor and the Xenopus 5S RNA gene. *Cell* 31:395-405.

Sancar, G. B., Sancar, A., Little, J. W., and Rupp, W. D. (1982). The uvrB gene of *Escherichia coli* has both lexA-repressed and lexA-independent promoters. *Cell* 28:523-530.

Shenk, T. (1981). Transcriptional control regions: Nucleotide sequence requirements for initiation by RNA polymerase II and polymerase III. *Current Topics Microb. Immunol.* 93:25-46.

Shimatake, H., and Rosenberg, M. (1981). Purified lambda regulatory protein cII positively activates promoters for lysogenic development. *Nature* 292: 128-132.

Talkington, C., and Pero, J. (1979). Distinctive nucleotide sequences of promoters recognized by RNA polymerase containing a phage coded "sigmalike" protein. *Proc. Natl. Acad. Sci. USA* 76:5090-5094.

Tegtmeyer, P., Anderson, B., Shaw, S. B., and Wilson, V. G. (1981). Alternative interactions of the SV40 A protein with DNA. *Virology* 115:75-87.

Tenen, D. G., Haines, L. L., and Livingston, D. M. (1982). Binding of an analog of the simian virus 40 T antigen to wild type and mutant viral replication origins. *J. Mol. Biol.* 157:473-492.

Tsurimoto, T., and Matsubara, K. (1981). Purified bacteriophage lambda O protein binds to four repeating sequences at the lambda replication origin. *Nucl. Acids Res.* 9:1789-1799.

Valentin-Hansen, P. (1982). Tandem CRP binding sites in the deo operon of *Escherichia coli* K-12. *Embo. J.* 1:1049-1054.

Van Dyke, M. W., Hertzberg, R. P., and Dervan, P. B. (1982). Map of distamycin, netropsin and actinomycin binding sites on heterogenous DNA: DNA cleavage-inhibition patterns with methidiumpropyl-EDTA. FeII. *Proc. Acad. Sci. USA* 79:5470-5474.

Vonka, V., and Hirsch, I. (1982). Epstein-Barr virus nuclear antigen. *Prog. Med. Virol.* 28:145-179.

Wallace, R. G., Lee, N., and Fowler, A. V. (1980). The araC gene of *Escherichia coli*: Transcriptional and translational start points and complete nucleotide sequence. *Gene* 12:179-190.

Wickens, M. P., and Laskey, R. A. (1981). Expression of cloned genes in cell-free systems and in micro-injected Xenopus oocytes. *Genet. Eng.* 1:103-167.

9

Transcription Initiation and Termination Signals

THOMAS SHENK

PROKARYOTIC PROMOTERS

Promoters for bacterial RNA polymerase have been relatively well defined (e.g., Gilbert, 1976; Brown et al., 1978; Simpson, 1979; Scherer, Walkinshaw, & Arnott, 1978; Reznikoff & Abelson, 1978; Rosenberg & Court, 1979; Siebenlist, Simpson, & Gilbert, 1980). Comparison of large numbers of bacterial promoters (Scherer, Walkinshaw, & Arnott, 1978; Rosenberg & Court, 1979; Siebenlist, Simpson, & Gilbert, 1980; Soberon, et al., 1982) has identified two consensus sequences (Fig. 9-1): the so-called Pribnow Box (Pribnow, 1975) whose consensus sequence is TATAATG, located approximately 11-6 nucleotides before the transcription initiation site, and a "−35 Box," with the consensus sequence TGTTGACA beginning at approximately −36 nucleotides before the transcription initiation site. The distance between the −35 and Pribnow boxes can vary by 1 or 2 nucleotides in various promoters but is generally 17 or 18 base pairs. Stephens and Gralla (1982) have constructed variant "*lac*" promoters with either 16, 17, or 18 base pairs between the −35 Box and the Pribnow Box and studied promoter efficiency as measured by open complex formation. A separation of exactly 17 base pairs resulted in almost an order of magnitude increase in rate over separations of 16 or 18 bases.

The −35 and Pribnow consensus sequences are clearly essential components of prokaryotic promoters. Most promoter mutations fall within these sequences. The sequences contain contact points for RNA polymerase (reviewed in Siebenlist, Simpson, & Gilbert,

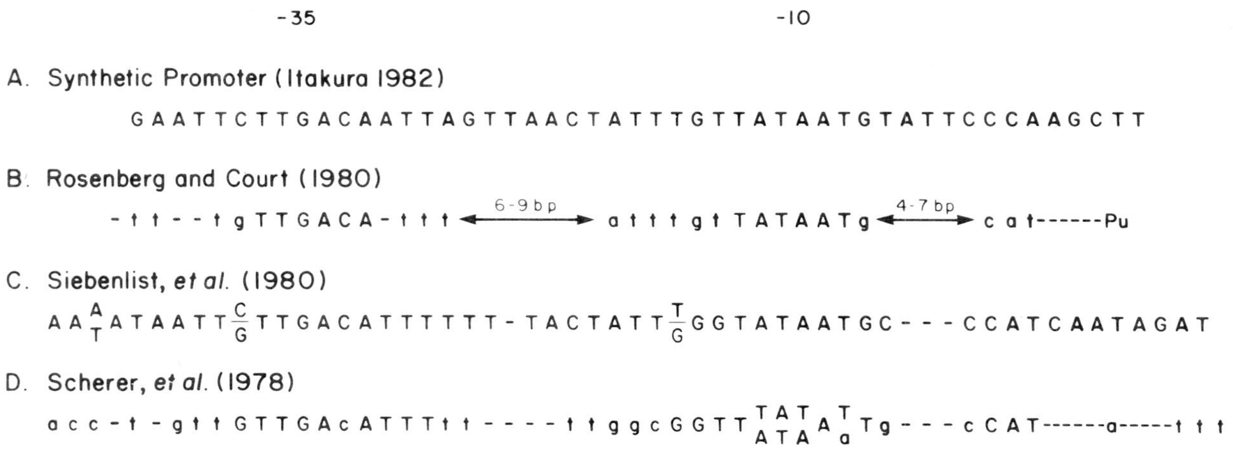

-35   -10

A. Synthetic Promoter (Itakura 1982)

G A A T T C T T G A C A A T T A G T T A A C T A T T T G T T A T A A T G T A T T C C C A A G C T T

B. Rosenberg and Court (1980)

- t t - - - t g T T G A C A - t t t $\xrightarrow{6-9\,bp}$ a t t t g t T A T A A T g $\xrightarrow{4-7\,bp}$ c a t - - - - - -Pu

C. Siebenlist, *et al.* (1980)

A A ^A_T A T A A T T ^C_G T T G A C A T T T T T T - T A C T A T T ^T_G G G T A T A A T G C - - - C C A T C A A T A G A T

D. Scherer, *et al.* (1978)

a c c - t - g t t G T T G A c A T T T t t - - - - t t g g c G G T T $^{T A T A}_{A T A A}$ ^T_a T g - - - c C A T - - - - - a - - - - t t t

Figure 9-1. Comparison of biologically active synthetic *E. coli* promoter and three forms of promoter consensus sequence (taken from Soberon et al., 1982).

1980). Recently, Soberon et al. (1982) synthesized a DNA fragment containing -35 and Pribnow consensus sequences and found it to function *in vivo*.

Additional sequences may also play some role in transcription initiation. For example, there appears to be a nonrandom distribution of bases located both 3 and 4 nucleotides upstream from the -35 Box and at the transcription initiation site. Some strong promoters are embedded in AT-rich sequences, and in one case a deletion of an AT-rich segment more than 70 b.p. upstream from the promoter decreased the rate of initiation of transcription (Wells, 1982). The strong binding of RNA polymerase at low temperatures suggests that there may be effects of flanking sequence on the ease of transcription initiation of nearby promoters. However, these sequences probably do not interact directly with RNA polymerase, since no promoter mutants have been found more than 38 b.p. upstream from the transcription initiation site. Furthermore, studies in which RNA polymerase is cross-linked to its template or in which polymerase/promoter complexes are subjected to chemical modification do not provide evidence of interaction more than 38 b.p. upstream of the transcription initiation site.

EUKARYOTIC RNA POLYMERASES

Three chromatographically separable RNA polymerases have been isolated from the nuclei of a wide variety of eukaryotic cells (reviewed by Roeder, 1976). They have been designated polymerases I, II, and III or A, B, and C. The polymerases are large, multisubunit complexes, and they can be differentiated on the basis of their sensitivity to the fungal toxin, α-amanitin. RNA polymerase I is responsible for the synthesis of ribosomal RNAs, polymerase II is responsible for heterogeneous nuclear RNA which is processed to mRNAs, and polymerase III transcribes genes encoding 5S ribosomal RNA, tRNAs, and a variety of small cellular and viral RNAs of uncertain function.

CELL-FREE TRANSCRIPTION BY EUKARYOTIC POLYMERASES

Until recently, the nucleotide sequences involved in the initiation of transcription by these polymerases in eukaryotic cells remained a mystery. Now, thanks to a variety of technological advances, many of which are described in this volume, it has become possible to identify these sequences and dissect their functions in the transcription process. A major breakthrough in our understanding of transcription by eukaryotic polymerases has come with the development of cell-free systems.

Polymerase III

Polymerase III present in nuclear (Birkenmeier, Brown, & Jordan, 1978; Korn, Birkenmeier, & Brown, 1979) and whole-cell (Ng, Parker, & Roeder, 1979) *Xenopus* oocyte extract selectively transcribes cloned, *Xenopus* 5S RNA genes. Cytoplasmic extracts from human KB cells (Wu, G.-J., 1978; Weil, Segall, et al., 1979), mouse plasmacytoma cells, and *Xenopus* oocytes (Weil, Segall, et al., 1979) accurately transcribe cloned 5S, tRNA, and VA RNA genes. These extracts contain the bulk (65–90%) of cellular polymerase III activity (Weil, Segall, et al., 1979). The polymerase and other components leak out of the nucleus during the fractionation procedure. Multiple factors are required for accurate transcription by polymerase III. Segall, Matsui, and Roeder (1980) have separated a KB cell extract into four different fractions by chromatography on phosphocellulose. Two fractions plus polymerase III are required for transcription of VAI and tRNA genes. An additional fraction must be included for transcription of 5S RNA genes.

Polymerase II

Shortly after cell extracts capable of polymerase III transcription became available, similar extracts were developed for polymerase II. Weil, Luse, et al. (1979) supplemented a cytoplasmic kB cell extract with calf-thymus polymerase II and obtained accurate initiation at the adenovirus 2 major late control site. Matsui et al. (1980) fractionated the kB cell extract and identified four fractions required for accurate and specific initiation by polymerase II. Manley et al. (1980) and, more recently, Weingartner and Keller (1981) prepared concentrated HeLa cell extracts that initiate specifically at a variety of adenovirus polymerase II control regions. These extracts do not require supplementation with polymerase as does the extract utilized by Weil, Luse, et al. (1979). The extracts have proved capable of transcribing a wide variety of RNA polymerase II-type genes.

Polymerase I

Most recently, extracts of Ehrlich ascites cells supplemented with polymerase I have been shown to accurately transcribe a mouse ribosomal DNA gene (Grummt, 1981a). Transcription was resistant to α-amanitin, and only extracts derived from exponentially growing cells were active.

5' ENDS AND INITIATION SITES

To discuss transcriptional control regions, it is necessary to know the sites at which transcripts are initiated. Available evidence indicates that 5' ends which correspond to transcriptional initiation sites have been identified for all three eukaryotic RNA polymerases.

Polymerase III

tRNAs are processed at their 5' ends after transcription, but 5S and VA RNAs do not undergo posttranscriptional cleavage. Tetraphosphate residues are released from the 5' ends of 5S RNA (Denis & Wegnez, 1973) and VA RNAs (Price & Penman, 1972; Celma, Pan, & Weissman, 1977), and VA RNAs can be labeled with (β-^{32}P)-nucleoside triphosphates in isolated nuclei (Söderlund et al., 1976). Also, 5S and VA RNAs are synthesized *in vitro* with no evidence of a precursor transcript (Birkenmeier, Brown, & Jordan, 1978; Wu, G.-J., 1978; Ng, Parker, & Roeder, 1979; Weil, Segall, et al., 1979; Fowlkes & Shenk, 1980). Transcription of 5S and VA RNA genes should, therefore, be initiated at the position to which their 5' ends have been mapped.

Polymerase II

The 5' ends of eukaryotic mRNAs usually occur in the form of a cap structure with the general formula m^{7}GpppNm (Shatkin, 1976). A variety of biochemical experiments using defined viral and cellular templates indicate that cap sites generally are in close proximity to transcription initiation sites. The initiation site of the adenovirus 2 late transcriptional unit has been located at 16.5 map units on the viral genome by a variety of analyses, including nascent chain mapping (Weber, Jelinek, & Darnell, 1977; Ziff & Evans, 1978), UV mapping (Goldberg et al., 1977), and mapping by means of DRB (5,6-dichloro-1-β-D-ribofuranosylbenzimidazole)-induced termination of transcription (Fraser, Sehgal, & Darnell, 1978, 1979). The capped 5' terminus of the late transcript was also mapped to 16.5 map units by aligning T_1 ribonuclease-generated oligonucleotides from nuclear RNA with the DNA sequence of this region (Ziff & Evans, 1978). Similar results were obtained when the capped 5' end was synthesized in isolated nuclei (Manley, Sharp, & Gefter, 1979; Baker & Ziff, 1980) or in cell-free extracts (Weil, Luse, et al., 1979; Manley et al., 1980). No oligonucleotides were evident corresponding to RNA sequences immediately upstream of the cap site. Similar experiments indicate that initiation sites lie in close proximity to capped 5' termini for early adenovirus transcripts (Berk & Sharp, 1977, 1978; Evans et al., 1977; Baker et al., 1979; Chow, Broker, & Lewis, 1979; Sehgal, Fraser, & Darnell, 1979; Wilson, Frazer, & Darnell, 1979; Baker & Ziff, 1981), SV40 transcripts (Ford & Hsu, 1978; Laub et al., 1979; Honda et al., 1980), and the mouse β-globin gene (Konkel, Tilgham, & Leder, 1978; Luse & Roeder, 1980; Hofer & Darnell, 1981). Further evidence for identity of initiation and cap sites has been provided by analysis of RNA synthesized in the presence of (β-^{32}P)-labeled nucleoside triphosphates. This precursor specifically labeled SV40 (Contreras & Fiers, 1981; Gidoni et al., 1981), adenovirus, and mouse β-globin (Hagenbuchle & Schibler, 1981) cap cores (Gpp$N$) in the β-position, permitting the conclusion that the β-phosphate in the cap is derived from the 5'-terminal polyphosphate of the RNA.

Polymerase I

Ribosomal genes are generally present in a repeating unit which includes a preribosomal RNA gene and a spacer. The preribosomal RNA is processed into mature 18S, 5.8S, and 28S RNA species. Eukaryotic preribosomal RNA transcripts range in size from 8–13 kB, and UV mapping indicates that each unit has its own promoter (Hackett & Sauerbier, 1975). The preribosomal RNA transcript has been shown to contain 5'-di- or triphosphate residues in *Xenopus* (Reeder, Sollner-Webb, & Wahn, 1977), *Drosophila* (Levis & Penman. 1978), *Dictystelium* (Batts-Young & Lodish, 1978), yeast (Klootwijk, de Jonge, & Planta, 1979; Nikolaev et al., 1979), and mouse (Grummt, 1981b). Thus, one can conclude that the 5' end of the preribosomal RNA molecule marks the initiation site for polymerase I transcription.

TRANSCRIPTIONAL CONTROL REGIONS

The broad outlines of eukaryotic RNA polymerase control regions are now evident. It appears that each polymerase recognizes a different set of primary sequences, and the critical sequences are located in different positions relative to the 5′ ends of the transcripts.

Polymerase III

Telford et al. (1979) isolated a *Xenopus laevis* DNA segment containing the coding sequence for a tRNA^{met} gene with only 22 base pairs preceding the 5′ end of the transcript. This DNA was actively transcribed when injected into centrifuged oocytes. Similarly, a cloned *Bombyx mori* tRNA^{ala} gene which contained only 6 base pairs preceding the transcriptional initiation site functioned as a template in *Xenopus* geminal vesicles (Garber & Gage, 1979). Thus the boundary of the sequence required for initiation of transcription by RNA polymerase III could be as little as 6 base pairs before the 5′ end of a tRNA coding region. When the *Xenopus* tRNA^{met} gene was cut into halves (Kressmann et al., 1979), neither the 5′ nor the 3′ portion was transcriptionally active in oocytes. Clearly, something other than the tRNA^{met} gene 5′-flanking sequence was required for its transcription.

D. D. Brown and his co-workers demonstrated that the polymerase III transcriptional control region for a *Xenopus* 5S RNA gene was contained within its coding region (Bogenhagen, Sakonju, & Brown, 1980; Sakonju, Bogenhagen, & Brown, 1980). Deletion mutations were generated by *in vitro* manipulation of a 5S RNA gene which was cloned in pBR322. The altered genes were then used as templates in *Xenopus* cell-free extracts to assess the effects of the deletions. One set of deletions extended from the 5′ side of the gene, through its 5′-flanking sequences, and into the coding region. Deletions within the flanking sequences did not prevent transcription of the gene, although several altered the specific initiation site. It was possible to delete to position $+50$ ($+1$ is the first nucleotide of the coding region; -1 is the first nucleotide of the 5′-flanking region) without preventing transcription. When the 5′ end of the coding region was removed, transcription initiated within plasmid sequences, producing a hybrid pBR322-5S transcript of about normal size. When the deletion extended to $+55$ or further, little or no RNA was synthesized. Thus the 5′ boundary of the polymerase III control region lies between positions $+50$ and $+55$

within the coding region of the 5S RNA gene. A similar set of deletions extending from the 3' side of the gene fixed the other boundary of the control region between +80 and +83. The location of the control region determined by deletion mutagenesis was confirmed by excising the DNA segment encoding positions +41 to +87 within the 5S transcript and recloning it into pBR322. This small segment directed specific initiation by polymerase III at an upstream site within plasmid sequences. Therefore, the 5'-flanking sequences, the normal initiation site, and the first 50 base pairs of the coding region are not essential. The critical segment within the control region of the 5S RNA gene was termed an intragenic control region.

Engelke et al. (1980) identified and purified a 40K dalton polypeptide from *Xenopus* oocytes which is required for transcription of the 5S gene. The activity of the factor was monitored by its ability to facilitate transcription of exogenous 5S RNA genes in unfertilized *Xenopus* egg extracts which are otherwise incompetent for 5S RNA transcription. The purified factor binds specifically to the 5S RNA gene. "Footprinting" analyses (method of Galas & Schmitz, 1978) demonstrated that the factor interacts with nucleotides +45 to +96, almost precisely the location of the intragenic control region defined by deletion mutagenesis. Furthermore, the factor failed to bind to transcriptionally inactive 5S RNA templates carrying deletions extending from their 3' side to points beyond +83 (Sakonju et al., 1981). The factor's function in the initiation of transcription is not yet clear. Since it recognizes and binds to the intragenic control region of the 5S RNA gene, its function might be to guide the polymerase to an appropriate initiation site in a manner analogous to the sigma factor for *E. coli* RNA polymerase. Its availability might also regulate the expression of 5S genes that require the factor independently of transcription units that do not require it.

In contrast to oocytes, *Xenopus* somatic cells contain very little of the 40K factor (Honda & Roeder, 1980; Pelham, Wormington, & Brown, 1981). Nevertheless, transcription of cloned 5S RNA genes in somatic cell extracts is inhibited by polyclonal antibodies raised against the oocyte factor, and the antibodies precipitate from somatic cell extracts a polypeptide whose migration in polyacrylamide gels suggests that it is slightly larger than the 40K factor. The relationship of the larger somatic polypeptide to the oocyte factor is, as yet, unclear.

VA RNA and tRNA genes do not require the 40K polypeptide to be transcribed (Engelke et al., 1980). Nevertheless, these genes also contain intragenic control regions. Koski et al. (1980) have isolated a mutant containing a single base-pair change within the coding region of the yeast SUP4 tRNA^{tyr} gene which prevents its transcription *in vitro*. Internal control regions have been identified by deletion mutagenesis within the adenovirus 2 and 5 VAI RNA genes (Fowlkes & Shenk, 1980; Guilfoyle & Weinmann, 1981), *Xenopus* tRNA^{met} (Hofstetter, Kressmann, & Birnstiel, 1981) and tRNA^{leu} genes (Galli, Hofstetter, & Birnstiel, 1981), and a *Drosophila* tRNA^{arg} gene (Sharp et al., 1981). Several of the tRNA control regions consist of two critical regions separated by a spacer sequence (Fig. 9-1). Hofstetter, Kressmann, and Birnstiel (1981) demonstrated that small insertions or substitutions within this spacer region did not alter either the site of initiation or the rate at which the *Xenopus* tRNA^{met} gene was transcribed. However, deletions within this spacer region inactivated transcription. Other tRNA genes tolerate both small insertions and deletions within this spacer region (Johnson et al., 1980; Wallace et al., 1980). Possibly, a minimum spacing must be maintained between the two segments of the control region for their function. The spacing might already be minimal in a gene that cannot tolerate a deletion and greater than the minimum in cases where a small deletion does not inactivate the gene.

Fowlkes and Shenk (1980) have identified two sequences within the adenovirus VAI RNA intragenic control region which are conserved in a variety of different genes transcribed by polymerase III. The first consensus sequence (derived from the genes listed in Fig. 9-2) is near the 5' end of the control region and reads 5'-GTGGC$^{G}_{C}$NANNGG-3'. The second is near the 3' boundary of the control region and reads 5'-G$^{A}_{G}$GTTCGANNCC-3'. These consensus sequences correlate well with the segments identified by functional tests as critical for transcription by RNA polymerase III.

Polymerase II

The control sequences that regulate transcription appear to reside in the 5'-flanking region of polymerase II-type genes. A good number of these genes have been sequenced, and the 5' ends of the mRNAs they encode have been located. The flanking sequences of a variety of viral and cellular genes are displayed in Figs. 9-3 and 9-4. Asterisks identify mRNA cap sites. It is quite common for a gene to encode multiple cap sites (e.g., early SV40 mRNAs, Fig. 9-3, and yeast iso-1-cyto *c* mRNAs, Fig. 9-4), and there is a clear hierarchy of cap-site preferences: $A > G \gg U > C$ (Baker & Ziff, 1981). Two short sequences that appear to be conserved in a variety of genes

```
         +1      +10       +20       +30       +40       +50       +60       +70       +80       +90

AD-2 VAI            GGGCACTCTTCCGTGGTCTGGTGGATAAATTCGCAAGGGTATCATGGCGGACGACCGGGGTTCGAACCCGGATCCGGCCGTCCGCCGTG

AD-2 VAII           GGCTCGCTCCCTGTAGCCGGAGGGTTATTTTCCAAGGGTTGAGTCGCAGGACCCCCGGTTCGAGTCTCGGGCCGGCCGGACTGCGGCGAA

EBV EBER-1          AGGACCTACGCTGCCCTAGAGGTTTTGCTAGGGAGGAGACGTGTGTGGCGTAGCCACCCGTCCCGGGTACAAGTCCCGGGTGGTGAGGA

EBV EBER-2          AGGACAGCCGTTGCCCTAGTGGTTTCGGACACACCGCCAACGCTCAGTGCGGTGCTACCGACCCGAGGTCAAGTCCCGGGGAGGAGAAG

XENOPUS tRNA^met    AATCAGCAGAGTGGGCGACGCGGAAGCGTGCTGGCCCATAACCCAGAGGTCGATGGATCGAAACCATCCTCTGCTAAAACTTTTGCCAG

DROSOPHILA tRNA^arg GTCAAGCGGTCCTGTGGGCAATGATAACGCGTCTGACTACGGATCAGAAGATTCCAGGTTCGACTCCTGGCAGGATCGAATTTTTTTG

XENOPUS tRNA^leu    ACAGTCAGGATGGCCGAGCGGTCTAAGGCGCTGCGTTCAGGTCGCAGTCTCCCCTGGAGGCGTGGGTTCGAATCCCACTTCTGACACTGA

YEAST tRNA^tyr-Sup4 AAAUACTCTCGGTAGCCAAGTTGGTTTAAGGCGCAAGACTTTAATTTATCACTACGAAATCTTGAGATCGGGCGTTCGACTCGCCCCCGG
```

MOUSE 4.5S RNA

GCCGGTAGTGGTGGCGCACGCCGGTAGGATTTGCTGAAGGAGGCAGAGGATCACGAGTTCGAGGCCAGCCTGGGCTACACATT

HUMAN A36 RNA

AGGCTGGGAGTGGTGGCTCACGCCTGTAATCCCAGAATTTGGGAGGCCAAGGCAGGCAGATCACCTGAGGTCAAGAGTTCAAGACCAAC

RAT 7S RNA

GCCGGGCGGTGCCGCACGCCTGTAGTCCCAGCTACTCGGGAGGCTGAGACAGGAGGATCGCTTGAGTCCAGGAGTTCTGGGCTGTAGTGCG

XENOPUS 5S RNA

GCCTACGGCCATACCACCCTGAAAGTGCCCGATATCGTCTGATCTCGGAAGCCAAGCAGGGTCGGGCCTGGTTAGTACTTGGATGGGAGA

CONSENSUS

GTGGC^G_CNANNGG   G^A_GGTTCGANNCC

Figure 9-2. Partial nucleotide sequence of a variety of genes transcribed by RNA polymerase III. Sequences are the same sense as the RNA encoded, and +1 marks the 5' end of the primary transcript. Conserved sequences are underlined, transcriptional control regions are underscored with a heavy, wavy line. The sources for the sequences and start points are as follows: Ad-2 VAI and VAII (Akusjarvi et al., 1980), EBER1 and EBER2 (Rosa et al., 1981), *Xenopus* tRNA^met (Hofstetter, Kriessmann, & Birnstiel, 1981), *Drosophila* tRNA^arg (Sharp et al., 1981), *Xenopus* tRNA^leu (Galli, Hofstetter, & Birnstiel, 1981), yeast tRNA^tyr-sup4 (Koski et al., 1980), mouse 4.5S (Harada and Kato, 1980), human A36 (Duncan et al., 1981), rat 7S (Li et al., 1981), *Xenopus* 5S (Korn & Brown, 1978).

359

```
          -120      -110      -100       -90       -80       -70       -60       -50       -40       -30       -20       -10        +1       +10

ADENOVIRUS-2
MAJ LATE     GTGAAGACACATGTGCCCCTTCTTCGGCATCAAGGAAGGTGATTGGTTTATAGGTGTAGGCCACGTGACCGGGTGTTCCTGAAAGGGGGTATAAAAGGGGGTGGGGGCGCGTTCGTCCTCACTCTCTTCC

ADENOVIRUS-2
E1A          GGAGACTGCCCAGGTGTTTTTCTCAGGTGTTTTCCGGCGTTCCGGGTCAAAGTTGGCGTTTATTATTATAGTCAGCTGACGTGAGTTCGTATTTATACCGGTGAGTTCCTCAAGAGGCCACTCTTGAGT

SV40
EARLY        CATCTCAATTAGTCAGCAACCATAGTCCCGCCCCTAACTCCGCCCATTCTCCGCCCATGGCTGACTAATTTTTTTTATTTATGCAGAGGCCGAGGCCGCCTCGGCCTCTGAGCTATTCC

HSV
TK           ATGATGACAAACCCCGCCCAGCGTCTTGTCATTGGCGAATTCGAACACGGAGATGCAGTCGGGGCGGCGGCCGGTCCACTTCGCATATTAAGGTGACGCGTGTGGCCTCGAACACCGAGGCGAC

ASV
LTR          GGTGGTACGATCGTGCCTTATTAGGAAGGCAACAGACAGGTCTGACATGGATTGGACGAACCACTGAATTCCGCATTGCAGAGATAATTGTATTTAAGTGCCTAGCTGATACAATAAACGCCATTTGAC

MLV
LTR          CCAAGGACCTGAAATGACCCTGTGCCTTATTTGAACTAACCAATCAGTTCGCTTCTGCTTCGCGGACTTCTGTTCCCGAGCTCAAAAAAAGAGCCCACAACCCCTCACTCGGGGCGCCAGTCC

MMTV
LTR          AAGGTTCTGATCTGAGCTCTGAGTGTTCTATTTCTATTTGTCTTTGGAATTTATCCAAATCTTATGTAAATGCTTATGTGTAAACCAAGATATAAAAGAGTGCTGATTTTTTGAGTAAACTTGCAACAGT

HUMAN
β-GLOBIN     NNNNNNNNNNNNNCCTCACCTGTGGAGCCACACCCTAGGGTTGGCCAATCTACTCCCAGGAGCAGGGAGGGCAGGAGCCAGGGCTGGGCATAAAAGTCAGGGCAGAGCCATCTATTGCTTACATTTGCTT

HUMAN
δ-GLOBIN     TGAAGGTTCATTTTCATTCTCACAAACTAATGAAACCCTGCTTATCTTAAACCAACCTGCTCACTGGAGCAGGCAGGAGGACAGCACCATAAAAGCAGGCAGAGTCGACTGTTGCTTACACTTTCTT

MOUSE
α₁-GLOBIN    TTTTACTGGGTAGAGCAAGCACAAACCAGCCAATGAGTAACTGCTCCAAGGGCGTGTCCACCTGCCTGAGGACAGCCCTTGGAGGGCATATAAGTCTACTTGCTGCGAGGTCCAAGACACTTCTGATT

MOUSE
β-GLOBIN     NNNNNNNNNNNNNCCTGATTCCGTAGAGCCACACCCTGGTAAGGGCCAATCTGCTCACAGGATAGAGAGGGACAGGAGCCAGGGCATATAAGGTGAGGTAGGATCAGTTGCTCCTCACATTTGCTT

RABBIT
β-GLOBIN     GTCATCACCCAGACCTCACCCTGCAGAGCCACACCCTGGTGTTGGCCAATCTACACACGGGGTAGGGGTTAGGGATTACATAGTTCAGGACTTGGGCATAAAAGGCAGAGCAGGGCAGCTGCTGCTTACACTTGCTT

MOUSE SALIVARY
α-AMYLASE    TGCATTACAGAGATTACCAGTGAAATATCATGCAAGTTTGTCCAAATCTAATGGCCAAAATAAGAACAATGTTTTTCTTTAGATGCTAATAAAATTGTCCTGCTCAGGTTAGAGCAGCCATTCTCAACG

MOUSE LIVER
α-AMYLASE MIN  ATATAAGTGATTTAGAGAAAGAGTTGAAAGTAGAAGAGAGGTGTTGTGTATTGAGAAAGCAAAGTGATGAATTTATGTTGCAATAGAGGAGGTTATAAAAGCAGCAGGGAGGAGGCAGTGGTTCCAAAGG
```

MOUSE IMMUNO-GLOBULIN V_H	TGTCACCCTAAGAGAAACTGAAACCTTGCTCATTGCTTCCTTTTATTCTCTCAGGAACCTCCCCAATGCAAAGCAGCCCTCAGGCAGGAGGATAAAAGCTCACACTAACTGAGAAGCTCC*ATCCTCTTCT
CHICK OVALBUMIN	AAAATCTAACCCAATCCCATTAAATGATTTCTATGGCGTCAAAGTCAAACTTCTGAAGGGAACCTGTGGGTGGGTCACAATTCAGGCTAT*ATATTCCCCAGGGCTCAGCCAGTGTCTGTACATACAGCT
CHICK CONALBUMIN	AGGGGGCAACTTGGGAGCTATTGAGAAACAAGGAAGGACAAACAGCGTTAGGTCATTGCTTCTGCAAACACAGCCAGGGCTGCTCCTCTCTATAAAAGGGGAAGAAAGAGGCTCCGCAGCCATCACAGACC
CHICK α_2-COLLAGEN	CAGGCGCTGGGAGCCGCGCCGGCCCCGGCCGCCGCGCCATTGCTGCAGCGCGCCGCGGTGCCCGCCGCGCGGGGACCCCCTGCGGTATAAATACGGCGGAGCGGGGCTTGATTAATTTAGCATCCCGG
SEA URCHIN HISTONE H2A	TCATGAAATCGAGGTGCGGGCAGCGTCCCGCTGATTGGACAATTGTCACAATGCCCTCGCTGACCGGTCTTCTCCGATCCCGACGTTTGGTATAAATAGCCAGCAAAAAGATTAGGTGTCAACCATTCAA
BOMBYX FIBRON	ATTAATTTCTATGATGTTGTATCTGTACAATACAATGTGTAGATGTTTATTCTATCGAAAGTAAATACGTCAAAACTCGAAAATTTCAGTATAAAAGGTTCAACTTTTCAAATCAGCATCAGTTCGG
DROSOPHILA 70K HEAT SHOCK	GCTCTCCACTCTGTCACCAGTAAACGGCATACTGCTCTGTTGGTTCGGAGAGCGCGCCTCGAATGTTCGCGAAAAGAGCGCCGGAGTATAAATAGAGGCGCTTCGTCTTACGGACGGCGACAATTCAAT
DROSOPHILA YOLK PROTEIN I	NNNNNNNNNNNNNNNNNNNNNNNNNNNNNNNNNNNCCTGAGCCAGCGAAAAGCAAGTCGGAAAAATGGGAAATGCTCAGCGTAAATTGTGGTATATAAACCACCATCGTTGGATTTGGAAGGCCAGTTCAA

Figure 9-3. Partial nucleotide sequence of a variety of genes transcribed by RNA polymerase II which contain consensus-related sequences. Sequences are the sense of the RNA encoded, and asterisks (+1) mark the capped 5′ ends of mRNAs. The first nucleotide of the 5′ flanking sequence is −1. Conserved sequences within the 5′-flanking region are underlined. The sources for the sequences and 5′-cap sites are as follows: adenovirus-2 major late (Ziff & Evans, 1978), adenovirus-2 E1A cap site (Baker & Ziff, 1981), adenovirus-5 sequence (Van Ormondt et al., 1978), SV40 early cap sites (Reddy et al., 1979; Thompson, Radonovich, & Salzman, 1979; Haegeman & Fiers, 1980), SV40 sequence (Fiers et al., 1978; Reddy et al., 1978), herpes simplex virus thymidine kinase (McKnight, 1980), avian sarcoma virus long terminal repeat (Yamamoto, deCrombrugghe, & Pastan, 1980; Yamamoto, Jay, & Pastan, 1980), murine leukemia virus long terminal repeat (Sutcliffe et al., 1980), mouse mammary tumor virus long terminal repeat (Donehower, Huang, & Hager, 1981), human beta-globin (Efstratiadis et al., 1980), human alpha-globin (Spritz et al., 1980), mouse alpha$_1$-globin (Nishioka & Leder, 1979), mouse beta-globin (Konkel, Tilghman, & Leder, 1978), rabbit beta-globin (Dierks et al., 1981), mouse salivary and liver minor alpha-amylase (Young, Hagenbuchle, & Schibler, 1981), mouse immunoglobulin V_H (Kataoka et al., 1981), chick ovalbumin (Gannon et al., 1979; Benoist et al., 1980), chick conalbumin (Cochet et al., 1979), chick alpha$_2$-collagen (Vogeli et al., 1981), sea urchin H2A cap sites (Hentschel et al., 1980), sea urchin H2A sequence (Busslinger et al., 1980), *Bombyx* fibroin (Tsujimoto & Suzuki, 1979), *Drosophila* 70K heat shock (Ingolia, Craig, & McCarthy, 1980), and *Drosophila* yolk protein I (Hovemann et al., 1981).

| | -120 | -110 | -100 | -90 | -80 | -70 | -60 | -50 | -40 | -30 | -20 | -10 | +1 | +10 |

SV40 EARLY AT LATE TIME
AGGCAGAAGTATGCAAAGCATGCATCTCAATTAGTCAGCAACCATAGTCCGCCCCTAACTCCGCCCATCTCCGCCCATGGCTGACTAATTTTTTT

SV40 LATE
CTTTGCATACTTCTGCCTGCTGGGGAGCCTGGGGACTTTCCACACCCTAACTGACACACATTCCACAGCTGGTTCTTTCCGCCTCAGCAGGGTACCTAACCAAGTTCCTCTTTCAGGCCGTTATTTCAGGCC

POLYOMA LATE
GCTTTTGCTTCCTCTTGCAAAACCACACGACCTCTGGAGGGCGGTGCCTAGCAACTAATTAAAAGAGGATGTGCACGGCCCAGCTGCGTCAGTTAGTCCACTTCCTGCTTAACTGACTTGACATTTTCTA

ADENOVIRUS-2 EIIA
GGCGCTGACGACGCTGCTGCGCCGGGTGTGGCCGCTGGAGATGACGTAGTTTTCGCGCTTAAATTTGAGAAAGGGCGCGAAACTAGTCCTTAAGAGTCAGCGCGCAGTATTTGTGAAGAGAGCCTCC

ADENOVIRUS-2 IVA2
TGCCGAAGAGGGCGACATGTGTCTTCACACCCTGGAGCGAGTGGACCCCCTAGTGGACAACGACCGCTACCCCTCCCACTTAGCCTCCTTCGTGCTGGCCTGGACGGCAGCCTTCGTCTCAGAGTGGTCC

VACCINIA VIRUS EARLY
TATACATAAACTGATCACTAATTCCAAACCACCCGCTTTTTATAGTAAGTTTTTCACCCATAAATAATAAATACAATAATTAATTTCTGTAAAAGTAGAAAATATATTCTAATTTATTGCACGGTAAG

MOUSE LIVER-α-AMYLASE MAJ
AAGAGGTTGTTGTGTATTGAGAAAGCAAAGTGATGAATTATGTTGCAATAGAGGAGGTTATAAAGCAGCCAAGGGAAGGCAGTGGTTTCCAAAGGAACACCAGGGTGGTGCCCAGTCCATCATA

MOUSE PSEUDO-α3-GLOBIN
NNNNNNNNGGGCTCTTAAAAGCCTGGCTGGGCAGAGCACAGGCCAGCCAATGAGGACGCACTCAAAGGAGGGTGTCTCAAAGAGCGTGTAGGCATAGAAGTGTGCTTATTTCAGGTCCAACACAGTTCTGA

SEA URCHIN HISTONE H4
CCCACCGGGCCGTCGCGGAGGGCGCACCTGTGCGGCAGGGCGCCACCTGTGCGGGGTCATCGGAGGCCGATCGGAGCCTCGTCATCCAAGTGCCATACGGGTGACAATACCCCGCTCACCGGGAGGGTTGGTCAATCGCTCAG

YEAST HIS 3
CATTTTTTTTTCCCCTAGCGGATGACTCTTTTTTTTTCTTAGCGATTGGCATTATCACATAATGAATTAAACTAATGTGATTTCTTCGAAGAATATACTAAAAATGAGCAGGCAAG

YEAST ISO-1-CYTO C
ACATGATCATATGGCATGCCATGTGCTCTGTATGTATATAAAACTCTTGTTTCTCTTCTTTCTCTAAATATTCTTTCTTACATTAGTCCTTTGTAGCATAATTACTATACTTCTATAGACACGCAA

Figure 9-4. Partial nucleotide sequences of a variety of genes transcribed by RNA polymerase II which do not contain consensus-related sequences. Sequences are the sense of the RNA encoded, and asterisks (+1) mark the capped 5' ends of the preribosomal RNA. The sources for the sequences and cap sites are as follows: SV40 early at late time cap sites (Ghosh & Lebowitz, 1981), SV40 late cap sites (Haegeman & Fiers, 1978; Reddy et al., 1978; Canaani et al., 1979), polyoma late cap sites (Flavell et al., 1979, 1980), polyoma sequence (Soeda et al., 1980), adenovirus-2 EIIA (Baker et al., 1979; Baker & Ziff, 1981), adenovirus-2 IVA2 (Baker & Ziff, 1981), vaccinia virus early (Venkatesan, Barondy, & Moss, 1981), mouse liver major alpha-amylase (Young, Hagenbuchle, & Schibler, 1981), mouse pseudo-alpha-3-globin (Nichioka, Leder, & Leder, 1980; this gene is not transcribed: Talkington, Nishioka, & Leder, 1980), sea urchin H4 (Hentschel et al., 1980), yeast his 3 (Struhl, 1981), and yeast iso-1-cyto c (Faye et al, 1981).

362

are underlined in Fig. 9-3. An A—T stretch which is generally found near positions −25 to −35 (preceding the transcriptional start site) was first recognized by M. Goldberg and D. Hogness. This sequence is generally referred to as the "TATA" box. It is remarkably similar to the Pribnow box (TATAAT) located about 10 base pairs upstream from the initiation site of prokaryotic transcription units (Pribnow, 1975), but the sequences likely serve quite different functions in prokaryotes and eukaryotes. A second homology occurs in the −70 to −80 region. The model sequence at this location is 5′-GG$_T^C$ CAATCT-3′, and it is sometimes referred to as the CAAT box (Benoist et al., 1980; Corden et al., 1980; Efstratiadis et al., 1980). The widespread occurrence of these two homologies and their comparatively constant locations relative to cap sites suggest that they probably play important roles in transcription. Some, but not all, of the genes contain regions of partial (hyphenated) twofold symmetry around the start sites for transcription (e.g., Adenovirus-2 EIA, −11 to +9, CTCAAGCGGC/CACTCTTGAG and chick ovalbumin, −7 to +7, TGTCTGT/ACATACA) (Gannon et al., 1979). Hyphenated symmetries are found in prokaryotic operators and at other regulatory sites which interact with proteins (e.g., Gilbert, 1976) and could serve similar functions in eukaryotic control regions.

A variety of experiments using cell-free extracts to monitor the effects of alterations constructed *in vitro* suggest that the TATA box and its surrounding sequences are the major critical components of the polymerase II control region. This has been shown to be the case for the chick conalbumin (Corden et al., 1980; Wasylyk, Kedinger, et al., 1980) and ovalbumin genes (Tsai et al., 1981a), the adenovirus 2 major late transcriptional unit (Corden et al., 1980; Hu & Manley, 1981), the murine leukemia virus control region (Ostrowski, Berard, & Hager, 1981), and the *Bombyx* fibron gene (Tsujimoto et al., 1981). Single-base-pair changes within the TATA box region can drastically reduce the efficiency of *in vitro* initiation. A TATAAAA to TAGAAAA change produced a 20-fold reduction and a TATAAAA to TAAAAAA change caused a 40-fold reduction in transcription of the chick conalbumin gene (Wasylyk, Derbyshire, et al., 1980; Wasylyk & Chambon, 1981). A TATATAT to TGTATAT change reduced transcription of the chick ovalbumin gene about 10-fold (Tsai et al., 1981b). Significantly, a cloned segment containing the adenovirus TATA box (−12 to −32) directed initiation by polymerase II in cell-free extracts (Sassone-Corsi et al., 1981). Removal of the TATA box does not always completely abolish *in vitro* activity of templates. Hu and Manley (1981) could detect residual activity (8% of wild-type) in a DNA segment carrying the adenovirus 2 major late control region lacking its TATA box, and the SV40 early gene is still transcribed in HeLa cell extracts after its TATA sequence has been deleted (Rio et al., 1980; Lebowitz & Ghosh, 1982; Mathis & Chambon, 1981).

Recent studies with SV40 showed that the set of three 21-b.p. repeated GC-rich sequences upstream from the TATA box play a role in the initiation of early transcription *in vivo* and *in vitro* at all early mRNA transcription initiation sites (Lebowitz & Ghosh, 1982). Deletion of two of the three 21-b.p.-repeats produces minimal change in transcription efficiency, but deletion of all three reduces the efficiency of the initiation *in vitro* by over 80%, even in the presence of the TATA box. The heterogeneous set of initiation sites downstream of the 21 b.p. repeats which remain active even in the absence of the TATA box no longer function actively in the absence of these repeats. Stimulation of *in vitro* transcription by sequences upstream of the TATA box have also been noted for the silk fibroin gene (Tsuda & Suzuki, 1981) and sea urchin histone genes (Grosschedl & Birnstiel, 1982).

In general, the TATA box is not essential for transcription *in vivo*. Templates carrying deletion mutations which remove the TATA box have been actively transcribed after being reintroduced to cells by a variety of methods (SV40 early transcription unit: Benoist & Chambon, 1980; polyoma virus early transcription unit: Bendig, Thomas, & Folk, 1980; sea urchin H2A transcription unit: Grosschedl & Birnstiel, 1980a; herpes virus thymidine kinase gene: McKnight et al., 1981). TATA box mutations can, however, reduce the level of *in vivo* transcription. This was most dramatically demonstrated by Grosschedl et al. (1981) who substituted a 34-base-pair segment from the chick conalbumin gene for the equivalent segment normally preceding the sea urchin histone H2A gene. The substituted segment contained either a normal TATA or mutated "TAGA" box. The template carrying the mutated "TAGA" box was expressed at a level 5-fold below that of the TATA box template upon injection into *Xenopus* oocyte nuclei. Similar results were obtained for the adenovirus E1A transcription unit. Osborne et al. (1981) cloned the E1A gene into pBR322, mutated the clone, and then tested its ability to complement an adenovirus 5 deletion mutant lacking this gene (*dl*312, Jones & Shenk, 1979a). The yield of *dl*312 after transfection of HeLa cells was increased about 50-fold by adding the plasmid. Deletion of sequences preceding position −38 in the E1A gene within the plasmid had no effect on its complementing activity, but there was a significant reduction in complementing activity if the deletion was extended to position −23, removing the TATA box.

If the TATA box is not essential for *in vivo* transcription, what is its function? Gluzman, Sambrook, and Frisque (1980) isolated several SV40 mutants ($\Delta 9$ and $\Delta 58$) which are relevant to this question. These mutants carry small deletions between the TATA box and initiation sites for early mRNAs. The mutants synthesize early mRNAs and can transform rat cells. Ghosh, Roy, et al. (1981) located the 5' ends of the early viral mRNAs synthesized by these mutants in transformed cells. The 5' ends were displaced downstream from the wild-type start sites. In fact, they were displaced by distances about equal to the size of each deletion. Another mutant, *dl*892 (Shenk, Carbon, & Berg, 1976; Subramanian & Shenk, 1978), which lacks 19 base pairs just upstream of the TATA box was found to generate wild-type mRNAs. Thus the measurement is made from a site located downstream from *dl*892 and upstream from $\Delta 9$ and $\Delta 58$. This region includes the TATA box. Apparently, the TATA box helps to specify the precise start site for transcription. Possibly the polymerase contacts this site, which serves to align it with a specific point at which to initiate transcription. Consistent with this notion, Grosschedl and Birnstiel (1980a) found that multiple initiation sites are utilized *in vivo* (instead of the normal unique site) when the TATA box is deleted from the sea urchin H2A gene. Similar observations have been made for the SV40 early gene (Benoist & Chambon, 1981). However, lack of a defined TATA box does not necessitate substantial heterogeneity in initiation sites, as is evidenced by several of the genes listed in Fig. 9-4.

What sequences are essential for initiation of transcription by polymerase II *in vivo*? One obvious candidate is the conserved sequence (CAAT box) located near -70 to -80 preceding the initiation site (see Fig. 9-3). The CAAT box and surrounding sequence appear to be important for expression of the rabbit β-globin gene. Dierks et al. (1981) constructed a series of deletions extending from a far-upstream point toward the β-globin initiation site, and assayed the mutant templates for expression in mouse L cells. Templates lacking sequences up to position -76 (CAAT box intact) functioned at near normal levels, but expression was dramatically reduced when the deletion was extended to -66 (CAAT box removed). However, the CAAT box sequence is clearly not essential in several other genes. Grosschedl and Birnstiel (1980a) reported that a 55-base-pair deletion which included this sequence increased the level of transcription of the sea urchin H2A gene *in vivo*. Also, P. Hearing and T. Shenk (unpublished results) have found that deletion of the CAAT sequence did not alter expression of the adenovirus E1A gene in infected cells, and McKnight and Kingsbury (1982) isolated two deletion substitution mutants of the herpes virus thymidine kinase gene in which several residues of the CAAT box were altered without any effect on transcriptional efficiency in injected oocytes.

A component of the early SV40 polymerase II control region essential for efficient *in vivo* transcription lies at least 112 base-pairs upstream from the mRNA cap site. The critical site lies within a 72-base-pair sequence which is tandemly duplicated. Gruss, Dhar, and Khoury (1981) found that one copy of this repeat could be removed with no effect on early transcription, but when both copies were deleted the early control region did not function. Gluzman, Sambrook, and Frisque (1980) and Benoist and Chambon (1981) have also isolated SV40 mutants that lack the 72-base-pair repeats and that fail to express early functions. Several workers independently noted that SV40 sequences containing these 72-base-pair repeat sequences could enhance transformation by a linked thymidine kinase gene, and/or expression of other linked genes when situated in either orientation relative to the enhanced gene or when situated downstream from the gene (Reddy et al., 1982; Banerji, Rusconi, & Schaffner, 1981; Capecchi, M., 1980). This effect has been shown to be a consequence of enhanced transcription. The mechanism of enhancement is unknown but occurs with a variety of genes, including bacterial and animal cell genes, as well as viral genes. The actual sequences involved have been narrowed down to linked 9- and 11-b.p. sequences (M. Botchan, personal communication). Mutants with similar phenotypes to the SV40 72-b.p.-deletion mutant have also been isolated far upstream of the polyoma virus cap site (M. Fried and E. Reeley, personal communication). Furthermore, polyoma mutants have been described that, unlike the wild-type virus, can grow in undifferentiated teratocarcinoma cells (Katina et al., 1980, 1981; Fujimura et al., 1981; Sekikawa & Levine, 1981). Infection of these cells with wild-type virus produces very small amounts of viral RNA; the mutants produce more RNA. They all contain alterations in a region about 200 base-pairs upstream from the major early mRNA initiation site, again suggesting that a critical element of the early transcriptional control region lies far upstream. A mutation that is located 184 base-pairs upstream from the sea urchin mRNA start site reduced its level of transcription 15-fold in *Xenopus* oocytes (Grosschedl & Birnstiel, 1980b) and inversion of these upstream sequences actually enhanced histone gene expression. Upstream sequences between -109 and -95 and between -61 and -41 are required for efficient expression of the herpes thymidine kinase gene (McKnight et al., 1981; McKnight & Kingsbury, 1982).

The −61 site contains the sequence GGGGCGGCGCGG, and the 105 site contains the partly complementary sequence CCCCGCCC.

There are several ways in which an upstream site might function in the initiation of transcription. Possibly, polymerase II spans a large region on the template and contacts regions far upstream as well as sites at −30 and +1. This is not unlikely when one considers that the *E. coli* polymerase contacts an 80-base-pair region (Schmitz & Galas, 1979). Polymerase II could very likely interact directly with sites on the template quite far removed from the initiation site if the template DNA is supercoiled in a chromatin structure (about 75–80 base pairs per superhelical turn of β-form DNA; Felsenfeld, 1978). Alternatively, a regulatory protein which is not part of the polymerase could interact with the far-upstream sequence or a transcription factor could initially interact at a distant site then migrate to the transcription initiation site.

It appears that cell-free extracts are not yet faithfully reflecting mechanisms operative *in vivo*. In several systems other than SV40, *in vitro* transcription is not influenced by sequences upstream of the TATA box region. The extracts may lack critical regulatory components. It is also likely that the superstructure of the DNA template is critical for its proper regulation. Normally, transcription units are "packaged" in chromatin, while naked DNA is supplied to extracts and may or may not be properly assembled into chromatin.

Several viral and cellular genes do not contain readily indentifiable TATA-box sequences in their −25 to −35 region (Fig. 9-4). Some of these genes nevertheless share short regions spread through positions −40 to −80, homologous both with each other and with a variety of genes that contain TATA boxes (Contreras & Fiers, 1981). Little is known about the transcriptional control regions of these genes. Some initiate transcription at relatively unique locations (e.g., adenovirus-2 E11A and IVA2: Baker et al., 1979; Baker & Ziff, 1981) and others initiate at a wide variety of sites which can be spread over several hundred base-pairs (e.g., SV40 and polyoma late transcriptional units: Haegeman & Fiers, 1978; Reddy, Ghosh, et al., 1978; Canaani et al., 1979; Flavell et al., 1979, 1980; Piatak et al., 1981). It is not yet clear which sequences constitute functional components of these control regions.

Some of the sequences specifying the localization of the principal initiation sites for SV40 late message have been studied by the analysis of RNA from cells infected with various deletion and substitution mutants (Haegeman et al., 1979; Villareal, White, & Berg, 1979; Ghosh, Roy, et al., 1981; Gosh et al., 1982; Piatak et al.,

1981). Mutants have been constructed in which the SV40 DNA sequences encoding the transcribed leader from a position 25 nucleotides beyond the major transcription initiation site at residue 243 have been altered. Substitution either of host sequences or of other virus sequences at this position depresses the efficiency of initiation at the major site 25 nucleotides upstream, even though all sequences surrounding and upstream from the initiation site are intact. At the same time, initiation at sites located further upstream is augmented. On the other hand, duplication of nucleotide sequences from -32 to $+25$ with respect to the major transcription initiation site at residue 243 leads to formation of principal RNA species that initiate at both the first and the second copy of the duplicated sequence, indicating that local sequences around the transcription initiation site also play an important role in localizing the site of transcription. Upstream deletions removing residues -30 to -15 still permit initiation at residue 243. Thus, no critical localizing sequence appears to reside at the site where a TATA box is present in other promoters. It has been suggested that palindromic sequences in SV40 DNA immediately downstream from the principal late initiation sites may be partly responsible for the location of these sites and that many of the effects on late initiation sites may be consequences of distorted structure in locally denatured DNA (Ghosh et al., 1982).

Polymerase I

The sequences surrounding the transcriptional initiation sites for several ribosomal RNA transcriptional units are shown in Fig. 9-5. No consensus sequences are evident among this limited group of genes. Cell-free transcription of the mouse ribosomal RNA is prevented by loss of a segment between positions -74 and -39 (Grummt, 1981b), while the region from -12 to $+16$ contains all of the sequence needed for transcription of the *Xenopus* ribosomal RNA gene upon injection into oocyte nuclei (B. Sollner-Webb & R. Reeder, personal communication).

REGULATION OF TRANSCRIPTION BY EUKARYOTIC POLYMERASES

A variety of mechanisms are apparent which serve to regulate transcription by eukaryotic polymerases at the level of initiation. One

	-70	-60	-50	-40	-30	-20	-10	$+1$	$+10$	$+20$

DROSOPHILA GCCCGTATGTTGGGTGGTAAATGGAATTGAAAAATACCCGCTTTGAGGACAGCGGGTTCAAAAACTACTAT*AGGTAGGCAGTGGTTGCCGA

XENOPUS GCCCGGCCTCTCGGGCCCCCCGCACGACGCCTCCATGCTACGCTTTTTTGGCATGTGCGGGCAGGAGGT*AGGGGAAGACCGGCCCTCGG

MOUSE CACTTTCCTCCCTGTCTCTTTATGCTTGTGATCTTTTCTATCTGTTCCTAATGGAACTTGGAGATAGGT*ACTGACACGCTGTCCTTTCC

HUMAN GGTGACGCGACCTCCCGGCCCCGGGGAGGTATATCTTTCGCTCCGAGTCGGCATTTTGGGCCGCCGGGTT*ATTGCTGACACGCTGTCCTC

YEAST GTAGATTGTTGTAATGAGAGGGGGTTAGTCATGGAGTACAAGTGTGAGGAAAAGTAGTTGGGAGGTACTTC*ATGCGAAAGCAGTTGAAGAC

Figure 9-5. Partial nucleotide sequence of several genes transcribed by RNA polymerase I. Sequences are the sense of the RNA encoded, and asterisks mark the 5′ end of the preribosomal RNA. The sources for the sequences and initiation sites are as follows: *Drosophila* (Long, Rebbett, & Dawid, 1981b), *Xenopus* (Sollner-Webb & Reeder, 1979), mouse (Grummt, 1981b; Miller & Sollner-Webb, 1981), human (R. Miesfield and N. Arnheim, personal communication), and yeast (sequence: Valenzuela et al., 1977; 5′ end: Klemenz & Geiduschek, 1980).

variable, which almost certainly serves to modulate the efficiency of transcription by all three polymerases, is the primary sequence at and surrounding the control region. Although consensus sequences are evident, control regions show considerable variation in their specific sequences. This variability probably influences the efficiency with which polymerase and control factors recognize and interact with individual genes. A variety of additional regulatory events are then superimposed on the basic transcriptional capability of the primary sequence.

Polymerase III

Several lines of evidence indicate that the 5'-flanking sequences of polymerase III-type genes serve to modulate the efficiency with which they are transcribed. DeFranco, Schmidt, and Soll (1980) isolated two *Drosophila* tRNA^{lys} genes with identical coding regions but different 5'-flanking sequences. The two genes were transcribed with very different efficiency in nuclear extracts of *Xenopus* oocytes. When the 5'-flanking sequences of the two tRNA genes were switched, the reconstructed genes displayed the characteristic of the gene from which the flanking sequences were derived. When the 5'-flanking region of the less-efficiency-transcribed gene was replaced by pBR322 sequences, the modified template directed the synthesis of substantially increased amounts of product. DeFranco, Schmidt, and Soll (1981) utilized deletion mutagenesis to identify the flanking sequence component responsible for downward modulation of tRNA^{lys} gene expression. The segment is located 13 base-pairs upstream from the mature tRNA coding sequence, and it reads 5'-GGCAGTTTTTG-3'. Expression of *Xenopus* tRNA^{met} genes is also modulated downward by 5'-flanking sequences (Clarkson et al., 1981). The flanking sequences of the adenovirus VAI RNA (Fowlkes & Shenk, 1980) and the *Xenopus* somatic 5S RNA (Wormington et al., 1981) genes produce the opposite effect to that documented for tRNA genes. VA and 5S genes are expressed less efficiently *in vitro* when they carry deletions in their 5'-flanking sequences. Finally, Sprague, Larson, and Morton (1980) found that a truncated silkworm tRNA^{ala} gene which contained only 11 nucleotides of its normal 5'-flanking sequences was not transcribed at detectable levels in silkworm extracts. Curiously, the truncated gene was transcribed as well as the wild-type gene in *Xenopus* oocyte extracts.

The mechanism by which flanking sequences alter transcriptional activity is not yet clear. Sequence similarities are evident in these regions when a variety of genes are compared (Korn & Brown, 1978), and it is quite likely that the polymerase interacts here during initiation of transcription since alterations can influence the precise start site (Thimmappaya, Jones, & Shenk, 1979; Fowlkes & Shenk, 1980; Sakonju, Bogenhagen, & Brown, 1980; Guilfoyle & Weinmann, 1981). Assuming the polymerase contacts this region during initiation of transcription, it is not surprising that its composition modulates function.

A second mechanism has been described which undoubtedly regulates transcription of the *Xenopus* 5S RNA gene (Honda & Roder, 1980; Pelham & Brown, 1980). 5S RNA is stored in the oocyte as a 7S ribonucleoprotein complex, and the major protein component of this particle is the 40K dalton polypeptide which also binds to the intragenic control region of the 5S gene. When 5S RNA is added to a *Xenopus* extract, transcription of the 5S gene is inhibited, presumably because the RNA competes with its template for the 40K polypeptide. As a result, the product inhibits its own transcription. This type of feedback inhibition might also function for other genes transcribed by polymerase III. For example, VA EBER and 4.5S RNAs exist in cells as ribonucleoprotein particles, and they can be precipitated by a particular class of sera, designated anti-La, from patients with systemic lupus erythematosis (Lerner et al., 1980, 1981; Hendrick et al., 1981). Conceivably, these ribonucleoprotein particles contain a common polypeptide that binds to their intragenic control regions and is required for their transcription.

Transcription by polymerase III may also be influenced by chromatin structure. Louis et al. (1980) demonstrated that nucleosomes in the 5S RNA gene cluster of *Drosophila* are positioned in two phases with respect to DNA sequence. In one phase, the center of the gene is exposed in the linker region (sensitive to micrococcal nuclease digestion); and in the second phase, the 5′-flanking sequences are located in this region. Both of these regions of the 5S RNA gene presumably contain key regulatory sequences. Perhaps the linker sequences that are accessible to micrococcal nuclease are equally accessible to transcriptional control factors, while sequences associated with the nucleosome are not. If this is true, nucleosome phasing could play an important role in the regulation of 5S gene expression.

Apparently, much of the regulation of polymerase III-type genes is accomplished by relatively simple means. These include gene copy number (e.g., there are many more copies of the 5S gene than individual tRNA genes), the efficiency with which the gene is transcribed (modulation by flanking sequences), and feedback inhibition (5S RNA binds the factor required for its transcription). Other levels of regulation, however, remain obscure. A case in point is the developmental switch from production of *Xenopus* oocyte-type to somatic-type 5S RNA (reviewed in Ford & Brown, 1976). This regulatory switch may be mediated by polypeptides which have yet to be identified (discussed in Honda & Roeder, 1980).

Polymerase II

Positive and negative regulatory polypeptides almost certainly play a significant role in regulation of polymerase II initiation. There is considerable *in vivo* evidence leading to the conclusion that the SV40 early gene product regulates its own synthesis (Tegtmeyer, 1974; Tegtmeyer et al., 1975; Reed, Stark, & Alwine, 1976; Alwine, Reed, & Stark, 1977; Khoury & May, 1977) and does so by binding to the control region for early mRNA synthesis (Reed et al., 1975; Jessel et al., 1976; Tjian, 1978). This autoregulation has recently been demonstrated to occur *in vitro* and to operate at the level of initiation (Rio et al., 1980). Several adenovirus early gene products also appear to regulate transcription. For example, an E1A gene product appears to be a positive regulator of other early regions (Berk et al., 1979; Jones & Shenk, 1979b; Nevins, 1981), and the EIIA gene product may be a negative regulator of EIV transcription (Nevins & Winkler, 1980). A wide variety of additional genes should prove to be interesting regulatory models. These include the herpes virus thymidine kinase gene (e.g., Zipser, Lipsich, & Kwoh, 1980), *Drosophila* heat-shock genes (e.g., Hackett & Lis, 1981), mouse α-amylase genes (e.g., Young, Hagenbuchle, & Schibler, 1981) and many others.

Glucocorticoid hormones regulate expression of certain genes, presumably at the level of transcriptional initiation. Steroid hormones associate with a specific soluble receptor in the target cell and this interaction increases the affinity of the receptor for binding sites in the nucleus (very likely DNA). Expression of the mouse mammary tumor virus (MMTV) genome is stimulated by glucocorticoid hormones. The target site for the stimulation event is on the viral genome (Buetti & Diggelmann, 1981; Hynes et al., 1981); more specifically, it is located within the MMTV long terminal repeat (LTR) sequences

(Huang et al., 1981; Lee et al., 1981). Payvar et al., (1981) have recently shown that purified glucocorticoid receptors bind specifically *in vitro* to cloned segments of the MMTV genome. In other hormonally controlled gene systems, Kurtz (1981) has shown that the rat $\alpha_{2\mu}$ globulin genes remain inducible after cloning and transfer to a mouse cell, and Mayo, Warren, and Palmiter (1982) have found that the mouse metallothionine gene remained responsive to heavy metals but not gluticocorticoids upon introduction into mouse cells.

Polymerase II transcription is also likely to be regulated at the level of chromatin structure (reviewed in Igo-Kemenis, Mbrz, & Zachau, 1982). Genes that are actively transcribed within a cell (e.g., globin genes in a reticulocyte) are preferentially transcribed when *E. coli* or eukaryotic RNA polymerases are added to cellular chromatin, suggesting that expressed genes are more accessible (Axel, Cedar, & Felsenfeld, 1973; Gilmour & Paul, 1973; Barrett et al., 1974; Steggles et al., 1974). The same genes are also more accessible to digestion by DNase I (Garel & Axel, 1976; Weintraub & Groudine, 1976). This type of experiment must be interpreted cautiously when building regulatory models, since the altered chromatin configuration could be either the cause or the effect of active transcription. Several groups (Wu, C., 1980; Keene et al., 1981) have demonstrated that the 5′ ends of *Drosophila* heat-shock genes are hypersensitive to DNase I digestion, focusing the altered chromatin structure to potential polymerase II initiation sites, and a nuclease hypersensitive site has been found 5′ to some globin genes (McGee et al., 1981). Recently, Larsen and Weintraub (1982) found that the globin hypersensitive site retains its nuclease sensitive nature in genes cloned in a plasmid, provided the plasmid is supercoiled. This indicates that the hypersensitive structure is latent in the DNA itself without any need for chromosomal proteins. There is also an "open" region on SV40 chromatin which, intriguingly, maps to the viral transcriptional control region (Scott & Wigmore, 1978; Varshavsky, Sundin, & Bohn, 1978; Waldeck et al., 1978; Sundin & Varshavsky, 1979; Varshavsky, Sundin, & Bohn, 1979; Jakobovits, Bratosin, & Aloni, 1980). Finally, there are strong indications that nucleosomes are not arranged randomly on DNA but exhibit a phase relationship with respect to DNA sequence (Musich, Maio, & Brown, 1977; Ponder & Crawford, 1977; Chao, Gralla, & Martinson, 1979; Wittig & Wittig, 1979; Levy & Noll, 1980; Samal et al., 1981). Phasing could represent an important means of regulating transcriptional initiation by positioning critical control sequences in locations within chromatin more or less accessible to polymerase and other regulatory factors.

Polymerase I

Little is known of the molecular mechanisms that serve to regulate transcription by RNA polymerase I. The ribosomal genes of *Drosophila*, however, provide a particularly attractive system for study of a polymerase I-related regulatory event. Several classes of these genes contain insertions within their coding region (Glover & Hogness, 1977; White & Hogness, 1977). Unlike normal ribosomal genes, those carrying insertions are rarely transcribed (Long & Dawid, 1979; Long, Rebbert, & Dawid, 1981a). Nevertheless, the sequence around the initiation sites of transcribed and insertion-carrying genes are identical (Long, Rebbert, & Dawid, 1981b). The basis for the differential transcriptional activity of these two classes of ribosomal genes is not yet clear and may well be related to the downstream insertions.

TERMINATION OF TRANSCRIPTION IN PROKARYOTES

A number of excellent reviews have discussed termination (Gilbert, 1976; Adhya & Gottesman, 1978; Pribnow, 1979; Rosenberg & Court, 1979; Platt, 1981) and attenuation (Yanofsky, 1981). Two classes of termination signals have been identified: rho-dependent and rho-independent.

Rho-independent termination signals function in a purified system containing only template DNA, RNA polymerase, and nucleoside triphosphates (Lebowitz, Weissman, & Radding, 1971). Terminators of this type exhibit two noteworthy sequence features (Fig. 9-6): a dyad symmetry potentially forming a hairpin loop in the newly transcribed RNA, and a stretch of uridylic acid residues (generally 6). Transcription terminates at the end of the run of U's. The exact sequences in the potential stem and loop structure vary, and, as pointed out by Gilbert (1976), they tend to be GC-rich. This would stabilize the potential hairpin loop structure. The stretch of U's might facilitate dissociation of the nascent template from the template, since rU-dA base pairs are quite unstable (Martin & Tinoco, 1980). Mutations in either the hairpin loop structure or the uridylic acid stretch have been shown to decrease the efficiency of termination.

Rho-dependent termination requires the presence of the rho factor (Roberts, 1969). Factor-dependent termination may occur in AT-rich regions that contain a dyad symmetry preceding the termi-

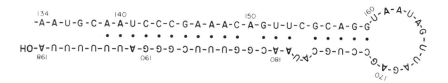

Figure 9-6.   Prototypic prokaryotic transcription termination signal. The 3′ OH of the transcript is at residue 198 (from Lebowitz, Weissman, & Radding, 1971). A small fraction of the RNA molecules contains additional A residues, presumably added posttranscriptionally (Rosenberg & Weissman, 1975).

nation site. The potential hairpin loop is generally less stable than that present at factor-independent sites. Rho-dependent termination is heterogeneous and can occur across a 50-nucleotide region (Platt, 1981).

Another cellular protein, nusA, can also facilitate termination (Greenblatt, McLimont, & Hanley, 1981; Ward & Gottesman, 1981). NusA has been shown to bind to core RNA polymerase (Greenblatt & Li, 1981) and may function through this association.

TERMINATION OF TRANSCRIPTION IN EUKARYOTES

Transcription termination has been studied best in the case of polymerase III. Here, transcription termination *in vitro* seems to reflect accurately situations that occur *in vivo* (Wu, G.-J., 1978; Birkenheimer, Brown, & Jordan, 1978). Deletion analysis of *Xenopus* 5S DNA has shown that termination occurred within clusters of four or more T residues in the noncoding DNA strand. The nucleotides preceding or following the T-run could be altered without abolishing termination (Bogenhagen & Brown, 1981). However, termination may be more efficient when the sequence flanking the T-run is GC-rich. Unlike prokaryotic termination signals, palindromic symmetries in the DNA immediately before the transcription termination site appear to be less important.

Transcription termination by polymerase II is a very much less clear situation because the majority of polymerase II transcripts have at their 3′ end a polyadenylic acid stretch that is added posttranscriptionally. It appears that, in many cases, polyadenylation occurs at sites where RNA has been cleaved posttranscriptionally, rather than at sites where primary transcription is terminated (Ford

& Hsu, 1978; Lai, Dhar, & Khoury, 1978; Nevins & Darnell, 1978; Nevins, Blanchard, & Darnell, 1980; Hofer & Darnell, 1981). In animal cells, the polyadenylation signal itself contains as an almost invariant feature the hexanucleotide AAUAAA, 12–21 b.p. preceding the poly-A additions (Proudfoot & Brownlee, 1974), although cases of single nucleotide deviation have been noted (Jung et al., 1980; Hagenbuchle, Bovey, & Young, 1980). This sequence is not in itself sufficient to specify strong polyadenylation, in that internal AAUAAA's have been found within messenger RNAs that do appear not to function at all *in vivo.* Extensive deletion analysis suggests that there are no specific sequences closely linked to the polyadenylation sites of SV40 that are necessary for efficient polyadenylation (Fitzgerald & Shenk, 1981). However, there are homologies downstream of polyadenylation sites of such heterogenous templates as those at SV40 late mRNA and rabbit β-globin mRNA or ovalbumin (Dhar et al., 1975; Benoist et al., 1980). So far, no function has been assigned to these sequences.

In yeast (Zaret & Sherman, 1982), some but not all polyadenylation sites are preceded by an AAUAAA. An alternative sequence, TAG . . . TA(T)GT . . . TTT, has been suggested to have some role in transcription termination or polyadenylation.

Histone mRNAs are generally not polyadenylated. They usually do not contain within them an AAUAAA signal, and the transcription termination signal may involve palindromic sequences within the DNA template (Birchmeier, Grosschedl, & Birnstiel, 1982). The inverted repeat GGCCCTTATCAGGGCC lies immediately upstream of the 3' end of mature RNA, but sequences downstream of the termination site also appear to be important for termination.

Presumptive transcription termination sites for polymerase I have been mapped by comparison of ribosomal DNA sequences with the sequences at the 3' end of the 45S ribosomal RNA precursor. These experiments suffer from the difficulty of correlation of 3' ends of RNA formed *in vivo* with transcription termination, rather than processing events. In mouse cells, the 3' end was localized to a site downstream of a dyad symmetry (Kominami et al., 1982). In tetrahymena transcription termination was determined to occur at the sequence TTTTTTACTTA (Din, Engberg, & Gall, 1982), almost identical to the sequence around the 3' end of yeast pre-RNA (Veldman et al., 1980). In tetrahymena there was also a dyad symmetry element in the DNA preceding the proposed termination site. A sea urchin protein factor necessary for polymerase I termination at the end of tetrahymena ribosomal genes has been described (Westergard, 1979).

ACKNOWLEDGMENTS

The author thanks Mary Fils-Aime for her capable assistance in the preparation of this manuscript. Thanks are also due to colleagues who kindly sent preprints of their most recent work. The author is an Established Investigator of the American Heart Association.

Editor's Note: This review was completed in November 1981, with partial updating in July, 1982.

REFERENCES

Adhya, S., and Gottesman, M. (1978). Control of transcriptional termination. *Ann. Rev. Biochem.* 47:967-996.

Akusjarvi, G., Matthews, M. B., Anderson, P., Vennstrom, B., and Petterson, U. (1980). Structure of genes for virus-associated RNA_I and RNA_{II} of adenovirus type 2. *Proc. Natl. Acad. Sci. USA* 77:2424-2428.

Alwine, C., Reed, S. I., and Stark, G. R. (1977). Characterization of the autoregulation of SV40 gene A. *J. Virol.* 24:22-27.

Axel, R., Cedar, H., and Felsenfeld, G. (1973). Synthesis of globin RNA from duck-reticulocyte chromatin in vitro. *Proc. Natl. Acad. Sci. USA* 70: 2029-2032.

Baker, C. C., and Ziff, E. B. (1981). Promoters and heterogeneous 5' termini of the messenger RNAs of adenovirus serotype 2. *J. Mol. Biol.* 149:189-221.

Baker, C. C., and Ziff, E. B. (1980). Biogenesis, structures, and sites of encoding of the 5' termini of adenovirus-2 mRNAs. *Cold Spring Harbor Symp. Quant. Biol.* 44:415-428.

Baker, C. C., Herisse, J., Courtois, C., Galibert, F., and Ziff, E. (1979). Messenger for the Ad2 DNA binding protein: DNA sequences encoding the first leader and heterogeneity at the mRNA's 5' end. *Cell* 18: 569-580.

Banerji, J., Rusconi, S., and Schaffner, W. (1981). Expression of a β-globin gene is enhanced by remote DNA sequences. *Cell* 27:299-308.

Barrett, T., Mayanka, D., Hamlyn, P. H., and Gould, H. J. (1974). Nonhistone proteins control gene expression in reconstituted chromatin. *Proc. Natl. Acad. Sci. USA* 71:5057-5061.

Batts-Young, B., and Lodish, H. F. (1978). Triphosphate residues at the 5′ ends of rRNA precursor and 5S RNA from *Dictyostelium discoideum. Proc. Natl. Acad. Sci. USA* 75:740–744.

Bending, M. M., Thomas, T., and Folk, W. R. (1980). Regulatory mutants of polyoma virus defective in DNA replication and the synthesis of early proteins. *Cell* 20:401–409.

Benoist, C., and Chambon, P. (1981). In vivo sequence requirements of the SV40 early promoter region. *Nature* 270:304–310.

Benoist, C., and Chambon, P. (1980). Deletions covering the putative promoter region of early mRNAs of simian virus 40 do not abolish T antigen expression. *Proc. Natl. Acad. Sci. USA* 77:3865–3869.

Benoist, C., O'Hare, K., Breathnach, R., and Chambon, P. (1980). The ovalbumin gene-sequence of putative control regions. *Nucl. Acids Res.* 8:127–142.

Berk, A. J., and Sharp, P. A. (1978). Structure of the adenovirus 2 early mRNAs. *Cell* 14:695–711.

Berk, A. J., and Sharp, P. A. (1977). Ultraviolet mapping of the adenovirus 2 early promoters. *Cell* 12:45–55.

Berk, A. J., Lee, F., Harrison, T., Williams, J., and Sharp, P. A. (1979). Pre-early Ad5 gene product regulates synthesis of early viral mRNAs. *Cell* 17: 935–944.

Birchmeier, C., Grosschedl, R., and Birnstiel, M. L. (1982). Generation of authentic 3′ termini of an H2A mRNA in vivo is dependent on a short inverted DNA repeat and on spacer sequences. *Cell* 28:739–745.

Birkenmeier, E. H., Brown, D. D., and Jordan, E. (1978). A nuclear extract of Xenopus laevis oocytes that accurately transcribes 5S RNA genes. *Cell* 15:1077–1086.

Bogenhagen, D. F., and Brown, D. D. (1981). Nucleotide sequences in *Xenopus* 5S DNA required for transcription termination. *Cell* 24:261–270.

Bogenhagen, D. F., Sakonju, S., and Brown, D. D. (1980). A control region in the center of the 5S RNA gene directs specific initiation of transcription: II. The 3′ border of the region. *Cell* 19:27–35.

Brown, K. D., Bennett, G. N., Lee, F., Schweingruber, M. E., and Yanofsky, C. (1978). RNA polymerase interaction at the promoter-operator region of the tryptophan operon of *E. coli* and *S. typhimurium. J. Mol. Biol.* 121: 153–177.

Buetti, E., and Digglemann, H. (1981). Cloned mouse mammary tumor virus DNA is biologically active in transfected mouse cells and its expression is stimulated by glucocorticoid hormones. *Cell* 23:335-345.

Burke, J. F., and Ish-Horowicz, D. (1982). Expression of *Drosophila* heat shock genes is regulated in rat-1 cells. *Nucl. Acids Res.* 10:3821-3829.

Busslinger, M., Portman, R., Irminger, J. C., and Birnsteil, M. (1980). Ubiquitous and gene-specific regulatory 5′ sequences in sea urchin histone protein variants. *Nucl. Acids Res.* 8:957-978.

Canaani, D., Kahana, C., Mukamel, A., and Groner, Y. (1979). Sequence heterogeneity at the 5′ termini of late simian virus 40 19S and 16S mRNAs. *Proc. Natl. Acad. Sci. USA* 76: 3078-3082.

Capechi, M. R. (1980). High efficiency transformation by microinjection of DNA into cultivated mammalian cells. *Cell* 22: 479-488.

Celma, M. L., Pan, J., and Weissman, S. M. (1977). Studies of low molecular weight RNA from cells infected with adenovirus 2. II. Heterogeneity at the 5′ end of VA-RNA I. *J. Biol. Chem.* 252:9043-9046.

Chao, M. V., Gralla, J., and Martinson, H. G. (1979). DNA sequence directs placement of histone cores on restriction fragments during nucleosome formation. *Biochem.* 8:1068-1074.

Chow, L. T., Broker, T. R., and Lewis, J. B. (1979). Complex splicing patterns of RNAs from the early regions of adenovirus-2. *J. Mol. Biol.* 134:265-303.

Clarkson, S. G., Koski, R. A., Corlet, J., and Hipskind, R. A. (1981). Influence of 5′ flanking sequences on tRNA transcription in vitro. *ICN-UCLA Symp. Mol. Biol.* 23:463-472.

Cochet, M., Gannon, F., Hen, R., Maroteaux, L., Perrin, F., and Chambon, P. (1979). Organization and sequence studies of the 17-piece chicken conalbumin gene. *Nature* 282:567-574.

Contreras, R., and Fiers, W. (1981). Initiation of transcription by RNA polymerase II in permeable, SV40-infected or noninfected CV1 cells: Evidence for multiple promoters of SV40 late transcription. *Nucl. Acids Res.* 9:215-236.

Corces, V., Pellicer, A., Axel, R., and Meselson, M. (1981). Integration, transcription and control of a *Drosophila* heat shock gene in mouse cells. *Proc. Natl. Acad. Sci.* 78:7038-7042.

Corden, J., Wasylyk, B., Buchwalder, A., Sassone-Corsi, P., Kedinger, C., and

Chambon, P. (1980). Promoter sequences of eukaryotic protein-coding genes. *Science* 209:1406–1414.

DeFranco, D., Schmidt, O., and Soll, D. (1980). The control regions of eukaryotic tRNA gene transcription. *Proc. Natl. Acad. Sci. USA* 77:3365–3368.

DeFranco, D., Sharp, S., and Soll, D. (1981). Identification of regulatory sequences contained in the 5′-flanking region of *Drosophila* lysine tRNA genes. *J. Biol. Chem.* 256:12424–12429.

Denis, H., and Wegnez, M. (1973). Récherches biochimiques sur l'oogenes. 7. Synthese et maturation du RNA 5S dans les petits oocytes de Xenopus laevis. *Biochimie* 55:1137–1151.

Dhar, R., Subramanian, K. N., Zain, B. S., Levine, A., Patch, C., and Weissman, S. M. (1975). Sequences in SV40 DNA corresponding to the "ends" of cytoplasmic mRNA. Les colloques de l'institut national de la sante et de la recherche medicale. *INSERM* vol. 47: 25–32.

Dierks, P., Van Ooyen A., Mantei, N., and Weissman, C. (1981). DNA sequences preceding the rabbit β-globin mRNA with the correct 5′ terminus. *Proc. Natl. Acad. Sci. USA* 78:1411–1415.

Din, N., Engberg, J., and Gall, J. G. (1982). The nucleotide sequence at the transcription termination site of the ribosomal RNA gene in Tetrahymena thermophila. *Nucl. Acids Res.* 10:1503–1513.

Donehower, L. A., Huang, A. L., and Hager, G. L. (1981). Regulatory and coding potential of the mouse mammary tumor virus long terminal redundancy. *J. Virol.* 37:226–238.

Duncan, C. H., Jagadeeswaran, P., Wang, R. R-C., and Weissman, S. M. (1981). Alu family RNA polymerase III transcriptional units interspersed in human β-like globin genes: structural analysis of templates and transcripts. *Gene* 13:185–196.

Efstratiadis, A., Posakony, J., Maniatis, T., Lawn, R., O'Connell, C., Spritz, R., deRiel, J., Forget, B., Weissman, S., Slightom, J., Blechl, A., Smithies, O., Baralle, F., Shoulders, C., and Proudfoot, N. (1980). The structure and evolution of the human β-globin gene family. *Cell* 21:653–668.

Engelke, D. R., Ng, S-Y., Shastry, B. S., and Roeder, R. G. (1980). Specific interaction of a purified transcription factor with an internal control region of 5S RNA genes. *Cell* 19:717–728.

Evans, R., Fraser, N., Ziff, E., Weber, J., Wilson, M., and Darnell, J. (1977). The initiation sites for RNA transcription in Ad2 DNA. *Cell* 12:733–739.

Faye, G., Leung, D. W., Tatchell, K., Hall, B. D., and Smith, M. (1981). Deletion mapping of sequences essential for in vivo transcription of iso-1-cytochrome c gene. *Proc. Natl. Acad. Sci. USA* 78:2258-2262.

Felsenfeld, G. (1978). Chromatin. *Nature* 271:115-121.

Fiers, W., Contreras, R., Haegeman, G., Rogiers, R., Van de Voorde, A., Van Heuverswyn, H., Van Herreweghe, J., Volckaert, G., and Ysebaert, M. (1978). Complete nucleotide sequence of SV40 DNA. *Nature* 273:113-120.

Fitzgerald, M., and Shenk, T. (1981). The sequence 5′-AAUAAA-3′ forms part of the recognition site for polyadenylation of late SV40 mRNAs. *Cell* 24:251-260.

Flavell, A. J., Cowie, A., Arrand, J. R., and Kamen, R. (1980). Localization of three major capped 5′ ends of polyoma virus late mRNAs within a single tetranucleotide sequence in the viral genome. *J. Virol.* 33:902-908.

Flavell, A., Cowie, A., Legon, S., and Kamen, R. (1979). Multiple 5′-terminal cap structures in late polyoma virus RNA. *Cell* 16:357-371.

Ford, J. P., and Hsu, M-T. (1978). Transcription pattern of in vivo-labeled late simian virus 40 RNA: Equimolar transcription beyond the mRNA 3′ terminus. *J. Virol.* 28:795-801.

Ford, P. J., and Brown, D. D. (1976). Sequences of 5S ribosomal DNA from Xenopus mulleri and the evolution of 5S gene-coding sequences. *Cell* 8: 485-493.

Fowlkes, D. M., and Shenk, T. (1980). Transcriptional control regions of the adenovirus VAI RNA gene. *Cell* 22:405-413.

Fraser, N. W., Sehgal, P. B., and Darnell, J. E. (1979). Multiple discrete sites for premature RNA chain termination late in adenovirus-2 infection: Enhancement by 5,6-dichloro-β-D-ribofuranosylbenzimidazole. *Proc. Natl. Acad. Sci. USA* 76: 2571-2575.

Fraser, N. W., Sehgal, P. B., and Darnell, J. E. (1978). DRB-induced premature termination of late adenovirus transcription. *Nature* 272:590-593.

Fujimura, F. K., Deininger, P. L., Friedman, T., and Linney, E. (1981). Mutation near the polyoma DNA replication origin permits productive infection of F9 embryonal carcinoma cells. *Cell* 23:809-814.

Galas, D. J., and Schmitz, A. (1978). DNAse footprinting: A simple method for the detection of protein-DNA binding specificity. *Nucl. Acids Res.* 5: 3157-3170.

Galli, G., Hofstetter, H., and Birnstiel, M. L. (1981). Two conserved sequence blocks within eukaryotic tRNA genes are essential for promotion of transcription. *Nature* 294:626–631.

Gannon, F., O'Hare, K., Perrin, F., LePennec, J. P., Benoist, C., Cochet, M., Breathnach, R., Royal, A., Garapin, A., Cami, B., and Chambon, P. (1979). Organization and sequences at the 5′ end of a cloned complete ovalbumin gene. *Nature* 278:428–434.

Garber, R. L., and Gage, L. P. (1979). Transcription of a cloned Bombyx mori tRNA$_2^{ala}$ gene: Nucleotide sequence of the tRNA precursor and its processing in vitro. *Cell* 18:817–828.

Garel, A., and Axel, R. (1976). Selective digestion of transcriptionally active ovalbumin gene from oviduct nuclei. *Proc. Natl. Acad. Sci. USA* 73: 3966–3970.

Ghosh, P. K., and Lebowitz, P. (1981). Simian virus 40 early mRNAs contain multiple 5′ termini upstream and downstream from a Hogness-Goldberg sequence: A shift in 5′ termini during the lytic cycle is mediated by large T antigen. *J. Virol.* 40:224–240.

Ghosh, P. K., Lebowitz, P., Frisque, R. J., and Gluzman, Y. (1981). Identification of a promoter component involved in positioning the 5′-termini of the simian virus 40 early mRNAs. *Proc. Natl. Acad. Sci. USA* 78:100–104.

Ghosh, P. K., Piatak, M., Mertz, J. E., Weissman, S. M., and Lebowitz, P. (1982). Altered utilization of splice sites and 5′ termini in late RNAs produced by leader region mutants of simian virus 40. *J. Virol.* 44:610–624.

Ghosh, P. K., Roy, P., Barkan, A., Mertz, J. E., Weissman, S. M., and Lebowitz, P. (1981). Unspliced functional late 19S mRNAs containing intervening sequences are produced by a late leader mutant of SV40. *Proc. Natl. Acad. Sci. USA* 78:1386–1390.

Gidoni, D., Kahana, C., Canaani, D., and Groner, Y. (1981). Specific initiation of transcription of SV40 early and late genes occurs at various cap nucleotides including cytidine. *Proc. Natl. Acad. Sci USA* 78:2174–2178.

Gilbert, W. (1976). In *RNA Polymerase*, eds. R. Losick and M. Chamberlin (Cold Spring Harbor, New York: Cold Spring Harbor Laboratories), pp. 193–205.

Gilmour, R. S., and Paul, J. (1973). Tissue-specific transcription of the globin gene in isolated chromatin. *Proc. Natl. Acad. Sci. USA* 70:3440–3442.

Glover, D. M., and Hogness, D. S. (1977). A novel arrangement of the 18S and

28S sequences in a repeating unit of *Drosophila melanogaster* rDNA. *Cell* 10:167-176.

Gluzman, Y., Sambrook, J. F., and Frisque, R. J. (1980). Expression of early genes of origin-defective mutants of SV40. *Proc. Natl. Acad. Sci. USA* 77:3898-3902.

Goldberg, S., Weber, J., and Darnell, J. E. (1977). Definition of a large transcription unit late in ad-2 infection of Hela cells: Mapping by effect as ultraviolet irradiation. *Cell* 10:617-621.

Greenblatt, J., and Li, J. (1981). Interaction of the sigma factor and the nusA gene protein of *E. coli* with RNA polymerase in the initiation-termination cycle of transcription. *Cell* 24:421-428.

Greenblatt, J., McLimont, M., and Hanley, S. (1981). Termination of transcription by the nusA gene protein of *E. coli. Nature* 292:215-220.

Grosschedl, R., and Birnstiel, M. L. (1982). Delimitation of far upstream sequences required for maximal *in vitro* transcription of an H2A histone gene. *Proc. Natl. Acad. Sci. USA* 79:297-301.

Grosschedl, R., and Birnstiel, M. L. (1980a). Identification of regulatory sequences in the prelude sequences of an H2A histone gene by the study of specific deletion mutants in vivo. *Proc. Natl. Acad. Sci. USA* 77:1432-1436.

Grosschedl, R., and Birnstiel, M. L. (1980b). Spacer DNA sequences upstream of the TATAAATA sequence are essential for promotion of H2A histone gene transcription *in vivo. Proc. Natl. Acad. Sci. USA* 77:7102-7106.

Grosschedl, R., Wasylyk, B., Chambon, P., and Birnstiel, M. L. (1981). Point mutation in the TATA box curtails expression of sea urchin H2A histone gene *in vivo. Nature* 294:178-180.

Grummt, I. (1981a) Specific transcription of mouse ribosomal DNA in a cell-free system that mimics control *in vivo. Proc. Natl. Acad. Sci. USA* 78:727-731.

Grummt, I. (1981b). Mapping of a mouse ribosomal DNA promoter by *in vitro* transcription. *Nucl. Acids Res.* 9:6093-6102.

Gruss, P., Dhar, R., and Khoury, G. (1981). The SV40 tandem repeats as an element of the early promoter. *Proc. Natl. Acad. Sci. USA* 78:943-947.

Guilfoyle, R., and Weinmann, R. (1981). The control region for adenovirus VA RNA transcription. *Proc. Natl. Acad. Sci. USA* 78:3378-3382.

Hackett, P. B., and Sauerbier, W. (1975). The transcriptional organization of the ribosomal RNA genes in mouse L cells. *J. Mol. Biol.* 91:235–256.

Hackett, R. W., and Lis, J. T. (1981). DNA sequence analysis reveals extensive homologies of regions preceding hsp 70 and αβ heat shock genes in *Drosophila melanogaster. Proc. Natl. Acad. Sci. USA* 78:6196–6200.

Haegeman, G., and Fiers, W. (1980). Characterization of the 5' terminal cap structures of early simian virus 40 mRNA. *J. Virol.* 35:955–961.

Haegeman, G., and Fiers, W. (1978). Localization of the 5' terminus of late SV40 mRNA. *Nucl. Acids Res.* 5:2359–2371.

Haegeman, G., van Heuverswyn, H., Gheysen, D., and Fiers, W. (1979). Heterogeneity of the 5' termini of late mRNA induced by a viable SV40 deletion mutant. *J. Virol.* 31:484–493.

Hagenbuchle, O., and Schibler, U. (1981). Mouse β-globin and adenovirus 2 major late transcripts are initiated at the cap sites in vitro. *Proc. Natl. Acad. Sci. USA* 78:2283–2286.

Hagenbuchle, O., Bovey, R., and Young, R. A. (1980). Tissue-specific expression of mouse α-amylase genes: Nucleotide sequence of isozyme mRNAs from pancreas and salivary gland. *Cell* 21:179–187.

Hall, B. P., Clarkson, S. G., and Tocchini-Valenti, G. (1982). Transcription initiation of eucaryotic transfer RNA gene. *Cell* 29:3–5.

Harada, S., and Kato, N. (1980). Nucleotide sequences of 4.5S RNA associated with polyA-containing RNAs of mouse and hamster cells. *Nucl. Acids Res.* 8:1273–1285.

Hendrick, J. P., Wolin, S. L., Rinke, J., Lerner, M. R., and Steitz, J. A. (1981). Ro scRNPs are a subclass of La RNPs: Further characterization of the Ro and La Small ribonucleoproteins from uninfected mammalian cells. *Mol. Cell. Biol.* 1:1138–1149.

Hentschel, C., Irminger, J. C., Bucher, P., and Birnstiel, M. (1980). Sea urchin histone mRNA termini are located in gene regions downstream from putative regulatory sequences. *Nature* 285:147–151.

Hofer, E., and Darnell, J. E. (1981). The primary transcription unit of the mouse β-major globin gene. *Cell* 23:585–593.

Hofstetter, H., Kressmann, A., and Birnstiel, M. L. (1981). *Cell* 24:573–585.

Honda, B. M., and Roeder, R. G. (1980). Association of a 5S gene transcription factor with 5S RNA and altered levels of the factor during cell differentiation. *Cell* 22:119–126.

Honda, H., Kaufman, R. J., Manley, J., Gefter, M., and Sharp, P. A. (1980). Transcription of SV40 in a HeLa whole cell extract. *J. Biol. Chem.* 256: 478–482.

Horn, G. T., and Wells, R. D. (1981). The leftward promoter of Bacteriophage λ. Structure, biological activity and influence by adjacent regions. *J. Biol. Chem.* 256:2003–2009.

Hoveman, B., Galler, R., Walldorf, U., Kupper, H., and Bautz, E. K. F. (1981). Vitellogenin in *Drosophila melanogaster*: sequence of the yolk protein I and its flanking regions. *Nucl. Acids Res.* 9:4721–4734.

Hu, S-L., and Manley, J. (1981). DNA sequence required for initiation of transcription in vitro from the major promoter of adenovirus-2. *Proc. Natl. Acad. Sci. USA* 78:820–824.

Huang, A. L., Ostrowski, M. C., Berard, D., and Hager, G. L. (1981). Glucocorticoid regulation of the HaMuSV p 21 gene conferred by the long terminal redundancy from mouse mammary tumor virus. *Cell* 27:245–255.

Hynes, N. E., Kennedy, N., Rahmsdorf, U., and Groner, B. (1981). Hormone responsive expression of an endogenous proviral gene of mouse mammary tumor virus after molecular cloning and gene transfer with cultured cells. *Proc. Nat. Acad. Sci. USA* 78:2038–2042.

Igo-Kemenis, T., Mbrz, W., and Zachau, H. G. (1982). Chromatin. *Ann. Rev. Biochem.* 51:89–122.

Ingolia, T. D., Craig, E. A., and McCarthy, B. J. (1980). Sequence of three copies of the gene for the major drosophila heat shock induced protein and their flanking regions. *Cell* 21:669–679.

Jakobovits, E. B., Bratosin, S., and Aloni, Y. (1980). A nucleosome-free region in SV40 minichromosomes. *Nature* 285:263–265.

Jessel, D., Landau, T., Hudson, J., Lalor, T., Tenen, D., and Livingston, D. M. (1976). Identification of regions of the SV40 genome which contain preferred SV40 T antigen-binding sites. *Cell* 8:535–545.

Johnson, J. D., Ogden, R., Johnson, P., Abelson, J., Dembeck, P., and Hakura, K. (1980). Transcription and processing of a yeast tRNA gene containing a modified intervening sequence. *Proc. Natl. Acad. Sci. USA* 77:2564–2568.

Jones, N., and Shenk, T. (1979a). Isolation of Ad5 host range deletion mutants defective for transformation of rat embryo cells. *Cell* 17:683–689.

Jones, N., and Shenk, T. (1979b). An adenovirus type 5 early gene function regulates expression of other early viral genes. *Proc. Natl. Acad. Sci. USA* 76: 3665–3669.

Jung, A., Sippel, A. E., Grez, M., and Schutz, G. (1980). Exons encode functional and structural units of chicken lysozyme. *Proc. Natl. Acad. Sci. USA* 77: 5759–5763.

Kataoka, T., Nikaido, T., Miyata, T., Moriwaki, K., and Honjo, T. (1981). The nucleotide sequences of rearranged and germline immunoglobulin V_H genes of a mouse myeloma MC101 and evolution of V_H genes in mouse. *J. Biol. Chem.* 257:277–285.

Katina, M., Vasseur, M., Montreau, N., Yaniv, M., and Blangy, D. (1981). Polyoma DNA sequences involved in control of viral gene expression in murine embryonal carcinoma cells. *Nature* 290:720–722.

Katina, M., Yaniv, M., Vasseur, M., and Blangy, D. (1980). Expression of polyoma early functions in mouse embryonal carcinoma cells depends on sequence rearrangements in the beginning of the late region. *Cell* 20:393–399.

Keene, M. A., Corces, V., Lowenhaupt, K., and Elgin, S. R. B. (1981). DNase I hypersensitive sites in Drosophila chromatin occur at the 5′-ends of regions of transcription. *Proc. Natl. Acad. Sci. USA* 78:143–146.

Khoury, G., and May, E. (1977). Regulation of early and late SV40 transcription: overproduction of early viral RNA in the absence of a functional T antigen. *J. Virol.* 77:167–176,

Klemenz, R., and Geiduschek, E. P. (1980). The 5′ terminus of the precursor ribosomal RNA of saccharomyces cerevisiae. *Nucl. Acids Res.* 8:2679–2689.

Klootwijk, J., de Jonge, P., and Planta, R. J. (1979). The primary transcript of the ribosomal repeating unit in yeast. *Nucl. Acids Res.* 6:27–39.

Kominami, R., Mishima, Y., Urano, Y., Sakai, M., and Muramatsu, M. (1982). Cloning and determination of the transcription termination site of ribosomal RNA gene of the mouse. *Nucl. Acids Res.* 10:1963–1979.

Konkel D., Tilghman, S., and Leder, P. (1978). The sequence of the chromosomal mouse β-globin major gene: Homologies in capping, splicing and polyA sites. *Cell* 15:1125–1132.

Korn, L. T., and Brown, D. D. (1978). Nucleotide sequence of *Xenopus borealis* oocyte 5S DNA: Comparison of sequences that flank several related eukaryotic genes. *Cell* 15:1145–1156.

Korn, L. J., Birkenmeier, E. H., and Brown, D. D. (1979). Transcription initiation of *Xenopus* 5S ribosomal RNA genes in vitro. *Nucl. Acids Res.* 7: 947–958.

Koski, R. A., Clarkson, S. G., Kurjan, J., Hall, B. D., and Smith, M. (1980). Mutations of the yeast SUP tRNA^{Tyr} locus: Transcription of the mutant genes *in vitro. Cell* 22:115–125.

Kressmann, A., Hofstetter, H., DiCapua, E., Grosschedl, R., and Birnstiel, M. L. (1979). A tRNA gene of *Xenopus laevis* contains at least two sites promoting transcription. *Nucl. Acids Res.* 7:1749–1763.

Kurtz, D. T. (1981). Hormonal inducibility of rate $\alpha_{2\mu}$ globulin genes in transfected mouse cells. *Nature* 291:629–631.

Lai, C-J., Dhar, R., and Khoury, G. (1978). Mapping the spliced and unspliced late lytic SV40 RNAs. *Cell* 14:971–982.

Larsen, A., and Weintraub, H. (1982). An altered DNA conformation detected by S1 nuclease occurs at specific sequences in active chick globin chromatin. *Cell* 29:609–622.

Laub, O., Bratosin, S., Horowitz, M., and Aloni, Y. (1979). The initiation of transcription of SV40 DNA at late time after infection. *Virol.* 92:310–323.

Lebowitz, P., and Ghosh, P. K. (1982). Initiation and regulation of simian virus 40 early transcription in vitro. *J. Virol.* 41:449–461.

Lebowitz, P., Weissman, S. M., and Radding, C. M. (1971). Nucleotide sequence of a ribonucleic acid transcribed *in vitro* from λ phage deoxyribonucleic acid. *J. Biol. Chem.* 246:5120–5139.

Lee, F., Mulligan, R., Berg, P., and Ringold, G. (1981). Glucocorticoids regulate expression of dihydrofolate reductase cDNA in mouse mammary tumour virus chimaeric plasmids. *Nature* 294:228–232.

Lerner, M. R., Andrews, N. C., Miller, G., and Steitz, J. A. (1980). Two small RNAs encoded by Epstein-Barr virus and complexed with protein are precipitated by antibodies from patients with systemic lupus erythematosus. *Proc. Natl. Acad. Sci. USA* 78:805–809.

Lerner, M. R., Boyle, J. A., Hardin, J. A., and Steitz, J. A. (1981). Two novel classes of small ribonucleoproteins detected by antibodies associated with lupus erythematosus. *Science* 211:400–402.

Levis, R., and Penman, S. (1978). Processing steps and methylation in the formation of the ribosomal RNA of cultured *Drosophila* cells. *J. Mol. Biol.* 121:219-238.

Levy, A., and Noll, M. (1980). Multiple phases of nucleosomes in the hsp 70 genes of *Drosophila melanogaster. Nucl. Acids. Res.* 8:6959-6968.

Li, W-Y., Reddy, R., Henning, D., Epstein, P., and Busch, H. (1982). Nucleotide sequence of 7S RNA: homology to Alu DNA and LA 4.5 RNA. *J. Biol. Chem.* 257:5136-5142.

Long, E. O., and Dawid, I. B. (1979). Expression of ribosomal DNA insertions in *Drosophila melanogaster. Cell* 18:1185-1196.

Long, E. O., Rebbert, M. L., and Dawid, I. B. (1981a). Structure and expression of ribosomal RNA genes of *Drosophila melanogaster* interupted by Type-2 insertions. *Cold Spring Harbor Symp. Quant. Biol.* 45:667-672.

Long, E. O., Rebbett, M. L., and Dawid, I. B. (1981b). Nucleotide sequence of the initiation site for ribosomal RNA transcription in *Drosophila melanogaster. Proc. Natl. Acad. Sci. USA* 78:1513-1517.

Louis, C., Schedl, P., Sanmal, B., and Worcel, A. (1980). Chromatin structure of the 5S RNA genes of *D. melanogaster. Cell* 22:387-392.

Luse, D. S., and Roeder, R. G. (1980). Accurate transcription initiation on a purified mouse β-globin DNA fragment in a cell-free system. *Cell* 22: 387-392.

Manley, J. L., Sharp, P. A., and Gefter, M. L. (1979). RNA synthesis in isolated nuclei in vitro: initiation of the Ad2 major late mRNA precursor. *Proc. Natl. Acad. Sci. USA* 76:160-164.

Manley, J. L., Fire, A., Cano, A., Sharp, P. A., and Gefter, M. L. (1980). DNA-dependent transcription of adenovirus genes in a soluble whole-cell extract. *Proc. Natl. Acad. Sci. USA* 77:3855-3859.

Martin, F. H., and Tinoco, I., Jr. (1980). DNA-RNA hybrid duplexes containing oligo(dA:rW) sequences are exceptionally unstable and may facilitate termination of transcription. *Nucl. Acids Res.* 8:2245-2249.

Mathis, D. J., and Chambon, P. (1981). The SV40 early region TATA box is required for accurate *in vitro* initiation of transcription. *Nature* 290: 310-315.

Matsui, T., Segall, J., Weil, P. A., and Roeder, R. G. (1980). Multiple factors required for accurate initiation of transcription by purified RNA polymerase II. *J. Biol. Chem.* 255:11992-11996.

Maquat, L. E., Thornton, K., and Reznikoff, W. (1980). *lac* promoter mutations located downstream from the transcription start site. *J. Mol. Biol.* 139: 537-549.

Mayo, H. E., Warren, R., and Palmiter, R. D. (1982). The mouse metallothionine gene is transcriptionally regulated by cadmium following transfection into human or mouse cells. *Cell* 29:99-100.

McGhee, J. D., Wood, W., Dolan, M., Engel, J. D., and Felsenfeld, G. (1981). A zoo base pair region at the 5' end of the chicken adult β-globin gene is accessible to nuclease digestion. *Cell* 27:45-55.

McKnight, S. L. (1980). The nucleotide sequence of the herpes simplex thymidine kinase gene. *Nucl. Acids Res.* 8:5949-5964.

McKnight, S. L., and Kingsbury, R. (1982). Transcriptional control signals of a eukaryotic protein-coding gene. *Science* 217: 316-324.

McKnight, S. L., Gabvis, E. R., Kingsbury, R., and Axel, R. (1981). Analysis of transcriptional regulatory signals of the HSV thymidine kinase gene: Identification of an upstream control region. *Cell* 25:385-398.

Miller, K. G., and Sollner-Webb, B. (1981). Transcription of mouse ribosomal RNA genes by RNA polymerase I: *In vitro* and *in vivo* initiation and processing sites. *Cell* 27:165-174.

Moreau, P., Hen, R., Wasylyk, B., Everett, R., Gaub, M. P., and Chambon, P. (1981). The SV40 72 base pair repeat has a striking effect on gene expression both in SV40 and other chimeric recombinants. *Nucl. Acids Res.* 9: 6047-6068.

Musich, P. R., Maio, J. J., and Brown, F. L. (1977). Interactions of a phas relationship between restriction sites and chromatin subunits in African green monkey and calf nuclei. *J. Mol. Biol.* 117:637-655.

Nevins, J. R. (1981). Mechanism of activation of early viral transcription by the adenovirus E1A gene product. *Cell* 26:213-220.

Nevins, J. R., and Darnell, J. E. (1978). Steps in the processing of Ad2 mRNA: PolyA$^+$ addition precedes splicing. *Cell* 15:1477-1493.

Nevins, J. R., and Winkler, J. J. (1980). Regulation of early adenovirus transcription: A protein product of early region 2 specifically represses region 4 transcription. *Proc. Natl. Acad. Sci. USA* 77:1893-1897.

Nevins, J. R., Blanchard, J-M., and Darnell, J. E. (1980). Transcription units of adenovirus type 2: Termination of transcription beyond the polyA addition site in early regions 2 and 4. *J. Mol. Biol.* 144:337-386.

Ng, S. Y., Parker, C. S., and Roeder, R. G. (1979). Transcription of cloned *Xenopus* 5S RNA genes by *X. laevis* RNA polymerase III in reconstituted systems. *Proc. Natl. Acad. Sci. USA* 76: 136-140.

Nikolaev, N., Georgiev, O. I., Venko, P. V., and Hadjiolov, A. A. (1979). The 37S precursor to ribosomal RNA is the primary transcript of ribosomal RNA genes in Saccharomyces cerevisiae. *J. Mol. Biol.* 127:297-308.

Nishioka, Y., and Leder, P. (1979). The complete sequence of a chromosomal mouse α-globin gene reveals elements conserved throughout vertebrate evolution. *Cell* 18:875-882.

Nishioka, Y., Leder, A., and Leder, P. (1980). Unusual α-globin-like gene that cleanly lost both globin intervening sequences. *Proc. Natl. Acad. Sci. USA* 77:2806-2809.

Osborne, T. F., Gaynor, R. B., and Berk, A. J. (1982). The TATA homology and the mRNA 5′ untranslated sequence are not required for expression of essential Adenovirus E1A function. *Cell* 29:139-148.

Osborne, T. F., Schell, R. E., Burch-Jaffe, E., Berget, S. J., and Berk, A. J. (1981). Mapping a eukaryotic promoter: A DNA sequence required for *in vivo* expression of adenovirus pre-early functions. *Proc. Natl. Acad. Sci. USA* 78:1381-1385.

Ostrowski, M. C., Berard, D., and Hager, G. L. (1981). Specific transcriptional initiation *in vitro* on murine type C retrovirus promoters. *Proc. Natl. Acad. Sci. USA* 78:4485-4489.

Palmiter, R. D., Chen, H. Y., and Brinstein, L. C. (1982). Differential regulation of metallothionine-thymidine kinase fusion genes in transgenic mice and their offspring. *Cell* 29:701-710.

Payvar, F., Wrange, O., Carlstedt-Duke, J., Lkret, S., Gustafsson, J-A., and Yamamoto, K. R. (1981). Purified glucocorticoid receptors bind selectively *in vitro* to a cloned DNA fragment whose transcription is regulated by glucocorticoids in vivo. *Proc. Natl. Acad. Sci. USA* 78:6628-6632.

Pelham, H. R. B., and Brown, D. D. (1980). A specific transcription factor that can bind either the 5S RNA gene or 5S RNA. *Proc. Natl. Acad. Sci. USA* 77:4170-4174.

Pelham, H. R. B., Wormington, W. M., and Brown, D. D. (1981). Related 5S RNA transcription factors in *Xenopus* oocytes and somatic cells. *Proc. Natl. Acad. Sci. USA* 78:1760-1764.

Piatak, M., Subramanian, K. N., Roy, P., and Weissman, S. M. (1981). Late mRNA production by viable SV40 mutants with deletions in the leader region. *J. Mol. Biol.* 153:589-618.

Platt, T. (1981). Termination of transcription and its regulation in the tryptophan operon of *E. coli. Cell* 24:10-23.

Ponder, B. A. J., and Crawford, L. V. (1977). The arrangement of nucleosomes in nucleoprotein complexes from polyoma virus and SV40. *Cell* 11:35-49.

Pribnow, D. (1979). Genetic control signals in DNA. In *Biological Regulation and Development, 1*, ed. R. F. Goldberger (New York: Plenum), pp. 219-227.

Pribnow, D. (1975). Bacteriophage T7 early promoters: Nucleotide sequences of two RNA polymerase binding sites. *J. Mol. Biol.* 99:419-443.

Price, P., and Penman, S. (1972). A distinct RNA polymerase activity, synthesizing 5.5S, 5S and 4S RNA in nuclei from Ad2-infected HeLa cells. *J. Mol. Biol.* 70:430-450.

Proudfoot, N. J., and Brownlee, G. G. (1974). Sequence at the 3′ end of globin mRNA shows homology with immunoglobulin light chain mRNA. *Nature* 252:359-362.

Reddy, V. B., Ghosh, P. K., Lebowitz, P., and Weissman, S. M. (1978). Gaps and duplicated sequences in the leaders of SV40 16S RNA. *Nucl. Acids Res.* 5:4195-4214.

Reddy, V. B., Ghosh, P., Lebowitz, P., Piatak, M., and Weissman, S. M. (1979). Simian virus 40 early mRNAs: Genomic localization of 3′ and 5′ termini and two major splices in mRNA from transformed and lytically infected cells. *J. Virol.* 30:279-296.

Reddy, V. B., Tevethia, S. S., Tevethia, M. J., and Weissman, S. M. (1982). Nonselective expression of simian virus 40 large tumor antigen fragments in mouse cells. *Proc. Natl. Acad. Sci. USA* 79: 2064-2067.

Reed, S. I., Stark, G. R., and Alwine, J. C. (1976). Autoregulation of SV40 gene A by T antigen. *Proc. Natl. Acad. Sci. USA* 73:3083-3087.

Reed, S. I., Ferguson, J., Davis, R. W., and Stark, G. R. (1975). T antigen binds to SV40 DNA at the origin of DNA replication. *Proc. Natl. Acad. Sci. USA* 72:1605-1609.

Reeder, R. H., Sollner-Webb, B., and Wahn, H. L. (1977). Sites of transcription initiation in vivo on *Xenopus* laevis ribosomal DNA. *Proc. Natl. Acad. Sci. USA* 74:5402-5406.

Reznikoff, W., and Abelson, J. (1978). The *lac* promoter. In *The Operon*, eds. J. Miller and W. Reznikoff (Cold Spring Harbor, New York: Cold Spring Harbor Laboratories), pp. 221–243.

Rio, D., Robbins, A., Myers, R., and Tjian, R. (1980). Regulation of simian virus 40 early transcription *in vitro* by a purified tumor antigen. *Proc. Natl. Acad. Sci. USA* 77:5706–5710.

Roberts, J. (1969). Termination factor for RNA synthesis. *Nature* 224: 1168–1174.

Robins, D. M., Paek, I., Seeburg, P. M., and Axel, R. (1982). Regulated expression of human growth hormone genes in mouse. *Cell* 29:623–631.

Roeder, R. G. (1976). Eukaryotic nuclear RNA polymerases. In *RNA polymerase*, eds. R. Losick and M. Chamberlin (Cold Spring Harbor, New York: Cold Spring Harbor Laboratories).

Rosa, M. D., Gottlieb, E., Lerner, M. R., and Steitz, J. A. (1981). Striking similarities are exhibited by two small Epstein-Bar virus-encoded RNAs and the adenovirus-associated RNAs VAI and VAII. *Mol. Cell. Biol.* 1: 785–796.

Rosenberg, M., and Court, D. (1979). Regulatory sequences involved in the promotion and termination of RNA transcription. *Ann. Rev. Gen.* 13:319–353.

Rosenberg, M., and Weissman, S. M. (1975). Termination of transcription in bacteriophage λ. Heterogeneous, 3′-terminal oligo-adenylate additions and the effects of *p* factor. *J. Biol. Chem.* 250:4755–4764.

Sakonju, S., Bogenhagen, D. F., and Brown, D. D. (1980). A control region in the center of the 5S RNA gene directs specific initiation of transcription: I. The 5′ border of the region. *Cell* 19:13–25.

Sakonju, S., Brown, D. D., Engelke, D., Ng, S-Y., Shastry, B. S., and Roeder, R. G. (1981). The binding of a transcription factor to deletion mutants of a 5S ribosomal RNA gene. *Cell* 23:665–669.

Samal, B., Worcel, A., Lous, C., and Schedl, P. (1981). Chromatin structure of the histone genes of *D. melanogaster. Cell* 23:401–409.

Sassone-Corsi, P., Corden, J., Kedinger, C., and Chambon, P. (1981). Promotion of specific *in vitro* transcription by excised "TaTA" box sequences inserted in a foreign nucleotide environment. *Nucl. Acids Res.* 9:3941–3958.

Scherer, G. E. F., Walkinshaw, M. D., and Arnott, S. (1978). A computer aided oligonucleotide analysis provides a model for RNA polymerase-promoter recognition in *E. coli. Nucl. Acids Res.* 5:3759–3773.

Schmitz, A., and Galas, D. J. (1979). The interaction of RNA polymerase and lac repressor with the *lac* control region. *Nucl. Acids Res.* 6:111-137.

Scott, W. A., and Wigmore, D. J. (1978). Sites in simian virus 40 chromatin which are preferentially cleaved by endonucleases. *Cell* 15:1511-1518.

Segall, J., Matsui, T., and Roeder, R. G. (1980). Multiple factors are required for the accurate transcription of purified genes by RNA polymerase III. *J. Biol. Chem.* 255:11986-11991.

Sehgal, P. B., Fraser, N. W., and Darnell, J. E. (1979). Early Ad2 transcription units: Only promoter-proximal RNA continues to be made in the presence of DRB. *Virol.* 94:185-191.

Sekikawa, K., and Levine, A. J. (1981). Isolation and characterization of polyoma host range mutants that replicate in multipotential embryonal carcinoma cells. *Proc. Natl. Acad. Sci. USA* 18:1100-1104.

Sharp, S., DeFranco, D., Dingerman, T., Farrell, P., and Soll, D. (1981). Internal control regions for transcription of eukaryotic tRNA genes. *Proc. Natl. Acad. Sci. USA* 78:6657-6661.

Shatkin, A. J. (1976). Capping of eukaryotic mRNAs. *Cell* 9:646-653.

Shenk, T. E., Carbon, J., and Berg, P. (1976). Construction and analysis of viable deletion mutants of SV40. *J. Virol.* 18:640-671.

Siebenlist, U., Simpson, R. B., and Gilbert, W. (1980). *E. coli* RNA polymerase interacts homologously with two different promoters. *Cell* 20:269-281.

Simpson, R. B. (1979). Contacts between *E. coli* RNA polymerase and thymines in the *lac* UV5 promoter. *Proc. Natl. Acad. Sci. USA* 76:3233-3237.

Soberon, X., Rossi, J. J., Larson, G. P., and Itakura, K. (1982). A synthetic consensus sequence, prokaryotic promoter is functional. In *Promoters: Structure and Function* (New York: Academic Press), pp. 407-431.

Söderlund, H., Pettersson, U., Venstrom, B., Philipson, L., and Mathews, M. B. (1976). A new species of virus-coded low molecular weight RNA from cells infected with Ad2. *Cell* 7:585-593.

Soeda, E., Arrand, J. R., Smolar, N., Walsh, J. E., and Griffin, B. E. (1980). Coding potential and regulatory signals of the polyoma virus genome. *Nature* 283:445-453.

Sollner-Webb, B., and Reeder, R. H. (1979). The nucleotide sequence of the initiation and termination sites for ribosomal RNA transcription in *X. laevis*. *Cell* 10:485-499.

Sprague, K. U., Larson, D., and Morton, D. (1980). 5'-Flanking sequence signals are required for activity of silkworm alanine tRNA genes in homologous *in vitro* transcription systems. *Cell* 22:171-178.

Spritz, R. A., deRiel, J. K., Forget, B. G., and Weissman, S. M. (1980). Complete nucleotide sequence of the human α-globin gene. *Cell* 21:639-646.

Stefano, J. E., and Gralla, J. D. (1982). Spacer mutations in *lac* P^S promoters. *Proc. Natl. Acad. Sci. USA* 79:1069-1072.

Steggles, A. W., Wilson, G. N., Kantor, J. A., Picciano, D. J., Flavey, A. K., and Anderson, W. F. (1974). Cell-free transcription of mammalian chromatin: Transcription of globin mRNA sequences from bone-marrow chromatin with mammalian RNA polymerase. *Proc. Natl. Acad. Sci. USA* 71: 1219-1223.

Struhl, K. (1981). Deletion mapping a eukaryotic promoter. *Proc. Natl. Acad. Sci. USA* 78:4461-4465.

Subramanian, K., and Shenk, T. (1978). Definition of the boundaries of the SV40 origin of DNA replication. *Nucl. Acids Res.* 5:3635-3642.

Sundin, O., and Varshavsky, A. (1979). Staphylococcal nuclease makes a single non-random cut in the simian virus 40 minichromosome. *J. Mol. Biol.* 132: 535-546.

Sutcliffe, J. G., Shinnick, T. M., Verma, I. M., and Lerner, R. A. (1980). Nucleotide sequence of murine leukemia virus: 3' end reveals details of replication, analogy to bacterial transposons and an unexpected gene. *Proc. Natl. Acad. Sci. USA* 77:3302-3306.

Talkington, C. A., Nishioka, Y., and Leder, P. (1980). *In vitro* transcription of normal, mutant and truncated mouse α-globin genes. *Proc. Natl. Acad. Sci. USA* 77:7132-7136.

Tegtmeyer, P. (1974). Altered patterns of protein synthesis in infection by SV40 mutants. *Cold Spring Harbor Symp. Quant. Biol.* 39:9-16.

Tegtmeyer, P., Schwartz, M., Collins, J. K., and Rundell, K. (1975). Regulation of tumor antigen synthesis by simian virus 40 gene A. *J. Virol.* 16:168-178.

Telford, J. L., Kressmann, A., Koski, R. A., Grosschedl, R., Muller, F., Clarkson, S. G., and Birnstiel, M. L. (1979). Delimitation of a promoter for RNA polymerase III by means of a functional test. *Proc. Natl. Acad. Sci. USA* 76:2590-2594.

Thimmappaya, B., Jones, N., and Shenk, T. (1979). A mutation which alters initiation of transcription by RNA polymerase III on the Ad5 chromosome. *Cell* 18:947–954.

Thompson, J. A., Radonovich, M. F., and Salzman, N. P. (1979). Characterization of the 5′-terminal structure of SV40 early mRNAs. *J. Virol.* 31: 437–446.

Tjian, R. (1978). The binding site on SV40 DNA for a T antigen-related protein. *Cell* 13:165–179.

Tsai, M-J., Tsai, S. Y., Kops, L. E., Schultz, T. Z., and O'Malley, B. W. (1981a). Elements required for initiation of transcription of the ovalbumin gene in vitro. *ICN-UCLA Symp. Mol. Cell. Biol.* 23:313–327.

Tsai, S. Y., Tsai, M-J., and O'Malley, B. W. (1981b). Specific 5′-flanking sequences are required for faithful initiation of *in vitro* transcription of the ovalbumin gene. *Proc. Natl. Acad. Sci. USA* 78:879–883.

Tsuda, M., and Suzuki, Y. (1981). Faithful transcription initiation of fibroin gene in a homologous cell-free system reveals an enhancing effect of 5′ flanking sequence far upstream. *Cell* 27:175–182.

Tsujimoto, Y., and Suzuki, Y. (1979). The DNA sequence of Bombyx mori fibroin gene including the 5′-flanking, mRNA coding, entire intervening and coding regions. *Cell* 18:591–600.

Tsujimoto, Y., Hirose, S., Tsuda, M., and Suzuki, Y. (1981). Promoter sequence of fibroin gene assigned by *in vitro* transcription system. *Proc. Natl. Acad. Sci. USA* 78:4838–4842.

Udvardy, A., Sümegi, J., Csorád-Toth, E., Gausz, J., and Gyurkovics, Y. (1982). Genomic organization and functional analysis of a deletion variant of the 87A7 heat shock locus of *Drosophila melanogaster*. *J. Mol. Biol.* 155: 267–286.

Valenzuela, P., Bell, G. I., Venegas, A., Sewell, E. T., Masiarz, F. R., DeGannaro, L., Weinberg, F., and Rutter, W. J. (1977). Ribosomal RNA genes of Saccharomyces cerevisiae II. Physical map and nucleotide sequence of the 5S ribosomal RNA gene and adjacent intergenic regions. *J. Biol. Chem.* 252:8126–8235.

Van Ormondt, H., Maat, J., DeWaard, A., and Van der Eb, A. J. (1978). The nucleotide sequence of the transforming HpaI-E fragment of adenovirus type 5 DNA. *Gene* 4:309–328.

Varshavsky, A. J., Sundin, O., and Bohn, M. (1979). A stretch of "late" SV40 viral DNA about 400 bp long which includes the origin of replication is specifically exposed in SV40 minichromosomes. *Cell* 16:453-466.

Varshavsky, A. J., Sundin, O. H., and Bohn, M. J. (1978). SV40 viral minichromosome: Preferential exposure of the origin of replication as probed by restriction endonuclease. *Nucl. Acids Res.* 5:3469-3478.

Venkatesan, S., Baroudy, B. M., and Moss, B. (1981). Distinctive nucleotide sequences adjacent to multiple initiation and termination sites of an early vaccinia virus gene. *Cell* 125:805-813.

Villareal, L., White, R. T., and Berg, P. (1979). Mutational alterations within the Simian Virus 40 leader segment generate altered 16S and 19S mRNAs. *J. Virol.* 29:209-219.

Vogeli, G., Ohkubo, H., Sobel, M. E., Yamada, Y., Pastan, I., and deCrombrugghe, B. (1981). Structure of the promoter for the chick alpha 2 type 1 collagen gene. *Proc. Natl. Acad. Sci. USA* 78:5334-5338.

Waldeck, W., Fohring, B., Chowdhury, K., Gruss, P., and Sauer, G. (1978). Origin of DNA replication in papovavirus chromatin is recognized by endogenous endonuclease. *Proc. Natl. Acad. Sci. USA* 75:5964-5968.

Wallace, R. B., Johnson, P. F., Tanaka, S., Schold, M., Itakura, K., and Abelson, J. (1980). Directed deletion of a yeast transfer RNA intervening sequence. *Science* 209:1396-1400.

Ward, D. F., and Gottesman, M. E. (1981). The nus mutations affect transcription termination in *E. coli*. *Nature* 292:212-215.

Wasylyk, B., and Chambon, P. (1981). A T to A bases substitution and small deletions in the conalbumin TATA box drastically decrease specific *in vitro* transcription. *Nucl. Acids Res.* 9:1813-1824.

Wasylyk, B., Derbyshire, R., Guy, A., Molko, D., Roget, A., Teolue, R., and Chambon, P. (1980). Specific *in vitro* transcription of conalbumin gene is drastically decreased by a single-point mutation in TATA box homology sequence. *Proc. Natl. Acad. Sci. USA* 77:7024-7028.

Wasylyk, B., Kedinger, C., Corden, J., Brinson, O., and Chambon, P. (1980). Specific *in vitro* initiation of transcription on conalbumin and ovalbumin genes and comparison with adenovirus-2 early and late genes. *Nature* 285: 367-373.

Weber, J., Jelinek, W., and Darnell, J. E. (1977). The definition of a large viral transcription unit late in Ad2 infection of HeLa cells: Mapping of nascent RNA molecules labeled in isolated nuclei. *Cell* 10:611-616.

Weil, P. A., Luse, D. S., Segall, J., and Roeder, R. G. (1979). Selective and accurate initiation of transcription at the Ad2 major late promoter in a soluble system dependent on purified RNA polymerase II and DNA. *Cell* 18:469–484.

Weil, P. A., Segall, J., Harris, B., Ng, S-Y., and Roder, R. G. (1979). Faithful transcription of eukaryotic genes by RNA polymerase III in systems reconstituted with purified DNA templates. *J. Biol. Chem.* 254:6163–6173.

Wells, R. D. (1982). Effects of neighboring DNA homopolymers on the biochemical and physical properties of the *Escherichia coli* lactose promoter (cloning and character studies). *J. Biol. Chem.* 257:12954–12961.

Weingartner, B., and Keller, W. (1981). Transcription and processing of adenoviral RNA by extracts from HeLa cells. *Proc. Natl. Acad. Sci. USA* 78: 4092–4096.

Weintraub, H., and Groudine, M. (1976). Chromosomal subunits in active genes have an altered conformation. *Science* 93:848–858.

White, R. L., and Hogness, D. S. (1977). R loop mapping of the 18S and 28S sequences in the long and short repeating units of *Drosophila melanogaster* rDNA. *Cell* 10:177–192.

Wilson, M. C., Fraser, N. W., and Darnell, J. E. (1979). Mapping of RNA initiation sites by high doses of UV irradiation: Evidence for three independent promoters within the left 11% of the Ad2 genome. *Virol.* 94:175–184.

Wittig, B., and Wittig, S. (1979). A phase relationship associates tRNA structural gene sequences with nucleosome cores. *Cell* 18:1173–1183.

Wormington, W. M., Bogenhagen, D. F., Jordan, E., and Brown, D. D. (1981). A quantitative assay for *Xenopus* 5S RNA gene transcription in vitro. *Cell* 24:809–817.

Wu, C. (1980). The 5′ ends of *Drosophila* heat shock genes in chromatin are hypersensitive to DNase I. *Nature* 286:854–860.

Wu, G-J. (1978). Adenovirus DNA-directed transcription of 5.5S RNA in vitro. *Proc. Natl. Acad. Sci. USA* 75:2175–2179.

Yamamoto, T., deCrombrugghe, B., and Pastan, I. (1980). Identification of a functional promoter in the long terminal repeat of Rous sarcoma virus. *Cell* 22:787–797.

Yamamoto, T., Jay, G., and Pastan, I. (1980). Unusual features in the nucleotide sequence of a cDNA clone derived from the common region of avian sarcoma virus mRNA. *Proc. Natl. Acad. Sci. USA* 77:176–180.

Yanofsky, C. (1981). Attenuation in the control of expression of bacterial operons. *Nature* 224:1168–1174.

Young, R. A., Hagenbuchle, O., and Schibler, U. (1981). A single mouse α-amylase gene specifies two different tissue-specific mRNAs. *Cell* 23:451–458.

Zaret, K. S., and Sherman, I. (1982). DNA sequence required for efficient transcription termination in yeast. *Cell* 28:563–573.

Ziff, E., and Evans, R. (1978). Coincidence of the promoter and capped 5' terminus of RNA from the adenovirus-2 major late transcription unit. *Cell* 15:1463–1475.

Zipster, D., Lipsich, L., and Kwoh, J. (1981). Mapping functional domains in the promoter region of the herpes thymidine kinase gene. *Proc. Natl. Acad. Sci. USA* 78: 6276–6280.

10

Signals for the Splicing of Eukaryotic Messenger RNA Transcripts

STEPHEN M. MOUNT and JOAN A. STEITZ

Exact colinearity between genes and their products came to be a fundamental tenet of molecular biology after several decades of work on prokaryotic organisms. Protein sequences were shown to correspond to RNA sequences as dictated by the genetic code, and RNA sequences indeed turned out to be exact replicas of the DNA from which they were transcribed. The discovery of intervening sequences in eukaryotic genes therefore came as something of a shock to the scientific world. Introns, or intervening sequences, can be defined as sequences in DNA which are absent from the mature RNA product specified by that DNA but which reside between sequences that are retained. The sequences that are retained are known as exons. Introns have now been found to occur in at least some genes of every eukaryotic species that has been examined. They appear in genes of all types. mRNA, tRNA, and rRNA genes with introns have all been identified. Even the genes of mitochondria and chloroplasts (both organelles that share features with prokaryotes) have been observed to contain introns. It is therefore reasonable to believe that the phenomenon of interrupted genes is one of great antiquity, and some have suggested (Darnell, 1978; Doolittle, 1978) that primordial organisms contained genes of this type. If true, then it seems likely that while the eukaryotic line of descent maintained a pattern of split genes, the prokaryotes lost their introns and presumably the ability to remove them as well.

The presence of introns in genes requires signals for their removal. Although "hopping polymerases" or DNA recombinational processes were first considered as possible mechanisms for producing RNA products lacking introns, early work showed that introns are dealt with on the RNA level after transcription of the entire gene into RNA (Horowitz et al., 1978; Tilghman et al., 1978). The cutting and religation process has come to be known as RNA splicing. Signals for RNA splicing can theoretically be divided into two types. One type must provide information about the precise location of a splice and identify a particular sequence as a splice site. Such a signal is mandated by the need for removal of introns to be accurate to within a single nucleotide. A second type of signal must dictate proper pairing of splice sites (and, incidently, whether or not certain splice sites will be used). That this second sort of signal exists is clear from the presence within many genes of more than one intron. Some mechanism must prevent the splicing out of one or more exons due to incorrect pairing of splice junctions at the 5' and 3' ends of two different introns. Since some transcripts are known to give rise to a number of different spliced products, it is clear that this second type of signal can be modulated in a surprisingly sophisticated manner.

The best-studied class of genes containing introns is that made up of nuclear and viral genes which encode proteins. In vertebrates, genes of this class so frequently contain introns that those which do not—histones (Kedes, 1978), adenovirus polypeptide IX (Aleström et al., 1980), interferon (Nagata, Mantei, & Weissmann, 1980), and a few artificial constructs (Carlock & Jones, 1981; Gruss et al., 1981)— must be considered the exceptions. Since fruit flies, yeast, and slime molds appear to splice a far smaller fraction of their protein-coding genes than do vertebrates, it is clear that splicing is not a universal attribute of these genes. As is the case for tRNA, rRNA, and organellar genes, a mosaic structure is optional for mRNA genes. Nonetheless, there is accumulating evidence that introns may be essential for the expression of most genes which contain them.

SPLICE JUNCTION SIGNALS

Because the sequences of so many protein-coding nuclear genes have been elucidated (Sharp, 1981) (Fig. 10-1), it is possible to articulate a consensus sequence for both donor splice junctions (those that occur at the 5' end of introns) and acceptor splice junctions (those that occur at the 3' end of introns). It is worthwhile to stress at the onset that these regularities apply as well to the genes of lower eukaryotes as to those of vertebrates, which dominate the roster in Fig. 10-1. The percent and number occurrence of each base in each position of the two consensus sequences is given in Fig. 10-2. Thus, the consensus ^C_AAG/GT^A_GAGT represents donors quite well, and the consensus $(^T_C)_{11}$N^C_TAG/G fits acceptors. The 139 donor sites included have an average agreement of 7.3 bases with the nine-base consensus, and the 130 acceptors show an average agreement of 3.4 bases with the four-base consensus ^C_TAG/G. The -14 to -5 region of acceptors (which can extend even farther into the intron, but always includes the sequence as far as -14) has regularities of its own—namely, the dinucleotide AG has never been seen to occur here. The overall frequency of occurrence of the four bases in the -14 to -5 region is T (51%) $>$ C (30%) $>$ A (10%), G (9%), and comparable frequencies can be discerned at each position. No acceptor sequence contains more purines than pyrimidines in this region.

Splice junction sequences as designated above have been shown to be important in that mutations within them result in defective processing. In one case, a donor sequence AGG/GTGAGA was mutated to AGG/GTGAAT with the result that the donor was no longer utilized (Solnick, 1981). The gene in question (adenovirus 2 early region 1a) has alternate donors, each spliced to the same acceptor, allowing three possible spliced products. Thus, it is unclear whether mutant transcripts that should have been spliced at this location simply utilize one of the two alternative splices or fail altogether to be spliced. This study also underscores the fact that some bases in the donor consensus (in this case, the G at position $+5$) are more important than others (the T at position $+6$, which fits the consensus in the mutant but not in the wild-type gene).

It was noticed early on that the GT at positions $+1$, $+2$ of donors and the AG at positions -2, -1 of acceptors are invariant; this rule remains valid for all the natural splice junctions now known. A mutation occurring in one form of β^0 thalassemia, a human anemia in which β-globin is absent, has shown by restriction enzyme analysis to involve one of the three nucleotides GGT at positions -1, $+1$, or $+2$ of a donor splice site (Baird et al., 1981). In another mutant with a splicing defect, one form of β^+ thalassemia, a human anemia in which β-globin is expressed well below normal levels, the only difference in over 1,600 nucleotide pairs of DNA is a single nucleotide change creating the dinucleotide AG from the dinucleotide GG at positions -21, -20 of an acceptor splice site. Although AG has been observed at -16, -15 in wild-type genes (see Fig. 10-1),

Gene and Organism	Line #	Pair #	D #	A #	Donor Sequence	Acceptor Sequence	D Fit	Match	AG	AG~ Ref.
Soybean leghaemoglobin	1	1	1	1	ATTCGTAAGT	AAATAGGAT	6	1	>5	1
Soybean leghaemoglobin	2	2	2	2	ATTGGTAAGT	TTGTAGGTG	7	3	>5	1
Soybean leghaemoglobin	3	3	3	3	CGTGGTAAGT	TGTAGGTG	7	3	>4	1
D. discoideum M4	4	4	4	4	TTAAGTTGTAT	TTTATTTATTTCAAGATT	5	0	112	2
D. discoideum M4	5	5	5	5	TAAAGTATGTTT	TAAATGATATATCAGGCA	7	1	92	2
French bean phaseolin	6	6	6	6	TCATGTACTGCC	ATGTTTGTCCTGTAGGAA	5	1	94	3
French bean phaseolin	7	7	7	7	CAATGTAAGAAA	GCATGATTTTATAGAGC	7	0	73	3
French bean phaseolin	8	8	8	8	AGAGGTAAATAC	TGTTGGCGGATTTAGGGA	7	3	40	3
Yeast actin	9	9	9	9	TCTGGTATGTTC	ATATTATATGTTTAGAGG	7	1	58	4
Sea urchin actin	10		10		ACAGGTAAGAAC		8			5
D. melanogaster actin	11	10	11	10	AATGGTGCGTGG	TTCTTTCCATTGCAGCTT	7	2	>15	6
D. melanogaster actin	12	11	12	11	CCAGGTGCGTAG	TGTGTTATCCTGCAGGCT	8	5	>14	6
D. melanogaster actin	13	12	13	12	CAAGGTGAGTAA	CTGTCCTGTTCAGGTA	9	2	>12	6
D. m. heat shock 83K	14	13	14	13	GAAGGTAACTAT	GTAAATCCATTGCAGATG	8	1	53	7
D. m. alcohol DH	15	14	15	14	GGCGGTAAGTTG	TGTATTCAATCCTAGAAC	7	1	>14	8
D. m. alcohol DH	16	15	16	15	GCAGGTGAGTTA	TTATAACACCTTTAGAAA	9	1	>15	8
B. mori fibroin	17	16	17	16	TCAGGTAAGTTT	AACATTTTGTTTCAGTAT	9	3	47	9
Silkmoth chorion	18	17	18	17	TCAGGTAAGGTA	ATATCCAAAACACAGTCC	8	3	23	10
Silkmoth chorion	19	18	19	18	CCAAGTGAGTTT	GTATGTCTTTTATAGTCT	8	2	35	10
Silkmoth chorion	20	19	20	19	CCAGGTAAGTGA	GTTTTTTTTTCTCAGTCT	9	0	14	10
Silkmoth chorion	21			20		CGCTTTCCTTTAAGAAT			147	10
Chick type 1 α2 collagen	22			21		TGTATTTAACAGGGT			>11	11
Chick type 1 α2 collagen	23	20	21	22	AGATGTAAGTCA	TGAATTTTTAGGGT	7	1	>11	11
Chick type 1 α2 collagen	24		22		TCAAGTAAGTAA		8			11
Chick type 1 α2 collagen	25			23		TTCTTTTTCTAGGGT			>11	11
Chick type 1 α2 collagen	26	21	23	24	CACTGTAAGTAC	TTTCACCTTCAGGGT	7	1	>11	11

Figure 10-1.

Gene and Organism	Pair #	D#	A#	Donor sequence	Acceptor sequence	D Fit	Match	AG-AG	Ref
Chick type 1 α2 collagen	27		24	ACGGGTACGTGG		7			11
Chick type 1 α2 collagen	28	25		CAAGGTGAGACTT	GTACTTCAACAGGGC	6	3	>11	11
Chick type 1 α2 collagen	29	22	26	TCCTGTAAGTTG	TTCCCTTGTTCACAGGGT	7	1	33	11
Chick type 1 α2 collagen	30	23	27	CGCTGTAAGTCT	TCTTTCCTGTTCCAGGGT	6	1	43	11
Chick type 1 α2 collagen	31		27	AACTGTAAGTCT		7			11
Chicken lysozyme	32	24	28	TCAGGTGAGCTC	ATGGAACTTCGACAGGGG	8	2	>15	12
Chicken lysozyme	33	25	29	CGTGGTAGGAGA	GCTTCCTTTCTTAAGCCC	5	2	>15	12
Chicken lysozyme	34	26	30	CCAGGTGAGTAA	TCTCCCTCCGCCCAGGGT	9	5	>15	12
Chicken ovomucoid	35	27	31	TGAGGTGAGAAA	TTTTTCCCCCAGATG	7	6	>11	13
Chicken ovomucoid	36	28	32	GCATGTGTGTAC	AACTTTGTCGAGGTG	7	0	>11	13
Chicken ovomucoid	37	29	33	TCCTGTAAGTGA	TTCCCTCTTCAGAGA	7	0	>11	13
Chicken ovomucoid	38	30	34	AAGTGTTATTGT	CTCCTTCCACAGATG	5	0	>11	13
Chicken ovomucoid	39	31	35	GAGTGTGAGTAG	TTTTCCTTTCAGAGA	7	2	>11	13
Chicken ovomucoid	40	32	36	TGCTGTGAGTGT	GTGCTTTTGCAGGTT	6	2	>11	13
Chicken ovomucoid	41	33	37	TCGTGTACGTAC		6			13
Chicken ovomucoid	42	34	38	AAAGGTAGGCAA	CCTCGCTTTCAGGGA	7	1	>11	13
Chicken ovalbumin	43	35	39	TAAGGTGAGCCT	TGCTGTTTGCTCTAGACA	8	2	42	14,15
Chicken ovalbumin	44	36	40	TCAGGTACAGAA	TATTTCAATTACAGGTT	6	4	22	14,15
Chicken ovalbumin	45	37	41	GCCAGTAAGTTG	TCTCTTTGTATTCAGTGT	7	4	33	14,15
Chicken ovalbumin	46	38	42		ACTCTTGCTTTACAGGAA		1	21	14,15

Figure 10-1. A compilation of splice junction sequences. The first column (Gene and Organism) designates the source of the sequences on that line. The third column (Pair #) indicates the number of the donor-acceptor pair, if one is given on that line. The fourth column (D#) indicates the number of the donor sequence reported on that line. The fifth column (A#) indicates the number of the acceptor sequence given on that line or used in a match reported on that line. The eighth column (D Fit) lists the number of bases the donor sequence on that line has in agreement with the consensus sequence. The ninth column (Match) lists the number of bases in the longest sequence shared by the donor and acceptor which is in registry and crosses or abuts the junction; this is also 1 less than the theoretical number of alternative splice points that could lead to the same RNA product. The tenth column (AG-AG) lists the distance (in bases) from the AG at the splice point (-2, -1) to the nearest AG 5' of -4. The eleventh column gives the number of the relevant reference, as listed on the final page of the figure. Reprinted with permission from Mount (1982).

Line #	Pair #	D#	A#	Gene and Organism	Donor Sequence	Acceptor Sequence	D Fit	Match	AG- AG	Ref.
47	39	43	42	Chicken ovalbumin	AATGGTAAGGTA	TATTCATTCTTAAAGGAA	7	2	>38	14
48	40	44	43	Chicken ovalbumin	TGAGGTATATGG	TTTTGGTTGCTCCAGCAA	6	2	>39	14
49	41	45	44	Chicken ovalbumin	GCAGGTATGGCC	CTCATTTCCTTGCAGCTT	7	4	>39	14
50			45	Chick ov-like gene X		GCTTCCTTCTTACAGGAC			28	16
51	42	46	46	Chick ov-like gene X	AAAGGTAGGGGA	GCTTCTGTTTTGCAGCAA	7	2	43	16
52	43	47	47	Chick ov-like gene X	GCGGGTACGGCC	TCCCTCTCTTTTCAGATT	6	1	28	16
53	44	48	48	Chick preproinsulin	GAAGGTCCATTT	GCTTCCTACCTCTAGGCC	6	3	52	17
54	45	49	49	Chick preproinsulin	CTAGGTAAGTCA	CTTTTCCCTTGGCAGTGA	8	2	45	17
55	46	50	50	Rat preproinsulin II	GCAGGTATGTAC	CTGACTATCTTCCAGGTC	8	5	23	18
56	47	51	51	Rat preproinsulin II	CAAGGTAAGCTC	TGACCTCCCTGGCAGTGG	8	2	83	18
57	48	52	52	Rat preproinsulin I	GCAGGTATGTAC	CTGCCTATCTTCCAGGTC	8	5	23	18
58	49	53	53	Human preproinsulin	GCAGGTCTGTTC	CTGCCTGTCTCCCAGATC	7	3	17	19
59	50	54	54	Human preproinsulin	CAGGGTGAGCCA	GGCACGTCCTGGCAGTGA	7	1	75	19
60	51	55	55	Human ε globin	GCAGGTAAGCAT	GGTGTCATTTCATAGACT	8	2	47	20
61	52	56	56	Human ε globin	CAAGGTGAGTTC	TCTTTTGCCTAACAGCTC	9	2	59	20
62	53	57	57	Human γ globin	GAAGGTAGGCTC	TTGTCAATCTCACAGGCT	7	3	48	21
63	54	58	58	Human γ globin	CAAGGTGAGTCC	CTTTCATCTCAACAGCTC	9	2	43	21
64	55	59	59	Human δ globin	GCAGGTTGGTAT	TTTTCCTACCCTCAGATT	7	3	41	22
65	56	60	60	Human δ globin	CAGGGTGAGTCC	ACCTCTTCTCCGCAGCTC	8	1	53	22
66	57	61	61	Human β globin	GCAGGTTGGTAT	TTTTCCCACCCTTAGGCT	7	3	43	23
67	58	62	62	Human β globin	CAGGGTGAGTCT	ATCTTCCTCCCACAGCTC	8	1	45	23
68	59	63	63	Rabbit β globin	GCAGGTTGGTAT	TTTTCATTTTCTCAGGCT	7	4	40	24
69	60	64	64	Rabbit β globin	CAGGGTGAGTTT	TTCTTTTTCCTACAGCTC	8	1	54	24
70	61	65	65	Mouse β globin minor	GCAGGTTGGTAT	TGTTTCCCTTTTTAGGCT	7	3	28	25
71	62	66	66	Mouse β globin minor	CAGGGTGAGTCT	TCTGTCTTCCCACAGCTC	8	1	53	25
72	63	67	67	Mouse β globin major	GCAGGTTGGTAT	TGTTTCCCTTTTTAGGCT	7	3	28	16

Name					Seq A	Seq B				
Mouse β globin major	73	64	68	68	CAGGGTGAGTCT	TCCATATTCCCACAGCTC	8	1	31	26
Mouse α globin	74	65	69	69	AAAGGTGAGAAAC	CTCTCCTTCTCCCAGGAT	8	3	52	27
Mouse α globin	75	66	70	70	CAAGGTATGCGC	ACTTTGTCTCCGCAGCTC	7	2	22	27
Frog α globin	76	67	71	71	ACAGGTAAATTA	ATTGTTTCTCTACAGGAT	8	6	46	28
Frog α globin	77		72		CCCAGTAAGTCC		7			28
Mouse α-amylase	78	68	73	72	CATGGTATGTAA	TATGTATTTTTCAGAAA	7	1	42	29
Mouse α-amylase	79	69	74	72	AATGGTGAGATT		7	1		29
Mouse α-amylase	80	70	75	73	GCAGGTGTGCAG	CTCATTTATTTGTAGGTC	7	4	35	29
Mouse metallothionein-I	81	71	76	74	ACCGGTAAGACT	TCTTTCTCCTCCCAGGCG	7	2	69	30
Mouse metallothionein-I	82	72	77	75	AAGAGTGAGTTG	TCCTCCTTCTTCTAGGCT	7	1	24	30
AAV 2	83	73	78	76	ACAGGTACCAAA	ATGGTTATCTTCCAGATT	6	3	27	31
Bovine ACTH-β-LPH	84	74	79	77	GCTGGTACGTGG	TGCCGCTCTCCGCAGGCG	7	2	>17	32
Influenza NS gene	85	75	80	78	TCAGGTAGACTG	TGCCTTCTCTTCCAGGAC	6	4	39	33
Influenza segment 7	86	76	81	79	GCAGGTAGATAT	CTTGAAAATTTGCAGGCC	7	5	23	34
Influenza segment 7	87	77	82	79	AAACGTATGTTC		7	1		34
Human glycoprotein α unit	88	78	83	80	AAAGGTAATATG	CATGTCTGTCTGCAGGAG	7	3	>47	35
Human glycoprotein α unit	89	79	84	81	CAGGGTGCGTGA	TTTTTCCCTGATAGATT	7	1	43	35
Human glycoprotein α unit	90	80	85	82	CAGGGTAAGAAAC	TTCCTTCCCCTTTAGGTC	7	3	27	35
Rat prolactin	91	81	86	83	AAAGGTATGTGC	CCTGAATTTCTTTAGCAG	8	2	32	36
Rat prolactin	92	82	87	84	ATTTGTAAGTAC	CTGTATAATTTCTAGGAT	6	1	33	36
Rat prolactin	93	83	88	85	CCCTGTGAGTCC	AACATTGTGGATTAGCCG	7	0	104	36
Rat prolactin	94	84	89	86	CCAGGTGAGCAT	TTGTTTTCTTATTAGGCC	8	3	46	36
SV40, late (D1-A2)	95	85	90	87	GAAGGTACCTAA	TGTCTTTATTTCACGTC	7	4	21	37
SV40, late (D1-A1)	96	86	90	88		TTGTGTTTGTTTTAGAGC		2	27	37
SV40, late (D3-A2)	97	87	91	87	ACTGGTAAGTTT		8	3		37
SV40, late (D3-A3)	98	88	91	89		TGCCTTTACTTCTAGGCC		2	22	37
SV40, late (D2-A2)	99	89	92	87	TAAGGTTCGTAG		7	4		37
SV40, late (D3-A1)	100	90	91	88				1		37

(Continued)

Gene and Organism	Line #	Pair #	D#	A#	Donor Sequence	Acceptor Sequence	D Fit	Match	AG–AG	Ref.
SV40, early (D4-A4)	101	91	93	90	TGAGGTATTTGC	GTTTGTGTATTTTAGATT	6	2	46	37
SV40, early (D5-A4)	102	92	94	90	TAAGGTAAATAT		8	2		37
SV40, Mouse β-globin	103	93	93	68				2		38
Polyoma, late (VP1)	104	94	95	91	TCAAGTAAGTGA	TTCCTTTAATTCTAGGGC	8	2	33	39
Polyoma, early (409-795)	105	95	96	92	CCAGGTAAGAAG	CCTATATTCTTACAGGGC	8	4	34	40
Polyoma, early (746-795)	106	96	97	92	CCAAGTAAGTAT		8	1		40
Polyoma, early (746-809)	107	97	97	93	GAGGGTGAGGAG	GGGCTCTCCCCCTAGAAC		0	14	40
Ad2, E1a	108	98	98	94	TACAGTAAGTGA	TGATTTTTTAAAAGGTC	7	3	27	41
Ad2, E1a	109	99	99	94	ATTTGTAAGTCC		7	2		41
Ad12, E1a	110	100	100	95	TACAGTAAGTGT	ATTTTGTTGTTTTAGGTC	6	2	26	42
Ad12, E1a	111	101	101	95	TACAGTAAGTGT		7	2		42
Ad12, E1a	111	101	101	95	TGGGGTGAGTAC		7	2		42
Ad2, late leader	112	102	102	96	AACGGTAAGAGC	CCTTTTTTCCACAGCTC	7	1	48	43
Ad2, late leader	113	103	103	97	CAAGGTAGGCTG	ACAATTGTTGTGTAGGTA	7	4	41	43
Ad2, late leader-fiber	114	104	104	98		TTCATATTGTTGCAGATG	7	2	>52	43
Ad2, leader 3-γ	115	105	104	99	GGAGGTGAGCTC	TGTAATTTACAACAGTTT		2	>14	43
Ad2, leader γ-fiber	116	106	105	98			7	2		43
Ad2, leader 3-hexon	117	107	104	100	AAAGGTAATTCA	CCATGTCGCCGCCAGAGG		2	>14	43
A human V_H gene	118	108	106	101	CCAGGTAAGGAT	TCTTTTTGTTTGCAGGTC	8	3	19	44
A human V_κ gene (HK101)	119	109	107	102	CCAGGTAAGGAA	TGTTTCCAATCTCAGGTG	8	5	63	45
A human V_κ gene (HK102)	120	110	108	103	AAACGTGAGTAG	TATTTCCAATCTCAGGTG	8	5	33	45
Human J_κ – C_κ	121	111	109	104	AAACGTAAGTGC	TGCTTCTTTCCTCAGGAA	8	1	43	46
Human J_κ – C_κ	122	112	110	104	ATTGGTGAGAGG		8	1		46
Mouse Cγ1	123	113	111	105	ACAGGTAAGTCA	GCTTTCTCTCCACAGTGC	6	1	14	47
Mouse Cγ1	124	114	112	106	AAAGGTGAGAGC	CTTCTTCATCCTTAGTCC	9	2	37	47
Mouse Cγ1	125	115	113	107		TTTACCCACCCACAGGCA	8	3	54	47

Mouse $C\gamma1$	126	116	114	108	ATTGGTGAGGAA	TACTTTTCTTGTAGCCA	6	1	27	47
Mouse $C\gamma2a$	127	117	115	109	CCAGGTAAGTCA	GCCTTCTCTCTGCAGAGC	9	3	14	48
Mouse $C\gamma2a$	128	118	116	110	AAAGGTGAGAGC	ATCTCTCCTCATCAGCAC	8	3	25	48
Mouse $C\gamma2a$	129			111		TTTCTACCCTCACAGGGT			33	48
Mouse $C\gamma2a$	130			112		CCTCCTCTCTTGCAGCCA			24	48
Mouse $C\gamma2b$	131	119	117	113	CTTGGTGAGAGA	GCCTTCTCTCTGCAGAGC	6	1	14	49,50
Mouse $C\gamma2b$	132	120	118	114	CCAGGTAAGTCA	ATCTCTCCTCATCAGCTC	9	3	35	49,50
Mouse $C\gamma2b$	133	121	119	115	AAAGGTGGGACC	TTTCTAACCCACCAGGGC	7	3	33	49,50
Mouse $C\gamma2b$	134			116		TCTCCTCTCTTGCAGCCA			24	49,50
Mouse $C\mu$	135	122	120	117	GCCAGTGAGTGG	TTCTCTTGACTGCAGGTC	7	2	17	51,52
Mouse $C\mu$	136	123	121	118	CCAGGTAAGAAC	GACCTTTCATTCCAGCTG	8	8	25	51,52
Mouse $C\mu$	137	124	122	119	AATGGTAGGTAT	TGTCTTCATTTACAGAGG	7	1	41	51,52
Mouse $C\mu$	138	125	123	120	ACTGGTAAACCC	TGTGTCCCTTCATAGAGG	6	1	37	51,52
Mouse $C\mu$	139	126	124	121	CAAGGTAGTATG	CTCTGTCACCTGCAGGTG	6	4	28	51,52
Mouse $C\mu$	140			122		GGTCCCTCAGAGA			>7	53
A mouse $V\kappa$ gene	141	127	125	123	ACAGGTAATGAA	ATTTTTCAATTGTAGGTG	7	4	87	54
A mouse $V\kappa$ gene (L6)	142	128	126	124	CCAGGTAAAATG	TGTCTCCATTCCTAGGTA	7	5	25	55
A mouse $V\kappa$ gene (L7)	143	129	127	125	CCAGGTATGACT	CTTCTATTTTTCCAGCCT	7	7	36	55
Mouse $V\lambda II$	144	130	128	126	TCAGGTCAGGAG	CTTACCTGTTTGCAGGAG	7	4	69	56
Mouse $V_{\kappa-41}$	145	131	129	127	CAAGGTTAAAAT	TGTCTCCACTCCTAGGTA	6	4	76	57
Mouse J5 – $C\kappa$	146	132	130	128	AAAACGTAAGTAG	TGCTTGCTTCCTCAGGGG	8	1	45	57
Mouse J4 – $C\kappa$	147	133	131	128	AAAACGTAAGTAG		8	1		57
Mouse J2 – $C\kappa$	148	134	132	128	AAAACGTAAGTAG		8	1		57
Mouse J1 – $C\kappa$	149	135	133	128	AAAACGTAAGTAG		8	1		57
Mouse J – $C\lambda$	150	136	134	129	CTAGGTGAGTCA	TACTTCATCCTGCAGGCC	8	2	>29	58
Mouse JH1 – $C\gamma1$	151	137	135	108	TCAGGTAAGCTG		8	2		59
Mouse JH1 – $C\gamma2a$	152	138	135	112				3		59
Mouse JH1 – $C\gamma2b$	153	139	135	116				3		59
Mouse JH1 – $C\mu$	154	140	135	122				7		59

(Continued)

Gene and Organism	Line #	Pair #	D#	A#	Donor Sequence	Acceptor Sequence	D Fit	Match AG	AG-AG	Ref.
Mouse JH2 - Cγ1	155	141	136	108	TCAGGTGAGTCC		9	2		59
Mouse JH2 - Cγ21	156	142	136	112				3		59
Mouse JH2 - Cγ2b	157	143	136	116				3		59
Mouse JH2 - Cμ	158	144	136	122				7		59
Mouse JH3 - Cγ1	159	145	137	108	GCAGGTGAGTCC		9	2		59
Mouse JH3 - Cγ2a	160	146	137	112				5		59
Mouse JH3 - Cγ2b	161	147	137	116				5		59
Mouse JH3 - Cμ	162	148	137	122				3		59
Mouse JH4 - Cγ1	163	149	138	108	TCAGGTAAGAAT		8	2		59
Mouse JH4 - Cγ2a	164	150	138	112				3		59
Mouse JH4 - Cγ2b	165	151	138	116				3		59
Mouse JH4 - Cμ	166	152	138	122				7		59
Mouse VH (M141)	167	153	139	130	AGCTGTAAGTGT	GTCACTTGTCACTAGGTA	6	3	37	59

(1) Jensen et al., 1981.
(2) Kimmel & Firtel, 1980.
(3) Sun et al., 1981.
(4) Gallwitz & Surex, 1980.
(5) Durica et al., 1980.
(6) Fryberg et al., 1981.
(7) Holmgren et al., 1981.
(8) Benyajati et al., 1981.
(9) Tsujimoto & Suzuki, 1979.
(10) Jones & Kafatos, 1980.
(11) Yamada et al., 1980.
(12) Jung et al., 1980.
(13) Stein et al., 1980.
(14) Breathnach et al., 1978.
(15) Benoist et al., 1980.
(16) Hellig et al., 1980.
(17) Perler et al., 1980.
(18) Lomedico et al., 1979.
(19) Bell et al., 1980.
(20) Baralle et al., 1980.
(21) Slighton et al., 1980.
(22) Spritz et al., 1980.
(23) Lawn et al., 1980.
(24) van Ooyen et al., 1979.
(25) Konkel et al., 1979.
(26) Konkel et al., 1978.
(27) Nishioka & Leder, 1979.
(28) Partington & Baralle, 1981.
(29) Young et al., 1981.
(30) Glanville et al., 1981.
(31) Green & Roeder, 1980.
(32) Nakanashi et al., 1980.
(33) Baez et al., 1980.
(34) Lamb et al., 1981.
(35) Fiddes & Goodman, 1981.
(36) Gubbins et al., 1980.
(37) Perricaudet et al., 1979.
(38) Chu & Sharp, 1981.
(39) Srivatsan et al., 1981.
(40) Treisman et al., 1981.
(41) Perricaudet et al., 1979.
(42) Perricaudet et al., 1980.
(43) Flint & Broker, 1980.
(44) Mattyssens & Rabbitts, 1980.
(45) Bentley & Rabbitts, 1980.
(46) Hieter et al., 1980.
(47) Honjo et al., 1979.
(48) Yamawaki-Kataoko et al., 1981.
(49) Yamawaki-Kataoka et al., 1980.
(50) Tucker et al., 1979.
(51) Kawakami et al., 1980.
(52) Early et al., 1980.
(53) Calame et al., 1980.
(54) Pech et al., 1981.
(55) Nishioka & Leder, 1980.
(56) Tonegawa et al., 1978.
(57) Max et al., 1979.
(58) Bernard et al., 1978.
(59) Sakano et al., 1980.

A. Donor Sequences

position	-4	-3	-2	-1	+1	+2	+3	+4	+5	+6	+7	+8
total	139	139	139	139	139	139	139	139	139	139	136	136
A	42	56	89	12	0	0	86	94	12	23	53	33
T	28	10	18	17	0	139	9	16	7	87	30	36
C	42	60	16	8	0	0	3	13	3	17	28	40
G	27	13	16	102	139	0	41	16	117	12	25	27
%A	30	40	64	9	0	0	62	68	9	17	39	24
%T	20	7	13	12	0	100	6	12	5	63	22	26
%C	30	43	12	6	0	0	2	9	3	12	21	29
%G	19	9	12	73	100	0	29	12	84	9	18	20

B. Acceptor Sequences

	-15	-14	-13	-12	-11	-10	-9	-8	-7	-6	-5	-4	-3	-2	-1	+1	+2	+3
total	113	113	114	126	126	126	127	127	127	129	130	130	130	130	130	130	130	130
A	17	11	11	19	8	19	14	24	15	4	13	33	5	130	0	29	22	25
T	58	50	57	67	75	62	62	57	57	73	75	38	40	0	0	11	48	37
C	21	28	35	27	30	38	42	35	46	46	36	28	84	0	0	23	28	42
G	17	24	11	13	13	7	9	11	9	6	6	31	1	0	130	67	32	26
%A	15	10	10	15	6	15	11	19	12	3	10	25	4	100	0	22	17	19
%T	51	44	50	53	60	49	49	45	45	57	58	29	31	0	0	8	37	28
%C	19	25	31	21	24	30	33	28	36	36	28	22	65	0	0	18	22	32
%G	15	21	10	10	10	6	7	9	7	4	5	24	1	0	100	52	25	20

Donor Consensus: $^{C}_{A} AG / GT ^{A}_{G} AGT$

Acceptor Consensus: $\left(^{T}_{C} \right)_{11} N \, ^{C}_{T} AG / G$

Figure 10-2. Tabulation of nucleotide frequencies by position. *A.* Donor sequences. The 139 donor sequences listed and numbered in Fig. 10-1 were used. Each value represents the number or percent occurrence of the nucleotide shown at left in the position numbered at the top. *B.* Acceptor sequences. The 130 acceptor sequences listed and numbered in Fig. 10-1 were used. Reprinted with permission from Mount (1982).

this proximity is rare; for the 97 acceptors, the average location of the nearest AG upsteam of −4 is −43, −42. Thus, the β^+ thalassemia mutation can be seen as an unacceptable change in the pyrimidine-rich stretch which is an essential component of all acceptor splice sites. In fact, the AG dinucleotide creates an alternate acceptor, and when the β°-thalassemia gene is expressed in an SV40 vector in HeLa cells a splice from the normal donor site to the alternate acceptor is observed (together with a much lower level of the normal splice: Spritz et al., 1981; Westaway & Williamson, 1981). The behavior of this mutant suggests that the universal absence of AG in the pyrimidine-rich segment of functional acceptors can be traced to potential problems with ambiguity.

ARE JUNCTION SIGNALS SUFFICIENT?

The question of whether junction sequences alone are sufficient for splicing does not have a simple answer. Several lines of evidence concur with the mutants mentioned above in supporting the sufficiency of these rather short signal regions. Deletions of exonic material to within 18 bases of the donor splice site for the mouse β-major globin gene's large intron does not interfere with the excision of that intron in an SV40 recombinant (Hamer & Leder, 1979). A deletion of intron sequence to within four bases of an SV40 late RNA donor splice site likewise has no effect (Piatak et al., 1981). Nor does the replacement of all exonic sequences 3′ to an acceptor splice site in rabbit β-globin or all but 25 bases of the large intron of the rabbit β-globin gene (A. Buchman & P. Berg, personal communication). Thus, additional signals embedded in large regions surrounding splice junctions cannot be required for proper splicing.

The fact that a chimeric gene containing a chimeric intron has the capacity to be properly spliced further excludes the possibility that sequences surrounding splice junctions provide information necessary for the proper pairing of specific donor and acceptor sites. This was demonstrated by Chu and Sharp (1981) who constructed a gene containing an SV40 T antigen exon 5′ and a mouse β-globin exon 3′ to a recombinant intron. Strikingly, the SV40 donor sequence shared only four bases (of the nine-base consensus region) with the β-globin donor it replaced. Yet it is spliced normally to the acceptor sequence present in the construct. Even more striking is the fact that the acceptors have identical bases in only 3 out of the 16 positions shown in the acceptor consensus of Fig. 10-1. Splicing thus appears to be an indiscriminate process by which any two properly positioned junctions can be joined.

The promiscuity of splice junctions is further substantiated by the splicing observed in adenovirus-SV40 hybrids. The adenovirus type 2-simian virus 40 hybrid virus Ad2$^+$ND$_4$ has been shown to synthesize an RNA which is spliced from an adenovirus donor splice site to an SV40 acceptor splice site (Westphal et al., 1979; Khoury et al., 1980). Similarly, Thummel, Tjian, and Grodzicker (1981) found that SV40 T antigen sequences were readily expressed in adenovirus constructs by virtue of their ability to participate in novel splicing events. Adenovirus promoters gave rise to transcripts that included copies of the inserted DNA, and anomalous splicing events allowed the synthesis of SV40 T antigen to be directed by mRNAs having adenovirus-encoded 5' and 3' termini. This interchangeability of splice sites, taken together with the lack of a requirement for signals in sequences surrounding splice sites, again suggests that the junctions themselves are a sufficient signal for splicing.

An apparent problem with the interchangeability of splice sites can be traced to the frequent repetition of a short sequence at the two intron termini. (Such a repeat also obscures the precise location of splicing.) The fact that nearly all introns have some repetition (see Fig. 10-1, especially the column labeled ("Match") suggests that this repetition may be a necessary attribute of a functional donor-acceptor pair, or might be an adaptation to an imprecise splicing mechanism. The latter hypothesis was carried to an extreme by Spritz et al. (1980), who pointed out that there are eight possible splice points (or splice frames) that would leave the human δ-globin amino acid sequence unchanged following excision of the small intron. However, the case of the ovomucoid gene's intron C brings this kind of thinking into question (Stein et al., 1980). Here, there is not only no ambiguity with respect to the splice point, but if the splice frame is moved one nucleotide in either direction amino acid changes result (ile → arg or ile → met). Therefore, the ovomucoid story leaves unaltered the impression that repetitious sequences at splice sites occur with a greater than random frequency, strongly suggesting that the sequence of a splice junction is responsive to the sequence of its proper mate and, vitiating the notion just elaborated, that splice junctions are fully interchangeable.

The resolution of this paradox lies in the similarity of the donor and acceptor consensus sequences. An occurrence of G in position −1 of both the donor and the acceptor is frequent because G is invariant in this position of acceptors and common in this position of donors (see Fig. 10-2). Similar logic applies to positions −3, −2,

Repeat length	0	1	2	3	4	5	6	7	8	9
Observed number	8	39	41	34	14	10	2	4	1	0
Expected number	20	33	43	32	17	6	2	1	0	0
Observed frequency	.05	.26	.27	.22	.09	.07	.01	.03	.01	.00
Expected frequency	.13	.21	.28	.21	.11	.04	.01	.00	.00	.00

Figure 10-3. Comparison of observed and expected donor-acceptor agreement. The observed numbers come from the 153 entries in the ninth column of Fig. 10-1; the observed frequencies were obtained by division by 153. The expected frequencies were calculated as described in Mount (1982); the expected numbers were calculated by multiplication by 153. See text for explanation. Reprinted with permission from Mount (1982).

+1, and +2. When the sequence overlap between the two consensus sequences was taken into account in the calculation of expected repeat lengths (Mount, 1982) the results shown in Fig. 10-3 were obtained. (The repeat length is the length of the longest sequence shared by a pair of donor and acceptor sites which crosses or abuts the junction, thereby creating ambiguity in the precise location of splicing.) What these results say is that if all known donor and acceptor sequences were randomly reassorted, the incidence of short repeats would not significantly change. In other words, the existence of these sequences does not bear on the interchangeability of splice sites and, as stressed above, interchangeability implies sufficiency.

There is, on the other hand, extensive data arguing against the sufficiency of junction sequences for specifying splicing. Most obvious is the presence of junction sequence analogs within exons. For example, mature mouse β-globin mRNA (Konkel, Tilghman, & Leder, 1978) contains the sequence AAG/*G*UGAAC (7 out of 9 bases in agreement with the donor consensus) 65 nucleotides upstream of the sequence CUGCUGGUUGUCUACCCUUGGACCC*AG*/C (a 23-base-long pyrimidine-rich stretch followed by CAG, displaying all known properties of acceptor splice sites). The lack of splicing between these two loci indicates that signals other than a simple match to the two consensus sequences are required for splicing. What those signals are remains the most elusive and perplexing question relating to the splicing of mRNAs.

THE VERSATILITY OF SPLICING

The complexity of rules governing splicing is best illustrated by a review of splicing patterns in transcription units where alternate splices are possible. Figure 10-4 shows a schematic representation of the splicing patterns observed in a number of characterized genes.

A illustrates the normal case. The number of donors equals the number of acceptors, donors and acceptors alternate along the sequence, and all the splice sites are utilized in the splicing of every transcript.

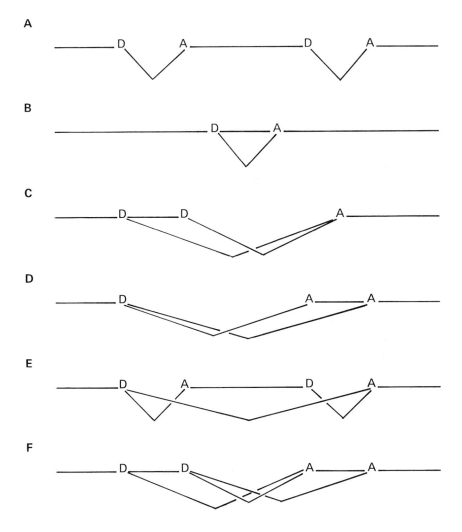

Figure 10-4. Schematic depiction of splicing patterns. *D* indicates a "donor" splice site; *A* indicates an acceptor splice site. See the text for a discussion of panels **A** through **F**.

B illustrates the case where a particular splice is optional. In some examples of this class, the optional splice is the only splice in the RNA. Examples are the 19S RNAs synthesized late in SV40 infection (Elder, Spritz, & Weissman, 1981; Ghosh et al., 1981) and the influenza virus segment 8 RNAs (Baez et al., 1980; Lamb & Lai, 1980). The optional splice of B can also be one of two or more splices, a situation occurring in adenovirus early region III transcripts (Chow, Broker, & Lewis, 1979; Berk & Sharp, 1978; Ziff, 1980).

C describes alternate donors. Examples of this type include the SV40 early region (Ziff, 1980) and adenovirus early region la genes (Chow, Broker, & Lewis, 1979; Berk & Sharp, 1978; Ziff, 1980). In the case of the SV40 early region splice, the ratio of the two products seems to be sensitive to a number of variables, including temperature (Alwine & Khoury, 1980) and the presence of deletions within the region between the two donors (Khoury et al., 1979).

D shows the case of alternate acceptors. Examples of this pattern abound in adenovirus late gene expression (Chow, Broker, & Lewis, 1979), and the distinction between the 16S form and some 19S forms of late SV40 mRNA is of this type (Piatak et al., 1981).

E illustrates the phenomenon of optional exons. The best-known example of this is the three small leader segments (known as x, y, and z) which are sometimes a part of adenovirus protein IV mRNA (Chow & Broker, 1978; Dunn et al., 1979). Another classic example is polyoma virus late mRNAs which are derived from giant transcripts generated by multiple circuits of the circular genome and which contain multiple copies of a 5′ leader sequence spliced to a single mRNA body (Acheson, 1978; Acheson et al., 1971; Birg, Favalora, & Kamen, 1977).

F illustrates one example (from the early region of polyoma virus) of the kind of complex pattern that can result from a combination of the other schemes. Here, three of the four conceivable splices have been observed, but the fourth (involving the outermost donor and acceptor) has not (Treisman et al., 1981). The splicing of SV40 late mRNA in SV40 mutants with deletions in the late leader region is extremely complex and includes an example in which deletion of sequences upstream of a donor prevent that site from being used in one of two alternate splices, but not the other (Elder, Spritz, & Weissman, 1981; Ghosh et al., n.d.).

One question that is raised by the existence of alternative splices is whether those junctions which have a better fit to the consensus will be used more often. This is certainly not a hard and fast rule. One example is the two closely spaced alternate donors at the 5' end of the chicken ovomucoid intron F (Stein et al., 1980). The two sites are GCT/GTGAGT (donor number 36 in Fig. 10-1; this is used 25% of the time) and AGT/GTGAGT (donor number 27, used 75% of the time). Since these two sequences have essentially the same match to the consensus it seems unlikely that the extent of match controls the ratio with which they are used. A clearer example is the production of the large T and small T polypeptides early in SV40 infection. Here, two donor sites (numbered 93 and 94 in Fig. 10-1) compete for pairing to a single acceptor site. The one with the better match to the consensus (donor number 94, used in the production of small T) gives rise to the less-abundant mRNA. Furthermore, the ratio of the two product mRNAs is extremely sensitive to the temperature at which the cells are grown and also to the presence of deletions well removed from the splice points (Alwine & Khoury, 1980; Khoury et al., 1979).

All the examples illustrated in Fig. 10-3 describe alternatives available to a single precursor RNA molecule, and it is generally observed that all the options ever used are utilized to some extent all the time. There is no well-documented example of a polyadenylated transcript being processed all one way under one set of physiological circumstances and all another way under another set of physiological circumstances, implying that splicing is not normally used as a means of regulating gene expression.

There are, however, examples of overlapping primary transcripts that are consistently spliced differently despite the fact that the transcripts differ only in regions some distance from the splice sites involved. For example, the splice that occurs between J5 and the κ constant region gene (Max, Seidman, & Leder, 1979) (see Fig. 10-5) is in some ways like the case of alternate splicing covered under C in Fig. 10-4. If DNA recombination has hooked a κ variable region to J5, then the donor sites at the 3' ends of J1, J2, and J4 are not used. Because the sequence around the J1 donor site is the same whether J2 or J5 is used as the DNA joining region, it is clear that sequences 5' to J1 (that is, over 30 bases upstream from this donor) must control its utilization. A more extreme case is the Amy-1^A gene in mouse. Transcription of this gene in liver cells begins at nucleotide 2,893 (in the nomenclature of Young, Hagenbüchle, & Schibler, 1981) and splicing occurs between a donor at 3,054 and an acceptor at 7,562. In the salivary gland, transcription begins at nucleotide 1 and splicing occurs between nucleotide 50 and nucleotide

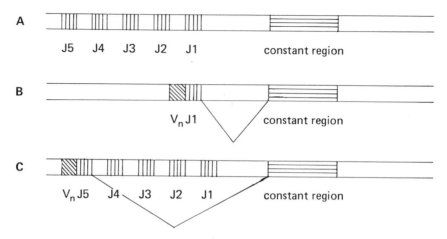

Figure 10-5. Schematic depiction of immunoglobulin κ gene rearrangements and splicing patterns. (These figures are not drawn to scale.) A. The unrearranged κ gene. Solid segments denote coding regions. Variable region coding blocks are not shown, but lie to the left. The unrearranged gene is not expressed. B. DNA rearrangement involving J1. In this case RNA splicing occurs between the J1 donor and the constant region acceptor. C. DNA rearrangement involving J5. In this case RNA splicing occurs between the J5 donor and the constant region acceptor, but not between the J1 donor and the constant region acceptor.

7,562; splices involving the donor at 3,054 are not seen. Thus, RNA sequences farther away than 150 nucleotides must be responsible for the failure of the donor at 3,054 to be used in the salivary gland.

Similar phenomena have been observed in cases where different transcript lengths provide new acceptors for splicing. The major late promoter of adenovirus gives rise to a series of long transcripts ending at one of five possible polyadenylation sites. mRNAs with a common polyadenylation site are referred to as members of a 3′ coterminal family. Strikingly, the acceptors used in one coterminal family are never used by RNAs of another coterminal family. Instead, exons of the leftward coterminal families are removed by splicing (reviewed in Ziff, 1980). A similar phenomenon is seen in the case of the C_μ immunoglobulin gene. This gene encodes both a secreted form and a membrane-bound form of the protein. The mRNA encoding the membrane-bound form is polyadenylated at a site far downstream from that where the secreted-form mRNA is polyadenylated. Here too, the acceptor closest to the polyadenylation site is used exclusively, and exons for the C-terminal region of the secreted form of the protein are excised as intronic material during splicing (Nakanishi et al., 1980).

In all these cases (J-C splicing, Amy-1^A, the adenovirus major late transcript unit, and the C_μ gene), the splice site closest to the variable end of the RNA transcript is used. Thus, what is observed is not truly differential splicing (in which a particular transcript is spliced differently at different times), but rather differential V-J joining, differential initiation of transcription, or differential poly-adenylation. How the splicing machinery responds faithfully to these changes with the appropriate splice pattern remains unknown.

Our inability to formulate rules governing the pairing of splice junctions (including the failure of potential splice junctions to be used at all) could be due to our ignorance of what intermediates are formed during the splicing process. It is quite possible that there are splicing events that remove only a portion of an intron, creating an intermediate that then goes on to be further spliced to form the mature RNA. The first report of the two-step excision of an intron was made by Kinniburgh and Ross, who observed an "intermediate" in the removal of the large intron from mouse β-globin major mRNA precursors (Kinniburgh & Ross, 1979). Similar intermediate-sized molecules have been reported for a particular intron in the chick α-2 collagen gene (Avvedimento et al., 1980). Interestingly, one of the proposed intermediate splices would regenerate an acceptor sequence that fits the consensus. However, it is not yet clear how general multistep splicing is or whether the intermediates observed are obligatory. Furthermore, proving that these forms are indeed intermediates, rather than aberrantly processed RNAs which turn over in the nucleus, threatens to be very difficult.

WHAT RECOGNIZES SPLICING SIGNALS?

Our current picture of mRNA splicing suffers further from the draw-back that, although a fair amount is known about the signals that govern splicing, virtually nothing is known about the biochemical machinery involved. Perhaps the best guess is a proposal that RNP complexes containing the small nuclear RNA U1 participate (Rogers & Wall, 1980; Lerner et al., 1980). This RNA has at its 5' end the sequence ^{m_3}GpppAmUmAC$\Psi\Psi$ACCUG, which is complementary to both the donor consensus and the region of the acceptor consensus immediately at the splice site (^C_TAG$|$G). It could therefore utilize

Watson-Crick base pairing to align donor and acceptor junctions for precise splicing as illustrated in Fig. 10-6. This hypothesis is supported by the experimental observation that when antibodies directed against the U1 containing RNP are included in a nuclear splicing system, splicing is greatly reduced (Yang et al., 1981). Note that U1 RNA exhibits exact complementarity to the consensus sequences (and is therefore partially complementary to any given splice junction). This means that U1 is likely to be involved in the recognition of junctions, but not in their proper pairing. Note also that the complementarity of U1 involves nucleotides that lie primarily within the intron. Sequences on the exonic sides of splice junctions are less constant, probably as a consequence of their involvement in protein-coding functions.

There has been much speculation in the literature that in addition to the involvement of U1 RNA, various other small RNAs might hybridize across specific splice junctions, thereby acting not in the recognition of junctions as junctions but in the correct pairing of junctions (Murray & Holliday, 1976; Harada, Kato, & Nishimura, 1980; Matthews, 1980). Such RNAs would be exon-bridging RNAs, since sequences that are specific to a particular donor-acceptor pair are typically exonic. In this regard, M. Matthews has found that the adenovirus small RNA VA$_I$ can hybridize to mRNA made late in infection, and to cDNA made from that RNA (Birg, Favalora, & Kamen, 1977). What is striking is that similar hybridization cannot be found for the genomic DNA specifying this mRNA. The most likely explanation for this phenomenon is homology between VA

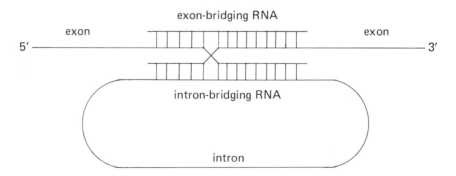

Figure 10-6. Models for the involvement of small RNAs in splicing. The precursor RNA is drawn so that sequences present in the mature RNA (exons) are on a single horizontal line; intervening sequences are drawn as a loop below. For simplicity, only a segment of the small RNA is shown. Short vertical lines indicate base pairs.

and exonic regions which span the splice junction. How this might work to facilitate proper splicing is shown in Fig. 10-6. It should be emphasized that exon-bridging models should be evaluated independently of intron-bridging models (such as that for U1). It is possible that both operate at the same time, that both operate but at different times, that only one operates, or that neither operate. It is also possible that small RNAs with exon-bridging capacity function in the transport of properly spliced mRNAs. The existence of their recognition sequence in spliced, but not in unspliced, RNAs makes this an attractive possibility. In conclusion, very little is known about the mechanism of mRNA splicing.

THE BIOCHEMISTRY OF SPLICING

There are three systems where specific information about the biochemistry of splicing is available. For various reasons, none of these three bear on the enigmatic question of how correct pairings are made between acceptor and donor junctions during mRNA processing.

The only system where the enzymes for splicing have been purified to any extent is yeast tRNA splicing. John Abelson and his colleagues have succeeded in separating a cutting and a ligating activity (Knapp et al., 1979; Peebles et al., 1979). The cleavages at the splice sites leave a 3' phosphate and 5' hydroxyl group (unusual for RNA processing enzymes). RNA has not been found as a major component of either enzymatic activity. All tRNA introns discovered so far occur in the same position in the common cloverleaf structure—between the two nucleotides that follow the anticodon. A compilation of exonic sequences about this site allows a short consensus to be discerned (intron sequences are indicated by "...": and the position of the anticodon is indicated by italic type):

X. laevis tRNA^{tyr}	U*GUA* . . . A*UCCU*	(Muller & Clarkson, 1980; Laski et al., 1982)
yeast tRNA^{tyr}	U*GUAGG* . . . AΨCUU	(Goodman, Olson, & Hall, 1977)
tRNA^{phe}	U*GAAY* . . . AΨCUG	(Valenzuela et al., 1978)
tRNA^{ser}	U*CGAA* . . . AΨCUC	(Ezcheverry, Colby, & Guthrie, 1979)

tRNA^{trp}	UCCAA . . . AUCGA	(Ogden et al., 1979)
Dm tRNA^{leu}	UCAAG . . . UUCUG	(Robinson & Davidson, 1981)
yeast tRNA^{leu_3}	UCAAG . . . CΨCAG	(Robinson & Davidson, 1981)
consensus	UCAAA . . . AUCUG	

Robinson and Davidson (1981) have noted an additional conserved sequence, AAAAUCUU, which appears in a particular position within the proposed secondary structures of tRNA introns. The similarity between this intronic sequence and the exonic consensus sequence (which is continuous in the completed tRNAs) may have some mechanistic significance. What is certainly meaningful is that all tRNA introns are confined to a specific location within the conserved tRNA secondary structure. This strongly suggests that the three-dimensional structure of the tRNA precursor (probably working in harmony with the conserved sequences described above) is primarily responsible for the recognition events required for removal of the intron. If so, the recognition system in mRNA splicing, which is capable even of joining splice junctions that have been paired arbitrarily by human investigators (Chu & Sharp, 1981), probably cannot be described by analogy to the tRNA system.

Another splicing event that has begun to be dissected *in vitro* is rRNA splicing by *Tetrahymena thermophila* (Zaug & Cech, 1980; Grabowski, Zaug, & Cech, 1981; Chech, Zaug, & Grabowski, 1981). In this case the intron is removed as a linear molecule which is subsequently converted to a circle of RNA also identifiable *in vivo*. Amazingly, the *in vitro* splicing reaction appears not to require any protein (Kruger et al., 1982). In the cell, circularized introns appear relatively stable, unlike the RNA removed from mRNA precursors. What function they might serve is unclear. The junction signals involved in rRNA splicing clearly do not fit those designated in Fig. 10-2 (Nomiyama, Sakakai, & Takagi, 1981).

A third system in which we have acquired some biochemical understanding of the splicing process is yeast mitochondria. The most intensively studied gene is called *cob* or *box*; it encodes the cytochrome *b* protein and in addition controls the synthesis of cytochrome oxidase. The *cob-box* gene possesses 6 exons and 5 introns (Slonimski et al., 1978; Grivell et al., 1979; Haid et al., 1979; Alexander et al., 1980; Nobreka & Tzagaloff, 1980; Jacq et al., 1981). Extensive genetic analysis has revealed that intron mutations not only arrest RNA splicing but lead to the accumulation of novel polypeptide chains (Church, Slonimski, & Gilbert, 1979; Haid et al.,

1980; Halbreich et al., 1980; Van Ommen et al., 1980; Claisse et al., 1978; Kreike et al., 1979; Solioz & Schatz, 1979). Abundant evidence supports the notion that these intron-encoded polypeptides play a role in the splicing both of the introns that specify them (in the process destroying their own mRNAs) and of intron(s) in other genes (Lazowski, Jacq, & Slonimski, 1980). However, the existence of nuclear mutations affecting mitochondrial splicing shows that the entire machinery is not specified by the mitochondrial genome (Jacq, 1981). Again, sequences at the splice junctions do not conform to the consensus sequences occurring in nuclear mRNA precursors.

Whether biochemical knowledge acquired in any of the three systems discussed above will be relevant to the mechanism of mRNA splicing in eukaryotic cell nucleic remains an open question. In each case, important differences can be identified which argue strongly that the identical machinery cannot be used. In tRNAs, the splice junction consensus sequences are not the same as those present in messenger precursors. Ribosomal RNA splicing likewise appears to utilize quite different junctions signals. Mitochondrial gene expression can be distinguished from that of nuclear genes by the coupling of transcription and translation. Potential protein-coding regions in the introns of nuclear mRNA precursors cannot possibly be functional since they are excised before reaching the ribosomes in the cytoplasm. These points suggest that at least four distinctly different sets of splicing machinery arose during the evolution of modern-day eukaryotes. Clearly, elements that recognize splice junction signals in these four systems must differ. On the other hand, some splicing functions could well be shared; for instance, the use of a common ligating enzyme is a possibility.

WHY SPLICING?

One question raised by the existence of introns is what advantage these interruptions in genes might confer on the cell. This question is closely related to the general problem of why higher eukaryotes have much more DNA than is required to encode their proteins; it may be that excess DNA is in fact not disadvantageous to these organisms. But what about the energy required to transcribe introns and then remove them by splicing? Surely there exists some compensating advantage.

Or does there? One suggestion is that introns exist because there is simply no mechanism for their removal from the genome.

This idea is not consistent with the case of the insulin gene. In chickens and humans the insulin gene has two introns, hinting that this situation prevailed in the primordial gene (Perler et al., 1980; Bell et al., 1980). In rats there are two insulin genes, both of which are functional and one of which has lost one of the two introns (Lomedico et al., 1979). Insulin, therefore, seems to be a gene from which an intervening sequence has been lost. If introns can be lost, it follows from their existence that they must be of some benefit. Gilbert (1978) has suggested an evolutionary advantage to being able to create new and useful proteins by recombination within introns. In other words, nature could produce new chimeric proteins in much the same manner as Chu and Sharp (1981) designed their experiment. This can be accomplished best if exons correspond to functional units or protein domains. Such a motif is most vividly illustrated in the case of the immunoglobulins, where exons do correspond strictly to recognizable domains in the three-dimensional protein structure (Sakano et al., 1981).

Another idea is that cells often use introns to expand coding potential much as viruses do, but that they do so more subtly. There are now two published cases of cellular genes—ovomucoid (Stein et al., 1980) and human growth hormone (Chapman et al., 1981; DeNotto et al., 1981)—where polymorphism in a protein (or cDNA) sequence has been explained by an alternate splicing scheme. Because few proteins have been sequenced with sufficient care to detect minor heterogeneities, the frequency of this phenomenon is hard to estimate. Two facts are relevant. One is that splicing makes it relatively easy to produce from a single gene two proteins which differ in only minor ways (such as by a small insertion, deletion, or substitution). The other fact is that protein polymorphism often confers a selective advantage. The latter is clear from a large body of data collected by population biologists in which heterozygote advantage has been used to explain allelism. What remains to be established is how frequently polymorphic proteins are produced by a single allele.

There is also a suggestion that introns and their removal might play some as yet undefined but crucial role in the process of messenger RNA maturation. The first evidence came from an SV40 deletion mutant precisely lacking the 16S late mRNA intron sequence (Gruss et al., 1979). In this mutant, the primary transcript of the late gene region should exactly correspond to the 16S mRNA coding for the protein VP1. However, no VP1 is made despite the fact that cells infected with the mutant virus express the early viral genes normally and show evidence of normal transcription of the late genes. Similar results were obtained by Hamer and Leder (1979), who made a series of constructs containing a portion of the mouse β major globin gene inserted into the SV40 late transcription unit (i.e., between the SV40 late promoter and the SV40 late polyadenylation site). Four possibilities were explored by independently varying the polarity of the β-globin insert and the presence or absence of a fragment of SV40 DNA containing an intron. The result was that only those constructs with at least one functional intron gave rise to stable mRNA.

But what about genes without introns? Histones (Horowitz et al., 1978), interferons (Keeds, 1978), the thymidine kinase gene of herpes virus (Wagner, Sharp, & Summers, 1981), and the gene for adenovirus polypeptide IX (Tilghman et al., 1978) are examples of mammalian genes without introns. Intron-free genes abound in yeast. Furthermore, intronless forms of genes that normally contain introns have recently been shown to be expressed. Gruss et al. (1981) saw expression of insulin encoded by an intronless rat preproinsulin gene which replaced the late region of SV40, and the mRNA involved was shown to be free of introns by cDNA sequencing. The tantalizing questions of what signals prevent the expression of some, but not all, intronless genes, and whether these signals are involved in the correct splicing of genes with introns, will probably continue to elude researchers for some time.

When the mechanisms for intron removal are finally elucidated, it may become clear what role is served by these interruptions in the continuity of eukaryotic genetic information. Perhaps they facilitate the maturation of mRNAs which otherwise could not be properly processed. Perhaps they function only in evolution, and their raison d'être will remain inaccessible to biochemical analysis. Whatever the case, the gathering of information concerning the signals that control splicing and the machinery that recognizes those signals rates high priority in today's molecular biology.

ACKNOWLEDGMENT

We would like to thank Drs. Randall Reed and Sherman M. Weissman for critically reading the manuscript. The writing of this review was made possible through a grant from the National Science Foundation.

REFERENCES

Acheson, N. (1978). Polyoma virus giant RNAs contain tandem repeat of the nucleotide sequence of the entire viral genome. *Proc. Natl. Acad. Sci. USA* 75:4754-4758.

Acheson, N., Buetti, K., Scherrer, K., and Weil, R. (1971). Transcription of the polyoma virus genome: Synthesis and cleavage of giant late polyoma-specific RNA. *Proc. Natl. Acad. Sci. USA* 68:2231-2235.

Aleström, P., Akusjärvi, G., Perricaudet, M., Mathews, M. B., Klessig, D. F., and Pettersson, U. (1980). The gene for polypeptide IX of adenovirus type 2 and its unspliced messenger RNA. *Cell* 19:671-681.

Alexander, N., Perlman, P., Hanson, D., and Mahler, H. (1980). Mosaic organization of a mitochondrial gene: Evidence from double mutants in the cytochrome b region of *Saccharomyces cerevisiae*. *Cell* 20:199-214.

Alwine, J., and Khoury, G. (1980). Control of simian virus 40 gene expression at the levels of RNA synthesis and processing: Thermally induced changes in the ratio of the simian virus 40 early mRNAs and proteins. *J. Virol.* 35:157-164.

Avvedimento, V., Vogeli, G., Yamada, Y., Maizel, J., Pastan, I., and de Crombrugghe, B. (1980). Correlation between splicing sites within an intron and their sequence complementary with U_1 RNA. *Cell* 21:689-696.

Baez, M., Taussig, R., Zazra, J., Young, J., Polese, P., Reisfeld, A., and Skalka, A. (1980). Complete nucleotide sequence of the influenza A/PR/8/34 virus N5 gene and comparison with the N5 gene of A/Udorn/72ano A/FPV/Rostock/34 strains. *Nucl. Acids Res.* 8:5845-5858.

Baird, M., Driscoll, C., Schreiner, H., Sciarratta, G., Sansone, G., Niazi, G., Ramirez, F., and Bank, A. (1981). A nucleotide change at a splice junction in the human β-globin gene is associated with $β°$-thalassemia. *Proc. Natl. Acad. Sci. USA* 78:4218-4221.

Baralle, F., Shoulders, C., and Proudfoot, N. (1980). The primary structure of the human ε-globin gene. *Cell* 21:621-626.

Bell, G., Pictet, R., Rutler, W., Cordell, B., Tischer, E., and Goodman, H. (1980). Sequence of the Human Insulin Gene. *Nature* 284:26-32.

Benoist, C., O'Hare, K., Breathnach, R., and Chambon, P. (1980). The ovalbumin gene-sequence of putative control regions. *Nucl. Acids Res.* 8: 127-142.

Bentley, D., and Rabbits, T. (1980). Human immunoglobulin variable region genes—DNA sequences of two V_κ genes and pseudogene. *Nature* 288: 730-733.

Benyajati, C., Place, A., Powers, D., and Sofer, W. (1981). Alcohol dehydrogenase gene of *Drosophila melanogaster*: Relationship of intervening sequences to functional domains in the protein. *Proc. Natl. Acad. Sci. USA* 78:2717-2721.

Berk, A., and Sharp, P. (1978). Structure of the adenovirus 2 early mRNAs. *Cell* 14:695-711.

Bernard, O., Hozumi, N., Tonegawa, S. (1978). Sequences of mouse immunoglobulin light chain genes before and after somatic changes. *Cell* 15: 1133-1144.

Birg, F., Favalora, J., and Kamen, R. (1977). Analysis of polyoma virus nuclear RNA by mini-blot hybridization. *Proc. Natl. Acad. Sci. USA* 74:3138-3142.

Breathnach, R., Benoist, C., O'Hare, K., Gannon, F., and Chambon, P. (1978). Ovalbumin gene: Evidence for a leader sequence in mRNA and DNA sequences at the exon-intron boundaries. *Proc. Natl. Acad. Sci. USA* 75:4853-4857.

Buchman, A., Burnett, L., and Berg, P. (1980). The SV40 nucleotide sequence. In *Molecular Biology of Tumor Viruses*, Part 2, ed. J. Tooze (Cold Spring Harbor, N.Y.: Cold Spring Harbor Laboratory), pp. 799-829.

Calame, K., Rogers, J., Early, P., Davies, M., Livant, D., Wall, R., and Hood, L. (1980). Mouse C_μ heavy chain immunoglobulin gene segment contains three intervening sequences separating domains. *Nature* 284:452-455.

Carlock, L., and Jones, N. C. (1981). Synthesis of an unspliced cytoplasmic message by an adenovirus 5 deletion mutant. *Nature* 294:572-574.

Chapman, G., Rogers, K., Brittain, T., Bradshaw, R., Bates, O., Turner, C., Cary, P., and Robinson, C. (1981). The 20,000 molecular weight variant of human growth hormone. *J. Biol. Chem.* 256:2395-2401.

Chech, T., Zaug, A., and Grabowski, P. (1981). In vitro splicing of the ribosomal RNA precursor of tetrahymena: Involvement of a quanosine nucleotide in the excision of the intervening sequence. *Cell* 27:487-496.

Chow, L. T., and Broker, T. (1978). The spliced structure of adenovirus 2 fiber message and other late mRNAs. *Cell* 15:497-510.

Chow, L. T., Broker, T., and Lewis, J. (1979). Complex splicing patterns of RNAs from the early regions of adenovirus 2. *J. Mol. Biol.* 134:265-287.

Chu, G., and Sharp, P. (1981). A gene chimaera of SV40 and mouse β-globin is transcribed and properly spliced. *Nature* 289:378-382.

Church, G. M., Slonimski, P. P., and Gilbert, W. (1979). Pleiotropic mutations within two yeast mitochondrial cytochrome genes block mRNA processing. *Cell* 18:1209-1215.

Claisse, M., Spyridakis, A., Wambier-Kluppel, M. L., Pajot, P., and Slonimski, P. O. (1978). Mosaic organization and expression of the mitochondrial DNA region controlling cytochrome *c* reductase and oxidase. II. Analysis of proteins translated from the *box* region. In *Biochemistry and Genetics of Yeast*, eds. M. Bacila, B. L. Horecker, and A. M. Stoppani (San Paulo: Academic Press), pp. 369-377.

Darnell, J. E., Jr. (1978). Implications of RNA-RNA splicing in evolution of eukaryotic cells. *Science* 202:1257-1260.

DeNoto, F., Moore, D., and Goodman, H. (1981). Human growth hormone DNA sequence and mRNA structure: Possible alternative splicing. *Nucl. Acids Res.* 9:3719-3730.

Doolittle, W. F. (1978). Genes in pieces: Were they ever together? *Nature* 272:581-482.

Dunn, A., Matthews, M., Chow, L., Sambrook, J., and Keller, W. (1979). A supplementary adenoviral leader sequence and its role in messenger translation. *Cell* 15:511-526.

Durica, D., Schoss, J., and Crain, W. (1980). Organisation of actin gene sequences in the sea urchin: Molecular cloning of an intron-containing DNA sequence coding for a cytoplasmic actin. *Proc. Natl. Acad. Sci. USA* 77:5683-5687.

Early, P., Rogers, J., Davis, M., Calame, K., Bond, M., Wall, R., and Hood, L. (1980). Two mRNAs can be produced from a single immunoglobulin μ gene by alternative RNA processing pathways. *Cell* 20:313-319.

Elder, J. T., Spritz, R. A., and Weissman, S. M. (1981). Simian virus 40 as a eukaryotic cloning vehicle. *Ann. Rev. Genet.* 15:295-340.

Ezcheverry, T., Colby, D., and Guthrie, C. (1979). A precursor to a minor species of yeast t-RNA[Ser] contains an intervening sequence. *Cell* 18:11-26.

Fiddes, J., and Goodman, H. (1981). The gene encoding the common alpha subunit of the four human glycoprotein hormones. *J. Molec. Applied Gen.* 1:3-18.

Flint, S., and Broker, T. (1980). Rytic infection by adenoviruses. In *Molecular Biology of Tumor Viruses*, Part 2, ed. J. Tooze (Cold Spring Harbor, N.Y.: Cold Spring Harbor Laboratory).

Fryberg, E., Bond, B., Hershey, D., Mixter, K., and Davidson, N. (1981). The actin genes of *Drosophila*: Protein coding regions are highly conserved but intron positions are not. *Cell* 24:107-116.

Gallwitz, D., and Surex, I. (1980). Structure of a split yeast gene: Complete nucleotide sequence of the actin gene in *Saccharomyces cerevisiae*. *Proc. Natl. Acad. Sci. USA* 77:2546-2550.

Ghosh, P., Piatak, M., Mertz, J., Weissman, S., and Lebowitz, P. (1981). Altered utilization of specific splices and $5'$-termini in late RNAs produced by leader region mutants of SV40. *J. Virol.* 44:610-624.

Ghosh, P., Roy, P., Barkar, A., Mertz, J., Weissman, S., and Leibowitz, P. (1981). Unspliced functional late 19S mRNAs containing intervening sequences are produced by a late leader mutant of SV40. *Proc. Natl. Acad. Sci. USA* 78:1386-1390.

Gilbert, W. (1978). Why genes in pieces? *Nature* 271:501.

Glanville, N., Durnam, D., and Palmiter, R. (1981). Structure of mouse metallothionein-*T* gene and its mRNA. *Nature* 292:267-269.

Goodman, H., Olson, M., and Hall, B. (1977). Nucleotide sequence of a mutant euckaryotic gene: The yeast tyrosine inserting ochre suppressor SUP4-0. *Proc. Natl. Acad. Sci. USA* 74:5453-5457.

Grabowski, P., Zaug, A., and Cech, T. (1981). The intervening sequence of the ribosomal RNA precursor is converted to a circular RNA in isolated nuclei of tetrahymena. *Cell* 23:467-476.

Green, M., and Roeder, R. (1980). Definition of a novel promoter for the major adenovirus-associated virus mRNA. *Cell* 22:231-242.

Grivell, L., Arnberg, A., de Boer, P., Borst, P., Bas, J., Groot, G., Hecht, N., Hengsens, L., Van Ommen, G., and Tabak, H. (1979). Transcripts of yeast mitochondrial DNA and their processing. In *Extra Chromosomal DNA*, eds. D. Cummings, P. Borst, I. David, S. Weissman, and C. Fox (New York: Academic Press), pp. 305-327.

Gruss, P., Efstradiadis, A., Karathanasis, S., König, M., and Khoury, G. (1981). Synthesis of a stable unspliced mRNA from intronless simian virus 40-rat preproinsulin gene recombinant. *Proc. Natl. Acad. Sci.* 6091-6095.

Gruss, P., Lai, C., Dhar, R., and Khoury, G. (1979). Splicing as a requirement for biogenesis of functional 16S mRNA of simian virus 40. *Proc. Natl. Acad. Sci. USA* 76:4317-4321.

Gubbins, E., Maurer, R., Lagrimini, M., Erwin, C., and Donelson, J. (1980). Structure of the rat prolactin gene. *J. Biol. Chem.* 255:8655-8662.

Haid, A., Grosch, G., Schomelzer, D., Schweyen, R. J., and Kaudewitz, C. (1980). Expression of the split gene COB in yeast mtDNA mutational arrest in the pathway of transcript splicing. *Current Gen.* 1:155-162.

Haid, A., Scheweyen, R., Bechman, H., Kaudewitz, F., Solioz, M., and Schatz, G. (1979). The mitochondrial *cob* region in yeast codes for apocytochrome b and is mosaic. *Eur. J. Biochem.* 94:451-472.

Halbreich, A., Pajot, P., Foucher, M., Grandchamp, C., and Slonimski, P. (1980). A pathway of specific splicing steps in cytochrome b mRNA processing revealed in yeast mitochondria by mutational blocks within the intron and characterization of a circular RNA derived from a complementable intron. *Cell* 19:321-329.

Hamer, D., and Leder, P. (1979). SV40 recombinants carrying a functional RNA splice junction and polyadenylation site from the chromosomal mouse beta [maj]-globin gene. *Cell* 17:737-747.

Harada, F., Kato, N., and Nishimura, S. (1980). The nucleotide sequence of nuclear 4.8S RNA of mouse cells. *Biochem. Biophys. Res. Comm.* 95: 1332-1340.

Hellig, R., Perrin, F., Gannon, F., Mandel, J.-L., and Chambon, P. (1980). The ovalbumin gene family: Structure of the X gene and evolution of duplicated split genes. *Cell* 20:625-637.

Heiter, P., Max, E., Seidman, J., Maizel, J., and Leder, P. (1980). Cloned human and mouse kappa immunoglobulin constant and J region genes conserve homology in functional segments. *Cell* 22:197-207.

Holmgren, R., Corces, V., Morimoto, R., Blackman, R., and Meselson, M. (1981). Sequence homologies in the 5' region of four *Drosophila* heat-shock genes. *Proc. Natl. Acad. Sci. USA* 78:3775-3778.

Honjo, T., Obata, M., Yamawaki-Kataoka, Y., Kataoka, T., Kawakami, T., Takahashi, N., and Mano, Y. (1979). Cloning and complete nucleotide sequence of mouse immunoglobulin γ 1 chain gene. *Cell* 18:559-568.

Horowitz, M., Laub, O., Bratosin, S., and Aloni, Y. (1978). Splicing of SV40 late mRNA is a post-transcriptional process. *Nature* 275:558-559.

Jacq, C., Pajot, P., Lazowska, J., Dujardin, G., Claisse, M., Groudinsky, O., de la Salle, H., Grandchamp, C., Labouesse, M., Gargouri, A., Guiard, B., Spyridakis, A., Dreyfus, M., and Slonimski, P. P. (1981). Role of introns in the yeast cytochrome b gene: Cis- and trans-acting signals, intron manipulation, expression and intergenic communications. In *Mitochondrial Genes*, eds. P. Slonimski, P. Borst, and G. Attardi (Cold Spring Harbor, N.Y.: Cold Spring Harbor Press).

Jensen, E. O., Paludan, K., Hyldig-Nielson, J., Jorgensen, P., and Marcker, K. A. (1981). The structure of a chromosomal leghaemoglobin gene from soybean. *Nature* 291:677-679.

Jones, W., and Kafatos, F. (1980). Structure, organization and evolution of developmentally regulated chorion genes in a silkmoth. *Cell* 22:855-867.

Jung, A., Sippel, A., Grez, M., and Schutz, G. (1980). Exons encode functional and structural units of chicken lysozyme. *Proc. Natl. Acad. Sci. USA* 77:5759-5763.

Kawakami, T., Takahashi, N., and Honjo, T. (1980). Complete nucleotide sequence of mouse immunoglobulin heavy chain genes. *Nucl. Acids Res.* 8:3933-3945.

Kedes, L. H. (1978). Histone gene and histone messengers. *Ann. Rev. Biochem.* 48:837-70.

Khoury, G., Alwine, J., Goldman, N., Gruss, P., and Jay, G. (1980). New chimeric splice junction in adenovirus 30 hybrid viral mRNA. *J. Virol.* 36:143-151.

Khoury, G., Gruss, P., Dhar, R., and Lai, C. (1979). Processing and expression of early SV40 mRNA. *Cell* 18:85-92.

Kimmel, A., and Firtel, R. (1980). Intervening sequences in a dictyostelium gene that encodes a low abundance class mRNA. *Nucl. Acids. Res.* 8: 5599-5610.

Kinniburgh, A., and Ross, J. (1979). Processing of the mouse β globin mRNA precursor: At least two cleavage ligation reactions are necessary to excise the larger intervening sequence. *Cell* 17:915-921.

Knapp, G., Ogden, R., Peebles, C., and Abelson, J. (1979). Splicing of yeast t-RNA precursors: Structure of the reaction intermediates. *Cell* 18: 37-45.

Konkel, D., Maizel, J., and Leder, P. (1979). The evolution and sequence comparison of two recently diverged mouse chromosomal β-globin genes. *Cell* 18:865-873.

Konkel, D., Tilghman, S., and Leder, P. (1978). The sequence of the chromosomal mouse β globin major gene: Homologies in capping, splicing and poly(A) sites. *Cell* 15:1125-1132.

Kreike, J., Bechmann, H., Van Hemert, F. J., Schweyen, R. J., Boer, P. H., Kaudewitz, F., and Groot, G. S. P. (1979). The identification of apocytochrome b as a mitochondrial gene product and immunological evidence for altered apocytochrome b in yeast strains having mutations in the COB region of mitochondrial DNA. *Eur. J. Biochem.* 101:607-614.

Kruger, K., Grabowski, P. J., Zaug, A. J., Sands, J., Gottschling, D. E., and Ceck, T. R. (1982). Self-splicing RNA: Autoexcission and autocyclization of the ribosomal RNA intervening sequence of tetrahymena. *Cell* 31: 147-157.

Lamb, R., and Lai, C. (1980). Sequence of interrupted and uninterrupted mRNAs and cloned DNA coding for the two overlapping nonstructural proteins of influenza virus. *Cell* 21:475-485.

Lamb, R., Lai, C.-J., and Choppin, P. (1981). Sequences of mRNAs derived from genome RNA segment 7 of influenza virus: Colinear and interrupted mRNAs code for overlapping proteins. *Proc. Natl. Acad. Sci. USA* 78:4170-4174.

Laski, F. A., Alzner-De Weerd, B., RajBhandary, U., and Sharp, P. (1982). Expression of *X. laevis* tRNA^{Tyr} genes in mammalian cells. *Nucle. Acids. Res.* 10:4609-4626.

Lawn, R., Efstradiadis, A., O'Connell, C., and Maniatis, T. (1980). The nucleotide sequence of the human β-globin gene. *Cell* 21:647-651.

Lazowski, J., Jacq, C., and Slonimski, P. P. (1980). Sequence of introns and flanking exons in wild type and box 3 mutants of cytochrome b reveals an interlaced splicing protein coded by an intron. *Cell* 22:333-341.

Lerner, M., Boyle, J., Mount, S., Wolin, S., and Steitz, J. (1980). Are snRNPs involved in splicing? *Nature* 283:220-224.

Lomedico, P., Rosenthal, N., Efstratiadis, A., Gilbert, W., Kolodner, R., and Tizard, R. (1979). The structure and evolution of the two non-allelic rat preproinsulin genes. *Cell* 18:545-558.

Matthews, M. (1980). Binding of adenovirus VA RNA to mRNA: A possible role in splicing? *Nature* 285:575-577.

Mattyssens, G., and Rabbitts, T. (1980). Structure and multiplicity of genes for the human immunoglobulin heavy chain variable region. *Proc. Natl. Acad. Sci. USA* 77:6561-6565.

Max, E., Seidman, J., and Leder, P. (1979). Sequences of five recombination sites encoded close to an immunoglobulin κ constant region gene. *Proc. Natl. Acad. Sci. USA* 76:3450-3454.

Mount, S. M. (1982). A catalogue of splice sequences. *Nucl. Acids Res.* 10: 459-473.

Muller, F., and Clarkson, S. (1980). Nucleotide sequence of genes coding for t-RNA^{phe} and t-RNA^{tyr} from a repeating unit of *X. laevis* DNA. *Cell* 19:345-353.

Murray, V., and Holliday, R. (1979). Mechanism for RNA splicing of gene transcripts. *FEBS Lett.* 106:5-7.

Nagata, S., Mantei, N., and Weissman, C. (1980). The structure of one of the eight or more distinct chromosomal genes for human interferon-alpha. *Nature* 287:401-408.

Nakanishi, S., Teranishi, Y., Noda, M., Notake, M., Wanatabe, Y., Kakadani, H., Jingami, H., and Numa, S. (1980). The protein-coding sequence of the bovine ACTH-β-LPH precursor gene is split near the signal peptide region. *Nature* 287:752-755.

Nishioka, Y., and Leder, P. (1980). Organization and complete sequence of identical embryonic and plasmacytoma κ V-region genes. *J. Biol. Chem.* 255:3691-3694.

Nishioka, Y., and Leder, P. (1979). The complete sequence of a chromosomal mouse α globin gene reveals elements conserved throughout vertebrate evolution. *Cell* 18:875-882.

Nobrega, F., and Tzagoloff, A. (1980). Assembly of the mitochondrial membrane system. DNA sequence and organization of the cytochrome b genes in *Saccharomyces cerevisiae* D273-10B. *J. Biol. Chem.* 255:9828-9837.

Nomiyama, H., Sakaki, Y., and Takagi, Y. (1981). Nucleotide sequence of a ribosomal RNA gene intron from slime mold *Physarum polycephalum*. *Proc. Natl. Acad. Sci. USA* 78:1376-1380.

Ogden, R., Beckman, J., Abelson, J., Kang, H., Söll, D., and Schmidt, O. (1979). In vitro transcription and processing of a yeast t-RNA gene containing an intervening sequence. *Cell* 17:399-406.

Oshima, Y., Itoh, M., Okada, N., and Miyata, T. (1981). Novel models for RNA splicing that involve a small nuclear RNA. *Proc. Natl. Acad. Sci. USA* 78:4471-4474.

Partington, C., and Baralle, F. (1981). Isolation of *Xenopus laevis* α-globin gene. *J. Mol. Biol.* 145:463-470.

Pech, M., Hochtl, J., Schnell, H., and Zachau, H. (1981). Differences between germ-line and rearranged immunoglobulin V_κ coding sequences suggest a localized mutation mechanism. *Nature* 291:668-670.

Peebles, C., Ogden, R., Knapp, G., and Abelson, J. (1979). Splicing of yeast t-RNA precursors: A two stage reaction. *Cell* 18:27-35.

Perler, A., Efstratiadis, A., Lomedico, P., Gilbert, W., Kolodner, R., and Dodgson, J. (1980). The evolution of genes: The chicken preproinsulin gene. *Cell* 20:555-566.

Perricaudet, M., Akusjärvi, G., Virtanen, A., and Petterson, U. (1979). Structure of two spliced mRNAs from transforming region of human subgroup C adenoviruses. *Nature* 281:694-696.

Perricaudet, M., Le Moullec, J.-M., Tiollas, P., and Petterson, U. (1980). Structure of two adenovirus type 12 transforming polypeptides and their evolutionary implications. *Nature* 288:174-176.

Piatak, M., Subramanian, K., Roy, P., and Weissman, S. (1981). Late mRNA production by viable SV40 mutants with deletions in the leader region. *J. Mol. Biol.* 153:589-618.

Robinson, R., and Davidson, N. (1981). Analysis of a *Drosophila* t-RBA gene cluster: Two t-RNA[Leu] genes contain intervening sequences. *Cell* 23: 251-259.

Rogers, J., and Wall, R. (1980). A mechanism for RNA splicing. *Proc. Natl. Acad. Sci. USA* 77:1877-1879.

Sakano, H., Maki, R., Kurosowa, Y., Roeder, W., and Tonegawa, S. (1980). Two types of somatic recombination are necessary for the generation of complete immunoglobulin heavy-chain genes. *Nature* 286:676-683.

Sakano, H., Rogers, J., Huppi, K., Brock, C., Traunecker, A., Maki, R., Wall, R., and Tonegawa, S. (1979). Domains and the hinge region of an immunoglobulin heavy chain are encoded in separate DNA segments. *Nature* 277:627-633.

Sharp, P. (1981). Speculations on RNA splicing. *Cell* 23:643-646.

Slighton, J., Blechl, A., and Smithie, O. Human fetal [G]γ- and [A]γ-globin genes: Complete nucleotide sequences suggest that DNA can be exchanged between these duplicated genes. *Cell* 21:627-638.

Slonimski, P., Claisse, M., Foucher, M., Jacq, C., Kochka, A., Kamouroux, A., Pajot, P., Perrodin, G., Spyridakis, A., and Wambier-Kluppel, M. (1978). Mosaic organisation and expression of the mitochondrial DNA region controlling cytochrome c reductase and oxidase. III. A model of structure and function. In *Biochemistry and Genetics of Yeast*, eds. M. Bacila, B. Horecker, and A. Stoppani (New York: Academic Press), pp. 391-426.

Solioz, M., and Schatz, G. (1979). Mutations in putative intervening sequences of the mitochondrial cytochrome b gene of yeast produce abnormal cytochrome b polypeptides. *J. Biol. Chem.* 254:9331-9334.

Solnick, D. (1981). An adenovirus mutant defective in splicing RNA from early region 1A. *Nature* 291:508-510.

Spritz, R., DeRiel, J., Forget, B., and Weissman, S. (1980). Complete nucleotide sequence of the human δ-globin gene. *Cell* 21:639-646.

Spritz, R. A., Jagadeeswaran, P., Choudary, P. V., Biro, P. A., Elder, J. T., de Riel, J. K., Manley, J. L., Gefter, M. L., Forget, B. G., and Weissman, S. M. (1981). Base substitution in an intervening sequence of a β^+ thalassemic human globin gene. *Proc. Natl. Acad. Sci. USA* 78:2455-2459.

Srivatsan, E., Deininger, P., and Freidman, T. (1981). Nucleotide sequence at polyoma VP_1 mRNA splice sites. *J. Virol.* 37:244-247.

Stein, J., Catterall, J., Kristo, P., Means, A., and O'Malley, B. (1980). Ovomucoid intervening sequences specify functional domains and generate protein polymorphism. *Cell* 21:681-687.

Sun, S., Slightom, J., and Hall, T. C. (1981). Intervening sequences in a plant gene—Comparison of the partial sequence of cDNA and genomic DNA of french bean phaseolin. *Nature* 289:37-41.

Thummel, C., Tjian, R., and Grodzicker, T. (1981). Expression of SV40 T antigen under control of adenovirus promoters. *Cell* 23:825-835.

Tilghman, S., Curtis, P., Tiemeier, D., Leder, P., and Weissman, C. (1978). The intervening sequence of a mouse globin β gene is transcribed within the 15S β globin mRNA precursor. *Proc. Natl. Acad. Sci.* 75:1309-1313.

Tonegawa, S., Maxam, A., Tizard, R., Bernard, O., and Gilbert, W. (1978). Sequence of a mouse germ-line gene for a variable region of an immunoglobulin light chain. *Proc. Natl. Acad. Sci. USA* 75:1485-1489.

Treisman, R., Novak, U., Favaloro, J., and Kamen, R. (1981). Transformation of rat cells by an altered polyoma virus genome expressing only the middle-T protein. *Nature* 292:595-600.

Tsujimoto, Y., and Suzuki, Y. (1979). The DNA sequence of Bombyx mori fibroin gene including the 5′ flanking, mRNA coding, entire intervening and fibroin protein coding regions. *Cell* 18:591–600.

Tucker, P., Marcu, K., Newell, N., Richards, J., and Blattner, F. (1979). Sequence of the cloned gene for the constant region of murine γ2b immunoglobulin heavy-chain. *Science* 206:1303–1306.

Valenzuela, P., Venegas, A., Weinberg, F., Bishop, R., and Rutter, W. (1978). Structure of yeast phe t-RNA genes: An intervening DNA segment within the region coding for the t-RNA. *Proc. Natl. Acad. Sci. USA* 75:190–194.

Van Ommen, G. J. B., Boer, P. H., Groot, G. S. P., De Haan, M., Roosendaal, E. and Grivvel, L. A. (1980). Mutations affecting RNA splicing and the interaction of gene expression of the yeast mitochondrial loci cob and oxi 3. *Cell* 20:173–183.

van Ooyen, A., van den Berg, J., Mantei, N., and Weissman, C. (1979). Comparison of total sequence of a cloned rabbit β-globin gene and its flanking regions with homologous mouse sequence. *Science* 206:337–344.

Wagner, M., Sharp, J., and Summers, W. (1981). Nucleotide sequence of the thymidine kinase gene of herpes simplex virus type 1. *Proc. Natl. Acad. Sci. USA* 78:1441–1445.

Westaway, D., and Williamson, R. (1981). An intron nucleotide sequence variant in a cloned β⁺-thalassaemia globin gene. *Nucl. Acids Res.* 9:1777–1788.

Westphal, H., Lai, S., Lawrence, C., Hunter, T., and Walter, G. (1979). Mosaic adenovirus SV40 RNA specified by non-defective hybrid virus AD2⁺. *J. Mol. Biol.* 130:337–351.

Yamada, Y., Avvedimento, V., Mudryi, M., Ohkubo, H., Vogeli, G., Meher, I., Pastan, I., and de Crombrugghe, B. (1980). The collagen gene: Evidence for its evolutionary assembly by amplification of a DNA segment containing an exon of 54 b.p. *Cell* 22:887–892.

Yamawaki-Kataoka, Y., Miyata, T., and Honjo, T. (1981). The complete nucleotide sequence of mouse immunoglobin γ2a gene and evolution of heavy chain genes: Further evidence for intervening sequence-mediated domain transfer. *Nucl. Acids Res.* 9:1365–1381.

Yamawaki-Kataoka, Y., Kataoka, T., Takahashi, N., Obata, M., and Honjo, T. (1980). Complete nucleotide sequence of immunoglobulin γ2b chain gene cloned from newborn mouse DNA. *Nature* 283:786–789.

Yang, V., Lerner, M., Steitz, J., and Flint, S. (1981). A small nuclear ribonucleoprotein is required for splicing of adenoviral early RNA sequences. *Proc. Natl. Acad. Sci. USA* 78:1371–1375.

Young, R., Hagenbüchle, O., and Schibler, U. (1981). A single mouse α-amylase gene specifies two different tissue-specific mRNAs. *Cell* 23:451–458.

Zaug, A., and Cech, T. (1980). In vitro splicing of the ribosomal RNA precursor in nuclei of tetrahymena. *Cell* 19:331–338.

Ziff, E. (1980). Transcription and RNA processing by the DNA tumour viruses. *Nature* 287:491–499.

INDEX

NAME INDEX

Abelson, J. N., 327, 349, 419
Acheson, N., 414
Adhya, S., 374
Agarwal, K. L., 1, 3, 24, 27, 34, 37, 51, 52
Agrawal, H. P., 179, 185, 199, 203, 210, 211, 212, 218
Aleström, P., 400
Alexander, M., 313
Alexander, N., 420
Aloni, Y., 373
Altenburger, W., 333
Altwegg, M., 235
Alvarado-Urbina, G., 5, 23
Alwine, C., 372
Alwine, J., 414, 415
Ames, B. N., 192, 193
Anderson, S., 74, 82, 95
Ansorge, W., 262
Armanath, V., 2
Arnott, S., 349
Astell, C. R., 23, 27, 29, 31, 37, 39, 44, 45, 54
Atkinson, M. R., 133
Avvedimento, V., 417
Axel, R., 373

Badek, T., 3
Baez, M., 414
Bahl, C. P., 23
Baird, M., 401
Baker, C. C., 354, 357, 367
Baldwin, B. L., 37
Bambara, R., 132
Banerji, J., 366

Bantle, J. A., 40
Barelle, F. E., 56
Barnes, W. M., 127, 129
Baron, F., 137, 138, 139, 265
Barrell, B. G., 69, 115, 121, 123, 126, 128, 130, 149
Barrett, T., 373
Bass, L. W., 137, 265
Bastia, D., 149
Batts-Young, B., 354
Beaucage, S. L., 5, 9, 10
Beaven, G. H., 115
Beck, E., 133
Beckwith, J. R., 327
Beers, R. J., 24, 25, 27
Behrens, K., 204
Bell, G., 422
Bendig, M. M., 364
Bennett, G. N., 321, 327
Benoist, C., 363, 364, 365, 366, 367
Benton, W. D., 174
Berard, D., 363
Berg, P., 53, 128, 174, 365, 367, 410
Berk, A. J., 354, 372, 414
Bernard, O. D., 56
Bernardi, G., 306, 314
Besmer, P., 23, 30, 38, 41, 54
Billeter, M. A., 123
Birchmeier, C., 376
Bird, R. E., 149
Birg, F., 414, 418
Birkenheimer, E. H., 375
Birkenmeier, E. H., 352, 353

SUBJECT INDEX

α-amatin, 351, 353
(α-^{32}P)ATP, 273,-74
acceptor splice junction, 400
acceptor splice site, 413
acetylation, 244
adenosine probing reaction, 289
adenosine reaction, 265
adenosines, 284–86
    limited RNA cleavage at, 278–79
adenovirus, 83, 145, 352, 353, 354,
        357, 363, 364, 367, 370,
        400, 401, 411, 412, 416
agarose gels, and detection of ^{3}H,
    176–78
allele, 422
amines, primary, 267
aniline-induced strand scission, 265,
    267–70, 275, 289
apyrimidinic acid, 138–39
araC protein, 308
ATP hydrolysis, 332

β-galactosidase, 146
β-globin, 401
β$^+$ thalassemia, 401
β°thalassemia, 401, 410
bacterial RNA polymerase, 349
bacteriophage
    G4 genome, 82
    T4 DNA, 37
bacteriorhodopsin gene, 49
BamH-I, 78
base analysis
    of mutagen/carcinogen-treated
        DNA, 197–98

of normal DNA, 196–97
of nucleic acids, 170
of RNA, by ^{3}H derivative method,
    178–87
base composition, effect of on oligo-
        nucleotide stability, 29
bases
    distinguishing, 115–16
    sequencing nucleic acids in, 114–23
base-specific
    cleavages, 122
    nested segment method, 132–44
base stacking interactions, 288
B. cereus RNase, 204, 207
bleomycin, 314
block overlap, for ordering nucleo-
        tides, 120–21
blunt-end cleavage enzyme, 78, 79
blunt-end cloning, and 5' sticky ends,
    95–96
blunt-end ligation, 76
Bombyx fibron gene, 363
bovine
    preproenkephalin mRNA, 47, 48
    proenkaphalin mRNA, 48
    prolactin mRNA, 57
−35 Box, 349–51
box gene, 420
B. subtilis tRNA, 243
Burton-Peterson reaction, 125

CAAT box, 363,365, 366
calf-thymus
    chromatin, 135
    DNA, 135, 139

450

rapid read-off sequencing procedure, 249
rapid sequencing, of RNA, 261
rat insulin mRNA, 49
reciprocal block cleavage, 120
recombinant clones, screening of, for size at inserted DNA, 105-6
recombinant DNA
  blotting protocol for, 102-5
  identification of, 102-6
restriction enzyme, 114, 334
  cleavage sites, 81
reverse transcriptase, primer for, 48-52
rho-dependent termination signals, 374
ribonucleoside analysis, radioiodination method for, 188-92
ribose-methylated dinucleoside triphosphates, 218-20
ribosubstitution, 123
  block technique, 123, 128-29
Rous sarcoma virus, 149
RNA
  aberrant sequencing patterns, 300-4
  base analysis of, 178-87
  base-specific chemical cleavage reactions, 275-86
  chemical sequencing, 262-70
  controlled hydrolysis of, 214
  copying DNA into, 126-28
  end-labeled, sequencing of, 261-303
  3'-end-labeling of, 270-73
  5'-end-labeling of, 273-74
  eukaryotic messenger, 399-423
  enzymatic degradation of, 179
  ^{3}H derivative method for sequence analysis of, 200-1
  ^{3}H/^{32}P derivative method for sequence analysis of, 201-2
  large, sequence analysis of, 223-4
  limited cleavage, at adenosines, 278-79
    at cytidines, 282-84
    at guanosines, 275-79
    at uridines, 279-81
  molecules, capped, 273
    uncapped, 273-74

polymerases, 351
preparation of 3' terminally ^{32}P-labeled, 206-7
preparation of 5' terminally ^{32}P-labeled, 205-6
5S, 353, 353
sequencing of end-labeled, 297-99
sequencing of, by ^{32}P gel readout method, 202-10
single-hit chemical cleavages of, 210-23
splicing, 400
U1, 417-18
RNase
  A, 204
    digest, 207
  Phy$_1$, 203, 204
    digest, 208
  Phy M digest, 208
  T1, 253
    digest, 207
  T2, 253
  U2 digest, 207
rRNA, 399, 400, 420

Sanger-Brownlee-Barrell sequencing method, 123
Sanger chain termination method
  and filamentous phage cloning vectors, 73-74
  and primed synthesis sequencing, 70-73
  for sequencing DNA, 69-108
Sanger fingerprinting technique, 245
Sanger two-dimensional electrophoresis, 125
S. aureus endonuclease, 204
S. cerevisiae, 41, 54
sea urchin
  H2A gene, 365
  histone gene, 364
sequence analysis, of nucleic acids, methods for, 170
sequencing gels, preparation of, 91-93
shotgun sequencing, 77
silk fibroin gene, 364
silkworm tRNA gene, 370

single-hit chemical cleavages of RNA, 210-23
single-stranded DNA, 40, 70, 307, 333
snake venom phosphodiesterase, 185, 200, 253
S1 nuclease, for removing 3′ overlaps, 96, 97
Southern blot experiment, 38
spleen exonuclease, 194, 197
splice junction signals, 400-10
splicing
 biochemistry of, 419-21
 of rRNA, 420
 versatility of, 412-17
 of yeast tRNA, 419
splicing signals, recognition of, 417-19
5S RNA gene, 356
stepwise degradation, 120
5′ sticky ends, filling in of, prior to blunt-end cloning, 95-96
strand scission, 120
 aniline-induced, 267-70
S. typhimurium, 246-52
supercoiling, 332, 334
support, synthesis of, 13-15
support, polymer
 HPLC-grade silica gel, 6-7
 silica gel, 6-7
SV40, 133, 366, 368, 372, 373, 376
 DNA, 128
 early gene, 364
 late mRNA, 414
 late polyadenylation site, 423
 late promoter, 423
 late RNA donor splice site, 410
 late transcription unit, 423
 recombinant, 410
 mutants, 365

TATA box, 363-68
5′ terminal analysis, 216-18
termination signals, 349-76
thin-layer chromatograms, for detection of ³H, 175-76
thin-layer chromatography, 272-73
thin-layer readout, 170, 172, 173
Tm, 25-27

tobacco mosaic virus RNA, 56
Torpedo acetylcholine receptor, 48
tract overlap, for ordering nucleotides, 120-21
transcription
 enhanced, 366
 initiation, 349-76
 regulation of, by polymerases, 368-74
 termination of, eukaryotes in, 375-76
 termination of, in prokaryotes, 374-75
transcriptional control regions, 355-68
trifluoroacetylation, 244
trimethylsilyation, 244
tritium-labeling method, 251
tRNA, 353, 356, 399, 400
 characterization of modified nucleosides in, 235-55
 carrier, 299
 introns, 420
 sequencing, and modified nucleotides, 245-51
 total, and modified nucleosides, 251-52
T4 RNA ligase, 270-73
two-dimensional chromatography, and modified nucleosides, 239-42
two-dimensional mobility shift analysis, 129

uncapped RNA molecules, 273-74
uridine reaction, 265
uridines, 284-86
 limited RNA cleavage at, 279-84
uridylic acid, 374
UV markers, 186
UV mapping, 354

VA RNA, 352, 353, 356
vector DNA, dephosphorylation of, prior to cloning, 100
viral mRNA, 273

wandering-spot method, 56, 129-32
Watson-Crick